THERMAL POWER CYCLES

G.H.A. COLE

Professor of Engineering Physics,
Department of Engineering Design and Manufacture,
University of Hull

Edward Arnold
A division of Hodder & Stoughton
LONDON   MELBOURNE   AUCKLAND

© 1991 G. H. A. Cole

First published in Great Britain 1991

British Library Cataloguing in Publication Data
Cole, G. H. A. (George Herbert Avery)
 Thermal power cycles.
 1. Heat engineering
 I. Title
 621.402

 ISBN 0-340-54522-4

All rights reserved. No part of this publication may be reproduced or transmitted in any form or by any means, electronically or mechanically, including photocopying, recording or any information storage or retrieval system, without either prior permission in writing from the publisher or a licence permitting restricted copying. In the United Kingdom such licences are issued by the Copyright Licensing Agency: 90 Tottenham Court Road, London W1P 9HE.

Typeset in 10 on 11pt Times by Wearside Tradespools, Fulwell, Sunderland
Printed and bound in Great Britain for Edward Arnold, a division of Hodder and Stoughton Limited, Mill Road, Dunton Green, Sevenoaks, Kent TN13 2YA by St Edmundsbury Press Limited, Bury St Edmunds, Suffolk, and bound by Hartnolls Limited, Bodmin, Cornwall.

PREFACE

Constant pressure shaft power plants convert energy into work on a large scale all over the world. The energy is usually derived by burning fossil fuels, although nuclear fuels have begun to be used, and the work is most often performed through electricity. The study of the practical characteristics and application of such plant remains a central feature of mechanical engineering. The study began essentially with the pioneering work of Sadi Carnot some 170 years ago and its development is an example of the application of the ideal to the real. This book offers a background to these studies. There is the restriction that automation of the plant and the associated environmental issues that must now be faced are not included, other than indirectly. To treat these topics would require a separate book for each and much of the practice has still to be established.

The substance of this book is based on material used for a core course that has been taught over a number of years to mechanical engineers reading for the BEng degree of the University of Hull. Apart from its value to mechanical engineers, it is hoped that the book will also prove useful as a text for option courses in other branches of engineering and applied physics: it is hoped that it will also be of use to readers outside the universities, polytechnics and colleges who need, or wish, to know more about these matters.

The book is not designed to be a treatise as such but is rather a collection of essays on the central theme of power production. This means that chapters need not be read in strict sequence although this would be the best approach. A wide range of examples is included for the reader and these form an important supplement to the text. They obviously provide the opportunity to practice the material treated in the text. But they do more. Many examples develop the arguments of the main text and occasionally offer new ideas. Even if the examples are not attempted the reader is advised at least to read them through after finishing the chapter.

Applied thermodynamics is a fascinating subject depending, as it does, on physical principles and practical application. It is necessary to optimise various factors that might, at first sight, seem to be at variance one with another. It is hoped that the reader will sense something of the flavour of this discipline from reading the book.

It is a pleasure to thank a number of people who have helped in the writing of the book. First those who cannot be properly named—past colleagues within the power industry and more recently students who have provided the background for me to become better acquainted with these various matters over the years. More recently I owe thanks to Dr. Alan Cummings of the University of Hull for his reading of proofs and for discussion at various stages. I must also thank the staff of Edward Arnold for their very professional work, courtesy and patience during the production of this book.

Although every effort has been made to exclude errors, it is unlikely that none remains and it would be most helpful to be told about any that are found. The many problems of energy facing the world at the present time make the matters discussed in the book of immediate relevance. If the book fires only one or two readers to become professionally involved it will have fulfilled a useful task.

G. H. A. Cole,
Hull, 1990

CONTENTS

Preface   iii

1. **THERMODYNAMICS: HEAT AND WORK**   1

   1.1 Some Preliminaries: Specifying the System   1
          Open and Closed Systems   1
          Flow and Non-flow Processes   2
          The Thermodynamic State   2
          Open and Closed Cycles   2
          Temperature and Heat   3
          Thermal Equilibrium   3

   1.2 Heat and Work   4

   1.3 Reversible and Irreversible Changes   5

   1.4 The First Law of Thermodynamics: The Enthalpy   5
          The Energy Equation   6
          Controlling the Flow   7
             The Throttle   7
             The Nozzle   7
             The Diffuser   8

   1.5 The Ideal Gas   8
          Equation of State   8
          Adiabatic Conditions   9
          The Entropy of the Gas   9
          With Heat Exchange   10
          Effect of Irreversibilities   11
          Cyclic Changes   11

   1.6 The Second Law of Thermodynamics   11

   1.7 Exercises   13

2. **ESSENTIAL CHARACTERISTICS OF THE CARNOT CYCLE: THE AVAILABILITY**   14

   2.1 The Carnot Engine   14

   2.2 The Carnot Cycle   15

   2.3 The Efficiency of the Carnot Engine: Temperature Scales   17

   2.4 The Carnot Refrigerator   18

   2.5 The Availability of a System   18
          Maximum Work from a Heat Reservoir   19
          Thermodynamic Potentials   19
             A Dual Transformation   19

vi *Contents*

	The Free Energies	19
	Interpretation of the Free Energies	20
	The Availability Function	21
	For a Closed System	21
	For an Open System	21
	Restrictions on the Availability	21
2.6	Summary	22
2.7	Exercises	23

Part I: GAS CYCLES

PROLOGUE

3. THE CARNOT ENGINE WITH GAS AS THE WORKING FLUID — 25

3.1	Work Transfer and an Ideal Gas	27
	The Validity of Using the Ideal Equation of State	27
	Reversible Adiabatic Conditions	28
	Expressions Describing Work Transfers	29
3.2	The Calculated Properties of the Carnot Engine	30
	Thermal Efficiency Using an Ideal Gas	30
	Specific Work Output	31
	Work Ratio	31
	Specific Flow of Gas	31
	Numerical Applications	32
3.3	Comment on the Equation of State for Real Gases	33
	The Virial Expansion	33
	Explicit Expressions for Thermodynamic Functions	34
	Empirical Equations of State	35
3.4	The Effects of Non-Ideal Components	36
	Isentropic Efficiencies	36
	The Cycle Thermal Efficiency	37
	The Work Output of the Cycle	38
3.5	Summary	38
3.6	Exercises	38

4. THE SIMPLE CONSTANT PRESSURE JOULE CYCLE — 41

4.1	Enthalpy Change and Work Transfer	41
4.2	Adiabatic Conditions	43
4.3	The Simple Joule Cycle	43
	Working Characteristics in Terms of Temperatures	44
	In Terms of Pressure and Temperature Ratios	45
	Characteristics of the Cycle	46
4.4	The Condition of Maximum Specific Work Output	49
	The Pressure Ratio	49
	A Numerical Comparison	50

Contents vii

	4.5	The Effects of Irreversibilities	51
		Compression and Expansion Efficiencies	51
		The Joule Cycle with Inefficiencies	52
		General Properties	52
		Maximum Specific Work Output	54
	4.6	Summary	54
	4.7	Exercises	55
5.	**REHEAT AND INTERCOOLING**		**56**
	5.1	The Principle of Reheat	56
	5.2	The Cycle with Reheat	57
		Specific Work Output is Maximised	58
		Thermal Efficiency and Work Transfer	59
	5.3	Reheat with Irreversibilities	60
		The Intermediate Pressure	63
		The Working Characteristics	63
	5.4	Intercooling	63
		Selecting the Intermediate Pressure	65
		The Working Characteristics	66
	5.5	Reheat and Intercooling	67
	5.6	Multi-Stage Cycles	70
	5.7	Isothermal Compression	73
	5.8	Summary	74
	5.9	Exercises	77
6.	**JOULE CYCLE WITH HEAT EXCHANGE**		**81**
	6.1	The Concept of Internal Heat Exchange	81
	6.2	Characteristics of the Cycle	82
		The Thermal Efficiency	83
		Efficiency at Maximum Work Output	83
	6.3	Less Than Perfect Heat Exchange	85
	6.4	A Less Than Perfect System	87
	6.5	Summary	89
	6.6	Exercises	90
7.	**THE COMPLETE SYSTEM**		**91**
	7.1	Total and Complete Systems	91
	7.2	Heat Exchange with Reheat	91
		Ideal Heat Exchange and Isentropic Change	91
		Inefficient Heat and Work Transfer	94

viii *Contents*

	7.3	Heat Exchange with Intercooling	96
	7.4	The Complete System	97
		n Complete Reheats and Intercoolings with Heat Exchange	97
	7.5	The Total System	100
	7.6	Comments on Practical Applications	104
		Marine Units	107
		Aircraft Propulsion	108
		Industrial Uses for Power	109
	7.7	Summary	109
	7.8	Exercises	110
8.	**Non-steady Flow Cycles for Low Power Output**		**112**
	8.1	The Turbine and Compressor Have Gone	112
	8.2	The Diesel Principle	112
		Thermal Efficiency and Work Output	113
		Comparison with the Joule Cycle	115
		Working Conditions	116
	8.3	Constant Volume Processes: The Otto Cycle	116
		Thermal Efficiency and Work Output	116
		The Cycle Characteristics	118
		Working Conditions	118
	8.4	A Mixed Constant Pressure/Constant Volume Cycle	118
	8.5	Summary	119

PART II: STEAM CYCLES

PROLOGUE

9.	**Some Properties of Steam**		**121**
	9.1	The T–V, p–T and p–V Diagrams	123
		The T–V Diagram	123
		The p–T Diagram	125
		The p–V Diagram	125
		Other Thermodynamic Variables	127
	9.2	Representing the Properties	127
		Steam Tables	127
		The Mollier Diagram	128
		The T–s Diagram	128
		Further Comment	129
	9.3	Summary	129
	9.4	Exercises	129

10.	**The Carnot Engine with Steam as the Working Fluid**	**131**
	10.1	Introduction	131
	10.2	The Working Characteristics	131
		A Specific Example	133
	10.3	Isentropic Efficiencies	136
		The Cycle Working Characteristics	136
		A Numerical Application	137
	10.4	Summary	137
	10.5	Exercises	137
11.	**The Simple Rankine Cycle**	**140**
	11.1	The Basic Cycle	140
	11.2	The Features of the Cycle	141
	11.3	The Compressor Work Transfer	142
	11.4	A Numerical Example	143
		A Low Pressure System	143
		The Compressor Work	144
		A System at Higher Boiler Pressure	144
	11.5	Imperfect Compression and Expansion	146
		Including Isentropic Efficiency	147
	11.6	The Mean Receptor Temperature	149
	11.7	Summary	150
	11.8	Exercises	150
12.	**Raising the Mean Receptor Temperature: Superheat, Reheat and Supercriticality**	**154**
	12.1	The Principle of Superheating	154
		The Enthalpy Changes	154
		The Effects of Irreversibilities	157
	12.2	Numerical Example of Superheating	158
		Ideal Cycle	158
		The Work of Compression	159
		General Results	159
		Inefficient Expansion	159
	12.3	Reheating	160
		General Criterion for Reheating	161
		Effect on Efficiency and Specific Steam Consumption	162
		Numerical Examples	167
	12.4	Supercritical Conditions	169
		Numerical Example	171
	12.5	Summary	173

x *Contents*

		12.6	Exercises	173
13.	**REGENERATIVE HEATING**			**177**
	13.1	Preliminary Thermodynamic Considerations		177
		Isothermal and Adiabatic Transfers		179
		Achieving the Carnot Efficiency		181
		The Availability Function		181
	13.2	Practical Implementation of Feed Heating		182
		Finite Number of Exchange Points		182
		Comments on the Heat Exchange		182
	13.3	A Possible Cycle Without Superheat		183
		Initial and Final Steam Conditions		185
		Selecting the Turbine Bleed Points		185
		Enthalpies and Entropies		185
		Energy Balance at the Bleed/Feed Point		186
		Work and Heat Transfers		186
		Two Feed Heaters		187
	13.4	The Thermodynamics of the Bleed Point Locations		187
		Steam and Feed Water Flows		187
		Optimum Feed Heat Transfer		189
	13.5	Feed Heating Using a Cascade		190
	13.6	Feed Heating in a Superheated System		192
		The Thermodynamics for Expanding the Bled Steam		192
		The Effect of Including Superheat		194
	13.7	Summary		195
	13.8	Exercises		197
14.	**THE FULL STEAM SYSTEM**			**200**
	14.1	Comments on the Simple Rankine Cycle		200
	14.2	Using the Flue-Gas Availability		200
		The Economiser		201
		The Effectiveness of the Economiser/Feed Heat System		202
		The Air Pre-Heater		203
	14.3	The Total Plant: Overall Plant Efficiency		203
		The Steam Cycle		204
		Working Characteristics (See also Table 14.2)		205
		Non-isentropic Behaviour		205
		Other Losses		206
		The Overall Efficiency		206
	14.4	Power and Heat Extraction Systems		207
	14.5	Combustion Products and their Control		210
		Impurities in the Fuel		210
		Features of Stack Emission		211
		Flue Gases from Coal		211

	Flue Gases from Oil	212
	Pollution Control	212
14.6	Comments on Metallurgical Limits	212
	Furnace Heat Transfer	213
	Ancillary Heat Transfer	213
	Limits are Set	213
14.7	The Usefulness of Exergy in Plant Assessment	214
14.8	Towards a Better Working Fluid	215
	Constant Temperature Furnace	215
	Varying Temperature Heat Source	217
14.9	Summary	217
14.10	Exercises	218

15. BINARY AND COMBINED CYCLES — 219

15.1	The Binary Vapour Cycle	219
15.2	Thermal Efficiency of the Dual Vapour/Steam Cycle	220
	Ideal Cycles	220
	Practical Cycles	221
15.3	The Mercury/Steam Cycle	223
15.4	Low Heat Cycles	225
15.5	A Combined Cycle	225
15.6	Summary	226
15.7	Exercises	226

16. NUCLEAR POWER CYCLES — 230

16.1	The Chain Reaction	230
16.2	Two Types of Reactor	232
16.3	Thermal Reactors	232
	The Gas Cooled Reactor	233
	The Dual Pressure Cycle	233
	Higher Gas Pressure	234
	The Water Cooled Reactor	235
	The Pressurised Water Reactor	236
	The Boiling Water Reactor	237
	The Heavy Water Reactor	237
	Other Possibilities	238
16.4	Reactors Using Fast Neutrons	239
16.5	Final Comment	239
16.6	Summary	240

Index — 241

1. THERMODYNAMICS: HEAT AND WORK

In the following chapters we shall be concerned with the conversion of energy into work and of work into energy using fluids (air and water particularly) which will be heated, cooled, compressed and expanded in various ways following particular sequences. The study of the transfer of heat from one body to another, the relation between heat and work and the consequent changes of state of the fluid is the concern of thermodynamics. Thermodynamics is, therefore, the basis for the study of the production of work by machinery of various kinds. Certain results of thermodynamics are reviewed in this chapter in so far as they refer to the production of work using constant pressure fluid cycles. Various definitions and conclusions are collected together for later use.

1.1 Some Preliminaries: Specifying the System

Very generally we will call a region of space which contains matter undergoing changes to produce work from energy (or vice versa) a thermodynamic system or simply a system. Because the chapters to follow are concerned with the use of fluids for the production of work, the present discussion will generally be restricted to fluid systems. The fluid may be a gas (Chapters 3 to 8 and part of Chapter 15) or it may be water in liquid or vapour form (Chapters 9 to 16). In all cases the system will involve a compression stage and an expansion stage, separated by a heating or a cooling stage.

Open and Closed Systems
If the fluid is permanently separated from its surroundings by a boundary through which it cannot pass, the system is said to be closed; otherwise the system is open. It follows that a closed system always involves the same fluid whereas in an open system the fluid can be changed. Although heat and work energy pass through a closed boundary, fluid may not pass through if the boundary is closed but does if the boundary is open. It is not required that the boundary of a closed system shall be rigid and it may change during the heat/work process (for instance, the material of the system might show thermal expansion or contraction). A closed fluid system can be represented as a collection of open systems. Thus, a compression or expansion stage can be studied individually as an open system with fluid passing through even though it is part of a total system which is closed. In another context the surroundings can be taken to be part of the system if this is useful, and an open system can be made closed in this wider sense.

The surroundings of a closed system are vast if the term is taken literally but usually only a very restricted part of the surroundings is affected by changes that might occur in the system. It is only this restricted part, in interaction with the system, that needs be taken into account and it is this

2   *Thermodynamics: Heat and Work*

region in thermodynamic contact with the system which will be meant when we talk of the surroundings of the system.

Flow and Non-flow Processes

The processes occurring in a closed system, when no fluid flows in or out of it, are generally described as non-flow processes (as viewed from the outside) while those in an open system, where fluid enters and leaves, are called flow processes. This is because there is no net loss or gain of fluid in a closed system (by definition) while there is in an open system. But in a closed system with several or many components, and involving a fluid, there will be fluid flow in the separate components which are then treated individually as open systems, even though there is no flow of matter to or from the system as a whole. That part of the total system considered in thermodynamic arguments can be marked by a hypothetical surface, called the control surface, to distinguish it from the remainder. It is necessary, of course, to specify the matter and energy flows across the control surface which link the material inside to that outside. In the following chapters we will not be concerned with the combustion process providing the heat transfer into the cycle nor the use of the work transfer (for example for generating electricity): these aspects lie outside the control surface enclosing the cycle.

The formal expressions of the Laws of Thermodynamics are usually made for closed systems but they are readily modified to describe open systems to which they also apply.

The Thermodynamic State

The state of a system at a given instant of time is determined by the total specification of its relevant properties at that time. What is relevant depends on the system under discussion.

For an engine acting as a closed system and using a fluid these are the thermodynamic properties of the fluid. It is known from experiment that a knowledge of two, and only two, properties of the fluid is sufficient to describe the state of the closed system, provided the relation is known (called the equation of state) between these independent properties and the other properties of the fluid. For example, for a very dilute gas (see Section 1.5) the equation of state is a relation between the pressure p, the volume V and the temperature T of the gas, and the knowledge of any two of these three quantities is sufficient to determine the other one. This specification of the state prescribes the thermodynamic state of the gas and the properties of the fluid are called the thermodynamic properties.

For an open system it is necessary to include information about the mechanical state of the fluid as well. This will involve the fluid velocity across the boundary together with any changes of elevation above ground level which might involve gravity. The effects of the mechanical properties may well be insignificant compared with the thermodynamic properties in particular cases, and generally, they are independent one of the other and so can be treated separately. This means that thermodynamic considerations can be made quite separate from the mechanical.

Open and Closed Cycles

An engine working steadily provides work and consumes energy at a constant rate. The condition of the fluid involved in the energy transfer, called the working fluid, is changed from some initial state where it absorbs energy to some final state where it transfers work. For steady work output the fluid must then be returned to its initial state so that the process can be repeated. The fluid is said to have passed round (or through) a cycle of operations, often simply called a cycle.

Because the working of the engine is steady it is clear that the properties of the system must, in the mean, remain constant at every internal point of the system, and the energy associated with each component must also be constant. The input of energy (through the fuel) and the work output (through the action of the engine) must also be constant in the mean otherwise the conditions inside would change and the system would not be working steadily. For a cycle undergoing a steady flow process the distinction between a closed and an open cycle is largely formal.

Temperature and Heat

There is no difficulty in using touch to distinguish a 'hot' object from a 'cold' one. In the first case heat enters our hand while it is in contact with the object (we might even get burned) while in the second case heat leaves it. We might say that the first object has a higher temperature than our hand while the temperature of the second object is lower. These physiological interpretations are made more precise through measurement. A gas can be used for this purpose.

Suppose a gas is contained in a fixed volume body at a chosen pressure, the boundary material being chosen not to impede the flow of heat. Through the equation of state the temperature of the gas is also fixed. Let it now be brought into direct contact with a second body. If the pressure and volume remain unaffected, then so does the temperature and we say that the gas and the second body have the same temperature. If, however, the pressure of the gas rises (correcting for any small change due to thermal expansion) the temperature of the second body is greater than that of the gas: by contrast, if the pressure falls, the second body is at a lower temperature. The actual difference in temperature between the gas and the second body can be assigned by introducing some numerical scale (perhaps, in this case, based on pressure measurements). Thermodynamics, being the study of the flow of heat, is able to provide a prescription for assigning a unique scale independent of the properties of a particular piece of matter, which can, in this sense, be regarded as absolute (called the Kelvin temperature scale. See Section 1.6).

Once a temperature scale has been set up the temperatures of any bodies in a group can be found by bringing the gas into contact with each separately, and in this way the gas behaves as a thermometer, though the gas can be replaced by more convenient and sophisticated equipment. The flow of heat involves a difference of temperature, and a difference of temperature is associated with a flow of heat.

Thermal Equilibrium

If several bodies are brought into contact with each other it is found by experience that the hotter will get cooler and the cooler will get hotter and each by a sufficient amount to reach a common temperature. It is not correct to suppose that they then each have the same energy content (whatever that may mean) and in general this will not be so. When the common temperature has been reached the bodies are said to be in thermal equilibrium. This is a general feature of matter and, in fact, forms the basis of measuring temperature. We may extend the argument to a continuous distribution of matter: if the temperature is not initially uniform throughout the volume it will become so after a sufficient length of time if the volume remains thermally isolated from its environment. It is also found from experience that an initially uniform distribution of temperature throughout a thermally isolated volume of matter will remain uniform and will not become non-uniform as time goes on. The natural tendency of the temperature towards uniformity is irreversible.

1.2 Heat and Work

Both heat and work are different manifestations of energy. Heat energy is transferred from one object to another if there is a temperature difference between them. We may say loosely that there is a flow of heat or that heat is transferred even though it is heat energy that is actually involved. We will often use the convenient terminology of heat transfer or heat flow in the following chapters on the understanding that it is really a shorthand for the flow of energy often associated with a difference of temperature. In SI units, the unit of heat flow is the joule per kilogramme per second (J/kg s).

Similarly, the ability of an engine to do work is also the ability to cause an energy flow able to move bodies in its surroundings. The movement of a body implies the action of a force through a distance and the transference of kinetic energy to the bodies in the surroundings to achieve the movement. There is, then, a mechanical equivalence between work and energy, and work is usually measured in terms of the energy equivalent. In SI units, work is measured in joules per kilogramme, like heat. Although we should really speak of the transfer of energy able to perform work we will find it more convenient to speak simply of a work transfer.

Heat and work transfer occur only when energy is being transferred within the system. In the transfer of heat to a body, the energy of the body is changed from an initial value U_1 to a final value U_2. The change of energy $dU(1,2) = U_2 - U_1$ is taken as being positive if heat enters the body and negative if it leaves. A change of energy will similarly result from the action of a force on the boundary of a closed body. This is taken as negative if the surroundings do work on (that is pass energy to) the body but positive if the energy flow is in the opposite direction.

dU depends upon the amount of energy entering or leaving the body and determines the energy state of the body relative to U_1 as the zero, datum state. There is no reference to the actual heat content of the body (which is given no meaning) but only to its change due to the transfer of energy. The quantity U is called the internal energy of the body and is a characteristic property: it can be defined only in relation to an additive constant which is assigned by specifying the zero state.

1.3 Reversible and Irreversible Changes

In special cases the change of state and the work done on the surroundings due to the addition of a quantity of heat to a body can be reversed if the same quantity of heat is taken away from the body. This is to be true even though the quantity of heat be indefinitely small and the changes brought about on the system are also infinitesimal. In such cases the body ends in precisely the same state and the environment ends in precisely the same condition as before the initial heat was transferred. Such a process is said to be reversible. If that is not the case, and the process cannot by any means be reversed precisely, the change is said to be irreversible. The main propositions of thermodynamics apply to reversible changes.

To obtain full thermal contact, the transfer of heat to the body must be made at a specific temperature T and this requirement has important consequences. First, the source of heat (which provides this temperature condition) must be of sufficient relative size to maintain its temperature against the loss to the body. The increment of loss by the source must be a vanishingly small proportion of its total heat content which means, in the limit of no incremental loss, that the heat source must be an energy reservoir of indefinite size. This is always assumed to be the case. Second, the heat transfer must be indefinitely small for the body receiving the heat to remain at the temperature of the reservoir and so to remain in thermal equilibrium with it. All changes, then, must be infinitesimal and thermal equilibrium is maintained within the body even though an infinitesimal change of state occurs. This requires any change to proceed indefinitely slowly. For a finite transfer of heat energy, when a finite amount of energy is involved and when the temperature is to rise by a finite amount, thermal equilibrium can be maintained at each stage only if the temperature is raised by an indefinitely large number of heat reservoirs spanning the finite temperature range and differing in temperature one from the next by an infinitesimal amount. This is the theoretical way that reversible changes must be presumed to come about. Although there is obviously no practical association with such requirements they can, in fact, be met surprising closely in practice under certain circumstances.

Irreversible effects in thermodynamics are supposed to arise from mechanical friction, as the mechanical parts move relative to each other, or in the general flow of a fluid by viscosity. They also arise by the conduction of heat in the presence of a temperature gradient, heat moving naturally down the temperature gradient. Any problems associated with the finite time available for heat energy to pass in practice are not accounted for.

1.4 The First Law of Thermodynamics: The Enthalpy

This concerns the energy content of a body. The formulation of the principle of the conservation of mechanical energy (a great achievement of the 18th Century) states that energy cannot be lost or gained in any closed system, but only transformed from one form into another. The First Law of Thermodynamics states that heat is a form of energy and is conserved. This means that any work energy that may be produced by an engine acting as a closed system can arise only from the conversion of existing energy into another form. Work

6   Thermodynamics: Heat and Work

cannot appear spontaneously without the transfer of energy; the production of work without the transfer of energy would provide perpetual motion 'of the first kind'. The First Law confirms generally, therefore, that such perpetual motion, which involves the creation of energy from nothing, cannot be achieved.

The Energy Equation

Suppose a quantity Q of heat is transferred to a closed body of mass M. As a result, the internal energy of the body is increased by the amount dU and the body performs the work W on its surroundings. The First Law expresses this relationship in the form

$$dU = Q - W$$

using the sign convention of Section 1.2. It will be convenient later to refer to unit mass of the body and we introduce the specific quantities u, q and w according to $U = Mu$, $Q = Mq$ and $W = Mw$, so that

$$du = q - w \tag{1.1}$$

This expression must be modified for an open system with material flows. Suppose the speed of the flow at the inlet is C_1, which is at a height z_1 above some zero elevation, and at the outlet is C_2 at the different height z_2. The acceleration of gravity is denoted by g and can be assumed constant for a given location. The sum of the kinetic and potential energies for unit mass flow of fluid at the inlet is $(C_1^2/2 + gz_1)$ while at the exit it is $(C_2^2/2 + gz_2)$: the difference represents the energy retained by the fluid. For the case of simple thermal expansion, the outward movement of the surface exerts the pressure p and expands through the distance l giving the volume $V = lA$ swept out by the surface. The work transfer is then 'force × distance' which is pV. If the entry pressure and specific volume of the fluid are respectively p_1 and v_1, while at the exit they are p_2 and v_2, the net work done on the surroundings by the fluid due to the energy change is the difference between that at the entrance and that at the exit. This can be shown to be $(p_2v_2 - p_1v_1)$. The energy de given to the system due to the flow is therefore

$$de = (1/2)(C_2^2 - C_1^2) + g(z_2 - z_1) + (p_2v_2 - p_1v_1) \tag{1.2}$$

The statement (1.1) for the closed system now becomes for the open system

$$du = q - w - de \tag{1.3}$$

with q being the heat received and w the mechanical work done on the surroundings. Eqn. (1.3) is called the steady flow energy equation and applies to many flow systems of engineering interest.

In the following chapters the inlet and outlet flows for the components taken as open systems will be the same and the difference of height between them will not be relevant. Then $de = (p_2v_2 - p_1v_1)$ and eqn. (1.3) reduces to the statement that, for the revesible change from state 1 to the state 2 due to the reception of heat q

$$u_2 - u_1 = q - w - (p_2v_2 - p_1v_1)$$

or

$$(u_2 + p_2v_2) - (u_1 + p_1v_1) = q - w \tag{1.4}$$

1.4 The First Law of Thermodynamics: The Enthalpy

The quantity $(u + pv)$ was called the total heat by Helmholtz but now is more usually called the enthalpy and is denoted by $h = u + pv$. With this notation

$$h_2 - h_1 = \mathrm{d}h = q - w \tag{1.5}$$

The enthalpy H for mass M of gas is $H = Mh$. Like the internal energy, the enthalpy is a characteristic of the state of the fluid and is defined relative to some state chosen as the standard.

Controlling the Flow

It is often necessary to channel the fluid in various ways and change its speed. Such processes can be described using the energy equation (1.2) with (1.3). There are three cases of special importance in thermal cycles.

The Throttle

This is a device for restricting the fluid flow: the simple screw valve of a tap is an immediate example. Such a restriction of the flow is valuable in low speed flows, such as for bled steam (see Chapter 13). No heat or work transfer occurs during the throttling process ($q = 0$; $w = 0$) and, since the flow is slow, there is essentially no change of kinetic energy of the flow. Again, the throttle is sufficiently small so that there is no change of gravitational energy on passing through. Eqn. (1.2) gives the change of energy of flow between the inlet and outlet of the throttle as $\mathrm{d}e = \mathrm{d}(pv)$, and this is combined with eqn. (1.3) to give $\mathrm{d}h = 0$. The enthalpy is constant throughout the throttling process so that the enthalpy h_1 on entry is equal to the enthalpy h_2 at the exit: the throttle is a constant enthalpy device and $h_1 = h_2$.

The Nozzle

It is often desirable in high speed flows to accelerate the fluid. A duct which is designed to accelerate the flow when a constant pressure drop is maintained across it is called a nozzle. The fluid acceleration is achieved by varying the cross-sectional area of the duct, and the simplest example is a convergent nozzle which is a duct of continuously decreasing cross-sectional area downstream. There is no work transfer across the boundary with the fluid ($w = 0$). Again, if the flow speed is sufficiently high there will be no heat transfer ($q = 0$) and the flow is adiabatic. The energy expression (1.2) becomes $\mathrm{d}e = (1/2)(C_2^2 - C_1^2) + \mathrm{d}(pv)$ where C_1 is the inlet speed and C_2 is the exit speed, which combines with eqn. (1.3) to give

$$(1/2)(C_2^2 - C_1^2) = (h_1 - h_2)$$

relating the exit conditions (2) to the inlet conditions (1).

A knowledge of the thermodynamic conditions at the inlet and outlet will allow C_2 to be deduced if C_1 is known. Usually only the outlet pressure is measured so the remaining thermodynamic information of the outlet side must be deduced in some other way. For this purpose it is usual to assume the flow to be reversible.

Irreversibilities (such as viscous friction) will reduce the speed of the fluid attained in passing through a given pressure drop. If C_2' is the exit speed in the presence of friction, the ratio $(C_2'/C_2)^2 < 1$ is often called the nozzle efficiency. This factor is typical of a given type of nozzle and can be found for that type by making measurements on a test nozzle: a typical nozzle efficiency will be about 0.95.

8   *Thermodynamics: Heat and Work*

The Diffuser
The flow of the fluid can be decreased by the inverse process of increasing the pressure through a duct, and this forms a diffuser. The largest pressure drop is required to provide a given decrease of the speed, and the simplest example is a diverging duct, where the cross-section continually increases downstream. The change of fluid kinetic energy is again expressed by the corresponding enthalpy change.

Irreversible processes also reduce the fluid speed so that the pressure drop is also reduced. The diffuser efficiency is often quoted as the ratio of the pressure drop associated with a given reduction of the speed with friction and without. The maximum efficiency is unity.

1.5   The Ideal Gas

Although the statements of thermodynamics are of wide generality it is often difficult to associate them with simple formulae because the equation of state of the matter involved is generally not amenable to being expressed in a simple and closed form. One exception is the indefinitely dilute gas, whose properties approach closely those of the ideal. This reason alone would make the expression of thermodynamics for the ideal gas of interest but there is another reason for us as well. Our main interest is in the application to gas and steam as working fluids and these are found to behave like the ideal to a first approximation.

Equation of State
The equation of state for the ideal gas is formed by combining Boyle's law (pressure p increases with decreasing volume at constant temperature, that is $pV = $ constant, where V is the volume of the gas) and Charles' law (pressure increases proportionally with the temperature T at constant volume). For a particular gas contained in volume V at temperature T, the pressure p is written

$$pV = R_1 T$$

where R_1 is a constant characteristic of that gas. The kinetic theory leads to the recognition of the universal gas constant $R = mR_1/M (= 8.314$ kJ/kg), where m is the molecular weight of the gas and M is the mass of gas present in the volume V. Then

$$pV = (M/m)RT \qquad (1.6)$$

For a mole of gas (a mass equal to the molecular weight in grammes: a kilomole refers to the molecular weight in kilogrammes) of volume $\tilde{V}$, $M = m$ (by definition) and the equation of state takes the universal form

$$p\tilde{V} = RT$$

This equation of state is sometimes called the Boyle's law equation of state or the ideal gas equation. The numerical value of R quoted above has been found experimentally from a range of real gases at vanishingly low pressures, when conditions closely approach the ideal.

An important property of an ideal gas (first recognised by Joule from experiments on real gases at very low pressures) is that the internal energy is independent of the volume

$$(\partial u/\partial v)_T = 0 \tag{1.7}$$

Adiabatic Conditions

The internal energy of a gas is dependent on temperature and volume which we can express by writing $u = u(T,v)$. For changes of the temperature at constant volume $du = (\partial u/\partial T)_v dT$. But, the specific heat capacity of the gas at constant volume C_v is $C_v = (\partial u/\partial T)_v$ so that

$$q = du + pdv = C_v dT + pdv \tag{1.8}$$

The specific heat capacity at constant pressure C_p is given by $C_p = (\partial h/\partial T)_p$.

If the reversible change undergone by a body does not involve the transfer of heat, so that $q = 0$, the change is said to be adiabatic. In this case, eqn. (1.8) gives

$$du = -w \qquad (q = 0)$$

and from eqn. (1.8)

$$0 = C_v dT + pdv \tag{1.9}$$

Introducing the equation of state (1.6) for the ideal gas we have instead, for unit mass of gas

$$0 = C_v dT + (M/m)RT(dv/v)$$

or, on dividing throughout by T,

$$0 = C_v(dT/T) + (M/m)R(dv/v) \tag{1.10}$$

This is an expression that can be easily integrated to provide information about the state of the gas before and after a reversible transition from the state 1 to the state 2. C_v is closely constant for changes of temperature and volume, and we find easily on integration that

$$C_v \log T + (M/m)R \log v = \text{constant} \tag{1.11}$$

Apparently, this expression remains constant during a reversible change of state without heat transfer. The value of this expression for the final state 2 is the same as that for the initial state 1.

The Entropy of the Gas

It is always of great importance to discover a quantity which remains constant during a transition of the state of a body and the expression (1.11) is one such. Clausius proposed that this constant should be called the entropy of the gas and be given a symbol: we choose s for the specific entropy (entropy per unit mass) and write

$$s = C_v \log T + (M/m)R \log v + \text{constant} \tag{1.12}$$

so that eqn. (1.10) becomes

$$ds = C_v(dT/T) + (M/m)R(dv/v) = 0$$

In moving reversibly from the initial state to the final state the entropy does not change. The entropy and the entropy change for the mass M of gas are then respectively $S = Ms$ and $dS = Mds$. According to eqn. (1.12), the entropy is itself defined up to an additive constant and so has a value determined entirely by the zero state chosen as the datum. In this respect it is the same as the internal energy and the enthalpy and, like them, can be called a state function because it does not depend upon the way the particular state has been reached. Eqn. (1.12) applies for adiabatic conditions at constant volume but an equivalent expression is derived later (Chapter 4, Section 4.1, eqn. (4.3)) for conditions of constant pressure.

By using eqns. (1.1) and (1.8), eqn. (1.12) can be rewritten in symbolic form

$$\int_1^2 \frac{dh + dw}{T} = \text{constant} \tag{1.13}$$

where w is the work done on the surrounding bodies per unit mass of the gas during the reversible change from state 1 to state 2. This allows the concept of entropy as a constant of a reversible adiabatic transformation, to be extended to any body even though its equation of state cannot be expressed in closed mathematical form.

With Heat Exchange

If unit mass of the gas now receives an infinitesimal quantity of heat q which causes an infinitesimal reversible change of state from state 1 to state 2, eqn. (1.1), using eqns. (1.8) and (1.13), is written:

$$\frac{dh + w}{T} = \frac{q}{T} = ds$$

and ds is the change of entropy of the gas in passing from the initial state 1 to the final state 2. Using eqn. (1.13) this is integrated to the form

$$ds(1,2) = \int_1^2 \frac{q}{T} = 0 \tag{1.14}$$

showing that the entropy is a constant of this reversible transition. This suggests defining more generally the infinitesimal change of entropy ds of a body associated with the move from an initial to a final state due to a transfer of heat q to the body as

$$ds = \frac{q}{T} \tag{1.15}$$

This definition of the entropy, through its change, was first proposed by Clausius. We can say that T is the integrating factor for q. q/T then becomes an exact differential, depending only on the initial and final states, whereas q (depending on the path linking these states) is not a complete differential.

While definition (1.15) is applicable to gases, it is not, in fact, generally applicable to all matter in this form. There is not a general thermodynamic expression which is valid in all cases and this makes the concept of entropy appear somewhat vague from this point of view. The situation is fully resolved, however, by statistical mechanical arguments describing the order/

disorder of the arrangements and motions of the atoms and molecules of which macroscopic matter is composed. This interpretation of entropy was made by Boltzmann who developed the associated concepts, which were extended later by Gibbs, but we cannot pursue this matter here. We need only realise that the arguments and formulae developed above are fully applicable to gases. In the more general case of condensed matter it is clear that the expression (1.13) has an integral whose value is constant for reversible changes of state but otherwise increases, so that the entropy can be defined there as well even though the describing expression cannot be written down explicitly.

Effect of Irreversibilities
Irreversibilities require the use of heat to overcome friction in the surrounding material so that the quantity of heat required to cause the state of a body to change from state 1 to state 2 is greater than if friction is not present. The work done on the surroundings depends on how the surroundings are changed and the entropy difference in passing from state 1 to state 2 depends upon the manner of the change. The entropy difference is not, now, a constant: indeed, the entropy difference is positive and we must write

$$(q/T)_{irr} > (q/T)_{rev}$$

Cyclic Changes
If, after a change of state, the thermodynamic state of a body returns exactly to its initial condition the change is said to be cyclic. If each element of the change is reversible, then so also is the total change. In this case the change of entropy is zero, according to eqn. (1.12), so that

$$\oint (q/T)_{rev} = 0$$

where the integration symbol denotes the cyclic nature of the change. If the change over the cycle is not reversible, not being between a succession of equilibrium states, the entropy change on moving from state 1 to state 2 is less than that involved with moving from state 2 to state 1. Consequently we have

$$\oint (q/T)_{irr} < 0$$

The cyclic nature of the integration can only apply to the body since the changes are not, in fact, reversible and the surroundings must remain affected by it.

1.6 The Second Law of Thermodynamics

Much that has been said so far applies to the special case of an ideal gas but the arguments have wider validity and apply quite generally to any matter. This is the content of the Second Law of Thermodynamics. It can be expressed in many equivalent forms. One form, originally due to Max Planck, is 'It is impossible to construct an engine working continuously that has no other effect than to receive heat from a source and raise a weight'.

Raising a weight is the generic expression for doing work on the surroundings and this means that it is not possible to provide work by simply cooling a body. If it were possible, an indefinite quantity of work could be obtained by cooling the oceans or cooling the Earth itself but no one has yet discovered how to achieve this dream, (or nightmare?): the Law says a method never will be discovered. If this was possible perpetual motion described as 'of the second kind' would be obtained: this is motion involving the annihilation of entropy with no other effect. The Second Law prohibits perpetual motion of the second kind.

There is a range of logical consequences that follows from the Second Law and we state a few now without proof. The proofs will be found in text books on thermodynamics. In general terms the proofs take the form of showing that if the inverse of a proposition were true then it would be possible to make an engine to provide perpetual motion of the second kind, which by the Second Law is impossible. Notice that the energy balance for such perpetual motion need not contravene the requirements of the First Law.

Consequence 1. An engine that produces work must reject heat energy as well as receive it: every engine requires a hot source and a relatively colder heat sink to function. This conclusion will be considered further in Chapter 2. The Second Law can then be stated in the form 'Every engine working continuously must involve an active exhaust pipe'. It seems there will always be at least a potential problem of pollution.

Consequence 2. While work energy can be completely converted into heat energy, heat energy cannot be completely converted into work energy.

Consequence 3. If the thermal efficiency of an engine is measured by the ratio (work output)/(energy input), it follows that no engine working continuously can have unit efficiency.

Consequence 4. The most efficient engine linking a given fixed upper temperature and a given fixed lower temperature is that moving reversibly between the linking heat stages. The working fluid is immaterial providing it is ideal in showing no mechanical or thermal inefficiencies. This engine was studied originally by Carnot and is explained in Chapter 2.

Consequence 5. All reversible engines working between the same two temperature reservoirs have the same thermal efficiency.

Consequence 6. It is possible to construct a scale of temperature which is independent of a particular thermometric material and is in this sense absolute. This provides also an absolute zero of temperature.

Another statement of the Second Law is that due to Clausius: 'The entropy of any closed system, thermally isolated from its surroundings, never decreases, remaining constant if the process undergone by the system is reversible or increasing if it is not'. All the consequences listed above follow equally from this statement as does the formulation of the Law given by Planck.

We will quote one further alternative statement of relevance to refrigeration:

'It is impossible for a self-acting machine, unaided by any external agency, to convey heat from one body to another at a higher temperature'.

This refers to the fact that heat energy, free to flow, moves naturally from a hotter to a colder body and can be made to move in the reverse direction only by the performance of external work.

Consequence 7. An ideal engine working backwards will absorb external work and cause energy to pass from a colder to a hotter body. This is the basis of the ideal refrigerator, although no real refrigerator is ideal.

It is often convenient to regard the First and Second Laws together as an expression of the statement that it is impossible to obtain perpetual motion of any form.

1.7 Exercises

1.1 By using the library, follow through the arguments that lead to the Consequences 1 to 7. The general approach is to accept the proposition that is to be disproved as true, and show that the second law is not then obeyed.

1.2 Water (50 kg) at the liquid saturation temperature is heated at 60 bar pressure to the vapour saturation condition (i.e. is vaporised). If the heat energy is added reversibly, calculate the entropy change of the water. Comment on your result.

1.3 The 50 kg of water of the last question is now cooled reversibly to 272 K. Calculate the change of entropy of the water.

1.4 Without referring to Chapter 4, derive the expression for the change of entropy of an ideal gas for a reversible change of state at constant volume, analogous to eqn. (1.12) of the main text for constant pressure. How might this be used to construct a graph showing the dependence of the entropy on the pressure at constant volume? (You can store this result for use later when you read Chapter 4.)

2. ESSENTIAL CHARACTERISTICS OF THE CARNOT CYCLE: THE AVAILABILITY

Sadi Carnot was a military engineer in early nineteenth century France who wondered which design would provide the most efficient engine. This set him on the road to discovering his now famous cycle and led to the founding of modern thermodynamics. Sadi Carnot's studies were first published in 1824 and it is interesting that the beginnings of thermodynamics, now one of the universal cornerstones of all science and technology, began with the study of the immediate problems of applying steam to produce work.

The thermodynamic study of the production of work is based on the consequences of the Second Law of Thermodynamics, set out in Chapter 1, Section 1.6. The best thermodynamic engine is the reversible heat engine described by Carnot, and although Carnot actually supported a theory of heat now known to be wrong, it was Clausius and, independently Thomson, who recognised the full generality of Carnot's arguments.

2.1 The Carnot Engine

Consider Carnot's arguments in modern terms of a simple engineering system, namely a piston moving in a sleeve. Working fluid (for instance, steam or air) is introduced into the piston volume and energy can also be fed in or removed. This is shown in Figure 2.1. We shall need to couple the piston system to a particular environment and it will be necessary sometimes to move the piston while keeping the temperature constant. This means we need to have available a very large energy reservoir of prescribed temperature. Ideally this will need to be of infinite capacity if the temperature of the piston is to be kept strictly constant but in practice it will be sufficient if it is simply very large.

How large is very large? Large enough to keep the temperature constant *within the accuracy of the measurements being taken*. The degree of error allowable in measurements is extremely important in practice because it allows idealised arguments to be applied to the real world within an acceptable degree of approximation. This acceptance of error is used widely in the application of the arguments of thermodynamics to the flow of fluids but we will not elaborate on this here.

In order to make clear the principles of the action of the piston system that we have, it is reasonable to suppose that irreversible effects such as friction and the conduction of heat other than at the reservoirs are negligibly small. Indeed, we will presume the ideal situation and suppose that such effects are absent altogether: there is no irreversibility. Again, we will suppose that the motion of the piston does not introduce significant random motions into the

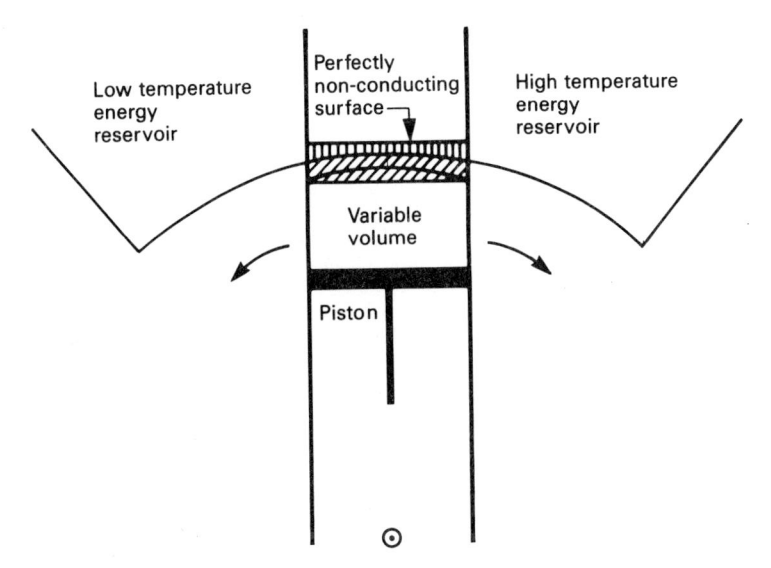

Fig. 2.1 The Piston System. The piston casing can be swung to be in contact with a high temperature reservoir, a low temperature reservoir or with a surface that does not conduct heat.

working fluid as it moves and that irreversible effects such as viscosity can be neglected. This is a further idealisation: we suppose that the working fluid behaves as an ideal fluid without irreversible effects. These various idealisations can often be approximately achieved in real life so that assuming that they are true here does not detract from the ultimate practical value of any deductions we might make—except that we must be careful in applying them to all cases.

Coming back to the Carnot engine, we have an ideal friction-free piston assembly, illustrated in Figure 2.1, containing an ideal working fluid and coupled to infinite energy reservoirs to control the temperature. We shall need one more facility: it will be necessary to isolate the piston system from time to time so that it neither gains nor loses heat through the walls from or to the outside. Again in practical terms this can often be done approximately but we will suppose now that we can do it exactly as part of our ideal apparatus.

2.2 The Carnot Cycle

Having got all this ideal system together let us make simple use of it. We will take the working fluid through a simple set of operations (called a cycle because the working field is in exactly the same state at the end as at the beginning). More specifically, we will raise and lower the temperature (using a hot and a cold heat reservoir) under conditions to be specified in a moment. It is convenient to represent the cycle in the form of a graph and there are various ways of doing this. For the moment we will plot the pressure against the volume, as in Figure 2.2.

Start from some initial setting of the piston so that the working fluid is at some specified low temperature achieved by coupling it to a cold reservoir of

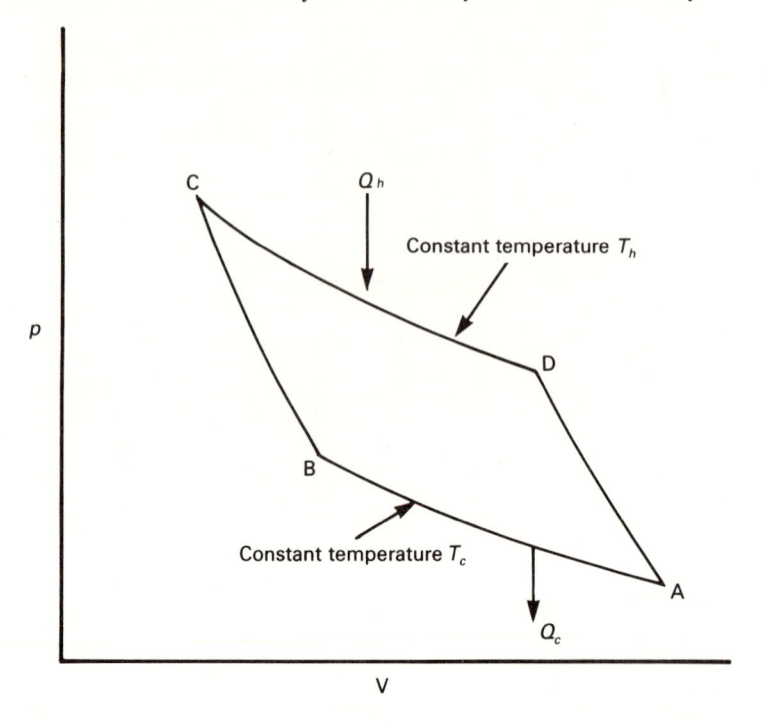

16 Essential Characteristics of the Carnot Cycle: The Availability

Fig. 2.2 The presssure-volume diagram for the Carnot cycle.

heat energy. The volume for a given mass of fluid is at its greatest. This initial condition of the fluid (called its 'state') is marked by point A in Figure 2.2 and is the initial 'state' of the working fluid.

The piston system is kept coupled to the heat reservoir and the piston pushed in so that the fluid is compressed at constant temperature. This must involve the use of a force external to the system which performs work on the system, and being at constant temperature, is called isothermal compression. Because work is done *on* the fluid, the fluid *rejects* heat to the external heat reservoir. The isothermal compression is continued until the fluid has reached a pre-arranged volume: the fluid is then in the state B in Figure 2.2.

The next stage is to compress the gas further but now without the passage of heat, and this is achieved by uncoupling it from the heat reservoir and instead insulating it totally from the outside. No heat is given up by the fluid during this process, which is consequently called adiabatic, and the gas heats up during this phase because energy of compression is added to it and no energy can leave it: the only thing the energy can do is heat the fluid. This stage of compression is continued until the temperature of the gas has been raised to some pre-arranged upper value. This is point C in Figure 2.2.

In relation to its initial state, the gas has now been compressed isothermally (when it rejected heat to a heat reservoir) and compressed adiabatically. To produce a cycle we must return the gas to its initial state and this means we must now allow it to expand. This is done in two stages.

The first expansion stage is to couple the piston system to a heat reservoir at the same temperature so that the compressed gas can expand at the constant higher temperature (that is isothermally). This is continued until the gas can be returned to its initial condition again by an adiabatic expansion, the piston system being uncoupled again from the heat reservoirs to allow this to happen. The isothermal expansion then moves from point C to point D in Figure 2.2. The final adiabatic expansion then is from point D to point A, the starting point.

The cycle we have described is the Carnot cycle. It can be started anywhere in the cycle, being brought back to the starting point always by one adiabatic expansion and one adiabatic compression (with no heat flow) together with one isothermal expansion (the fluid takes in heat from the external reservoir) and one isothermal compression (the fluid now expelling heat to an external reservoir).

2.3 The Efficiency of the Carnot Engine: Temperature Scales

The efficiency of the engine is measured by the amount of work that can be extracted per unit amount of energy used in the process. If energy Q_h is taken in at the hot end and energy Q_c is rejected at the cold end, the energy available to do work is $W = Q_h - Q_c$. The heat is put into the engine at the high temperature side so that the thermal efficiency η of the engine is defined as

$$\eta = W/Q_h = [Q_h - Q_c]/Q_h \tag{2.1}$$

This is alternatively written

$$\eta = 1 - Q_c/Q_h$$

Because there must always be some rejection of heat $Q_c > 0$ and $\eta < 1$. It is not possible, even as a matter of principle, to achieve a unit efficiency in the engine.

We have said earlier that the quantity of heat energy in a body is measured by its temperature and the heat flows can be used to define a temperature scale which is independent of the particular working fluid used (that is of the thermometer employed) and so is absolute in this sense.

The ratio of the energies absorbed and rejected in the Carnot cycle is equated to the ratio of the corresponding temperatures of the hot source T_h and the cold source T_c: then $Q_h/Q_c = T_h/T_c$. If the hot and cold ends are associated with some fixed thermometry points (for instance the triple point of water or the boiling and freezing points of pure water under standard conditions of pressure) there is, here, a method of determining the ratio of the hot and cold stages. Dividing this range up into a number of equal subranges (for the Celsius scale this is 100 stages between the boiling and melting points for pure water under standard conditions of pressure) we are able to establish a practical scale of temperature.

This allows us to rewrite the expression for the efficiency of the Carnot cycle in terms of the high and low temperatures

$$\eta = [T_h - T_c]/T_h = 1 - T_c/T_h \tag{2.2}$$

2.4 The Carnot Refrigerator

The Carnot engine is presumed to be without friction or irreversibility and so is entirely reversible, in the thermodynamic sense. If its action is then reversed work is *applied* to the engine with the result that heat passes from the cold reservoir to the hot one. The cold reservoir gets colder: this is the action of a refrigerator.

The effectiveness of the refrigerator is measured by the Carnot Performance Coefficient (*CPC*) defined as the flow of heat from the cold reservoir to the hot one per unit performance of work. This is

$$CPC = [Q_h - Q_c]/W = T_h/[T_h - T_c] \qquad (2.3)$$

It is seen that the *CPC* depends only on the temperatures of the hot and cold reservoirs and this is the performance coefficient of any reversible refrigerator working between the given upper and lower temperatures. No real refrigerator can have a lower *CPC* and generally will have a higher one.

The minimum amount of work necessary to remove a given quantity of heat Q_c from the cold reservoir is then given by

$$W = Q_c[T_h - T_c]/T_h \qquad (2.4)$$

A real refrigerator does not act entirely reversibly and so would require the expenditure of more work than this to remove the heat Q_c.

We see that the work required increases as the temperature difference increases and also increases as T_c decreases. We can also notice that an infinite amount of work would be required to reduce the cold reservoir to the absolute zero of temperature ($T_c = 0$), making the absolute zero unattainable. Indeed, such a statement is often quoted as an alternative statement of the Second Law of Thermodynamics.

The cooling process can be viewed alternatively as a heating process for the hot reservoir. In this case the temperature of the hot side is raised by transferring heat from the cold side by the performance of external work. Looked at in this way the Carnot engine working backwards acts as a heat engine warming the hot side. The *CPC* coefficient is also applicable here and the heating engine is called a heat pump.

2.5 The Availability of a System

The Carnot cycle provides the maximum thermal efficiency of any possible cycle but this does not mean that it also ensures the maximum possible work output from any engine based upon it. Indeed the relatively poor working performance of the Carnot cycle compared with that which can be achieved in practical plant is seen by comparing the various results of Chapters 3 and 7. We shall be concerned in the following Chapters with the useful work that can be derived from a variety of real engines in interaction with the environment, the heat acceptance and rejection being at constant pressure. It is important, in this context, to explore thermodynamic arguments to find the maximum amount of useful work that can be derived from a particular thermal system in interaction with its environment at the constant temperature T_e and the constant pressure p_e. Thermodynamics allows this to be done through a function called the availability, or sometimes the exergy.

2.5 The Availability of a System

Maximum Work from a Heat Reservoir
It is useful to remember (Section 2.3 dealing with the Carnot efficiency) that there is a maximum of work W_{max} obtainable by the extraction of a quantity of heat Q from a heat reservoir at temperature T_h in an environment at the lower temperature T_e. The heat and work transfers must be achieved reversibly and then $W_{max} = Q[(T_h - T_e)/T_h]$. This quantity is the theoretical availability of work within the reservoir. The available work of a real reservoir, involving irreversible heat and work transfer, is less than this.

A power plant can be regarded as a system involving the passage of heat from a hot reservoir to the environment acting as the cold reservoir: the associated work transfer is available for use. The action of the plant, then, is to reduce the initial thermodynamic state denoted by (p,T) to that of the environment, denoted by (p_e, T_e). Once the environmental condition is reached no further work can be obtained. Fuel is put into the plant in order to counter this move to the environmental conditions and the steady working is achieved when the rate of approach to the environmental state determined by the design of the engine is precisely balanced by the consumption of fuel. An estimate of the effectiveness of any chosen power plant can be made by comparing the work it actually produces with the maximum quantity that could be expected on the basis of reversible thermodynamics.

Thermodynamic Potentials
Conditions for stability (which must not apply for conditions of change) are expressed quite generally in terms of a minimum value of an appropriate quantity: in thermodynamics, the appropriate quantity is a thermodynamic potential.

A Dual Transformation
Suppose there is some quantity X (later we will interpret this as a thermodynamic state variable; u, v and w in this subsection are mathematical symbols only) which depends on two variables u and w (say): we write this $X = X(u,w)$. Let us change the variable u by the small amount du, keeping the other variable w fixed—it is natural to call u the active variable and w the passive variable. The result is the small change $dX = (\partial X/\partial u)_w du$ of the quantity X. The subscript w reminds us that w is a constant of the change. Let us write $v = (\partial X/\partial u)_w$ so that $dX = vdu$. Let there be a second quantity Y which depends on v and w, that is $Y = Y(v,w)$. A small change of v, again keeping w constant, gives $dY = (\partial Y/\partial v)_w dv$. Suppose, now, that v is chosen such that $(\partial Y/\partial v)_w = u$. This will be the case if Y is chosen so that $Y = uv - X$. To see that this is, indeed, the case let both u and v change to give $dY = udv + vdu - dX = udv + vdu - udv = vdu$, as we know is the case. This whole procedure is called a dual transformation.

The Free Energies
Let us apply this transformation to thermodynamic variables for a system in equilibrium at the temperature T. Begin with the specific internal energy u for the system. This will have a kinetic contribution (associated with the random motion of the atoms and molecules of which it is composed) and a potential contribution (associated with the interatomic forces). The random motion we shall specify by the specific entropy s, while the effect of the

inter-particle forces (depending on how close the atoms are on the average) will be specified by the volume v. Then, $u = u(s,v)$. Apply the dual transformation of the last sub-section, choosing s to be the active variable and v the passive variable. A small change of entropy ds results in a small change du of the internal energy so that

$$du = (\partial u/\partial s)_v ds$$

But $T = (\partial u/\partial s)_v$ so that $du = Tds$. Here u is to be identified with A in the sub-section so that $B = Ts - u$ is the second quantity. This is conventionally written $-F(T,v)$ and F is called the Helmholtz free energy and, notice, applies to constant volume. The dual transformation is completed when we note that $s = -(\partial F/\partial T)_v$ as required. This allows us to write

$$F = u - T(\partial F/\partial T)_v \tag{2.5}$$

The procedure can be repeated starting with $-F(T,v)$. Let us now choose v to be the active variable and T the passive variable and remember that $p = -(\partial F/\partial v)_s$. The result is the new quantity $G(T, p) = pv + F$, called the Gibbs free energy referring to constant pressure. Because $p = -(\partial G/\partial v)_s$ we can write alternatively

$$G = F - v(\partial G/\partial v)_s \tag{2.6}$$

Interpretation of the Free Energies

It was seen in Section 1.5 that if a material is transformed from an initial state 1 to a final state 2 at the temperature T, the change of entropy ds is related to the corresponding change of energy dq by $dq \leq Tds$. The equality sign applies if the change is entirely reversible. The statement of the First Law, eqn. (1.1), for a gas can then be rewritten in the form

$$Tds \geq du + pdv \tag{2.7}$$

and this controls the nature of the change of state from state 1 to state 2.

The Helmholtz free energy $F = u - Ts$ will change during a transformation from the initial state 1 to the final state 2 by amount $dF = du - Tds - sdT$. Introducing eqn. (2.7) this becomes

$$dF \leq -pdv - sdT$$

For an isothermal process ($dT = 0$) at constant volume ($dv = 0$) it follows that

$$dF \leq 0 \tag{2.8}$$

This means that the Helmholtz free energy cannot increase, and must either remain constant (reversible process) or decrease. Consequently, if F has a minimum value no change is possible and the thermodynamic system is stable.

An identical argument involving the Gibbs free energy G can be applied for an isothermal change ($dT = 0$) and at constant pressure ($dp = 0$) giving the condition

$$dG \leq 0 \tag{2.9}$$

The stable thermodynamic condition is that G should have a minimum value. We are familiar in rational mechanics with the result that a mechanical system is in stable mechanical equilibrium if the potential energy has a minimum value. Because the same condition obtains in thermodynamics for equilibrium, eqns. (2.8) and (2.9), the free energies F and G are sometimes referred to as the thermodynamic potentials. An engine linking an initial and a final state will provide work output only if the initial Gibbs free energy does not have a minimum value, so that we are particularly concerned with an initial condition of thermodynamic instability from this point of view.

The Availability Function

In the power cycles that will concern us in the chapters to follow, the heat transfer is achieved at constant pressure so it is the Gibbs function $G(p,T)$ that is of special interest in describing the system. We have seen that work is obtained from the system only if G decreases during the thermodynamic change and this requires u and p to be greater initially than finally. It is for this reason that an engine requires a fuel for combustion and some compression process. The work output increases as the initial enthalpy and pressure are (often independently) increased. The precise form taken by G depends upon whether the system is open or closed.

For a Closed System

The plant will use the environment for heat rejection and the environmental condition determines the availability. In particular, the pressure p_e and temperature T_e are the controlling variables. The Gibbs free energy now takes a special form, denoted by

$$A_e = u + p_e v - T_e s \tag{2.10}$$

and called the non-flow availability function, and is a property of both the system and of the environment. The change of the thermodynamic state of the system from the initial state 1 to the final state 2 requires that $A_1 > A_2$: if state 2 is the environment the difference $A_1 - A_e$ is often called the exergy for the change.

For an Open System

In an engine fuel is put in and work is taken out: this means that such a system is open even though its components may individually form closed systems. The open system differs from the closed system in that the final *pressure* is not necessarily that of the environment, in which case the Gibbs energy is called the steady-flow availability function, denoted by B, and is

$$B = u + pv - T_e s \tag{2.9}$$

B differs from A in the term involving the pressure. As for A so for B, the system will provide a work output in passing from state 1 to state 2 only if $B_1 > B_2$: again $B_1 - B_e$ is the exergy of the process.

Restrictions on the Availability

The maximum work that can be taken from an energy source is given by the algebraic differences between the various availabilities which characterise its various components. It is useful to notice here that the overall availability that can be used in a particular case will depend on the thermodynamic details employed. The thermodynamics of the combustion processes (not considered here) can themselves be the source of restrictions. For the Carnot cycle the availability of the hot reservoir is a determining factor in this connection because heat energy can only be received at the high temperature, and the combustion gases that are rejected to the environment contain considerable enthalpy that cannot be used. It is this feature that is largely responsible for the poor work output of the Carnot cycle. In practical engines regenerative heating processes will provide similar restrictions on the total usable availability (see Chapter 14).

2.6 Summary

1. The Carnot engine, working between an upper temperature T_h and a lower temperature T_c (being the temperature difference $dT_{Carnot} = T_h - T_c$) represents the (unattainable) ideal as a source of work. Energy Q_h is withdrawn from the high temperature source and energy Q_c is rejected to the low temperature source: the difference is available to provide work W, so that $Q_h = Q_c + W$. (The difference $(Q_h - Q_c)$ is sometimes referred to as the available energy of the engine.)
2. No engine can convert all the heat that it receives into work: there is always a prescribed quantity of energy rejected. The more efficient the engine, the less heat is rejected. But, every engine must have an active exhaust pipe.
3. The thermal efficiency η of the reversible engine is

    $$\eta = [Q_h - Q_c]/Q_h < 1$$

    between the same temperature difference dT_{Carnot}, irrespective of the working fluid provided the fluid is ideal in its properties.
4. The thermal efficiency of a reversible engine working across a given temperature difference cannot be exceeded by any real engine working through the same temperature difference.
5. The most efficient way of producing power continuously is to use a Carnot engine working continuously.
6. A refrigerator system is produced by working the Carnot engine backwards, heat now being forced from the colder to the hotter reservoir by the performance of external work W.
7. The performance of the refrigerator is measured by the Carnot Performance Coefficient $(CPC) = Q_c/W$.
8. No refrigerator can have a better performance coefficient than that of the Carnot refrigerator.
9. The Carnot cycle (which is closed—the final state coincides identically with the initial state) gives no net change in the entropy of the system. This is a characteristic of reversible processes.
10. The thermodynamic free energies (or potentials) are recognised as a means of specifying the conditions necessary for an engine to perform useful work. The usefulness of a modified form of the Gibbs free energy, and called the availability function, is recognised and defined separately for closed and open thermodynamic systems.
11. The difference between the availability of an initial state and that of the environment as the final state is called the exergy of the system.

2.7 Exercises

2.1 A Carnot heat engine receives 12.55 kW from a hot reservoir and rejects 3.35 kW to a cold reservoir. Show that the thermal efficiency of the engine is 0.73, and that the maximum useful work that can be extracted from the engine is 9.2 kW.

2.2 Heat is supplied to a Carnot engine at the rate of 10.46 kW and is rejected by the engine at the rate of 2.93 kW. Show that the efficiency of the engine is 0.72 and that the work output of the engine is 7.52 kW.

2.3 A heat engine absorbs 12.55 kW and rejects 3.35 kW. If the useful work output of the engine is 8 kW show that the rate of dissipation of mechanical energy in the engine is 1.2 kW. Show further that the thermal efficiency of the engine is 0.64.

2.4 The Carnot engine in Question 1 is run backwards so that 3.35 kJ are absorbed from the cold reservoir and 12.55 kJ are rejected to the hot reservoir by the expenditure of 9.2 kJ of external work. Show that the Carnot Performance Coefficient for the process is 1.37. If, in an actual refrigerator, 1.2 kJ were expended in overcoming the dissipative processes in the engine, deduce that the Performance Coefficient of the refrigerator then would be 1.55.

2.5 The hot reservoir of a Carnot engine is held at a temperature of 100 °C. What must the temperature of the cold reservoir be if the thermal efficiency of the engine is to be 0.75? If the temperature of the cold reservoir is changed to 0 °C, show that the efficiency of the engine is 0.268. What is the useful work output of the engine per kilogramme of working fluid (specific work output) if helium gas is used as the working fluid.

2.6 The Carnot engine of Question 2 is run backwards to provide refrigeration. How much ice at 0 °C can be produced at the cold reservoir? (The density of ice can be taken to be 1×10^3 kg/m^3 and the specific heat capacity to be 1 J/kg K.) What would be the amount of ice produced if the temperature were instead lowered to −2 °C?

2.7 The Carnot engine of Question 2 is run backwards to provide a heat pump for the hot end. The cold reservoir is held constant at 0 °C and heat is to be supplied to a body of mass 1 kg, of density 3×10^3 kg/m^3 and specific heat capacity 1 J/kg K. What is the temperature rise of the body?

2.8 A Carnot heat engine absorbs 12.55 kJ from the hot reservoir at temperature 100 °C and rejects 2.93 kJ at the cold reservoir. Calculate the temperature of the cold reservoir and the change of entropy for this transfer of heat.

2.9 It is required to increase the efficiency of a Carnot engine by *either* raising the temperature of the hot reservoir *or* by decreasing the temperature of the cold reservoir. Which temperature change will increase the efficiency more readily?

2.10 With the result of Question 9 in mind, can you see a reason for an airline choosing a polar route rather than an equatorial route for travel between a city in North America and a city in Europe? What other factors would be involved in the choice?

2.11 Unit mass of a fluid is held compressed in a cylinder by a piston at pressure p_1, volume v_1, temperature T_1 and internal energy u_1. The piston is now released and the initial compression in the fluid falls until the conditions of the constant environment are reached with pressure p_e, volume v_e, temperature T_e and internal energy u_e. If q is the quantity of heat transferred during the process, show:

(i) the work done by the fluid during the expansion is
$$W = -(u_e - u_1) + q$$
(ii) the work done on the environment is $p_e(v_e - v_1)$
(iii) the entropy change of the system is $s_e - s_1 > q/T_e$
(iv) the useful work done on the piston is
$$W \leqslant -(u_e - u_1) + T_e(s_e - s_1) - p_e(v_e - v_1)$$
(v) the *maximum* work that can be done by the system is
$$W_{max} = A_1 - A_e$$
where A is the non-flow availability function defined by eqn. (2.10) of the main text.

By drawing an appropriate *T–s* diagram, substantiate the statement that the path from state 1 to state e can be reversible *only* if the heat exchange between the system and the environment occurs (a) at the single point where the system has the temperature T_e, or (b) throughout the expansion if the system is linked to the environment through a series of Carnot engines working between the full range of intermediate temperatures $T \leqslant T_e$, and T_e.

PART I. CYCLES USING GAS AS THE WORKING FLUID

Prologue

In practice the transfer of heat to and from the working fluid of a gas cycle is achieved by heat conduction through extended boundary surfaces. This involves everywhere a temperature gradient at the surface and the heat transfer conditions cannot, therefore, be isothermal as would be required for a Carnot engine. For a reasonable rate of heat transfer the temperature gradient must not be too small (perhaps in excess of 10 °C) and there can be no question of whether the conditions are thermodynamically reversible. The heat transfer process will, on the other hand, be at constant pressure to a high degree of approximation. Such a cycle is called a constant pressure cycle and the gas turbine is an example of its application. The Carnot principle has no direct practical application for constant pressure cycles (and so to gas turbine engines) but it does provide a theoretical datum for the evaluation of the properties of a particular practical cycle. More than this, the Carnot analysis provides a guide for the modification of a practical cycle to improve its working characteristics in a systematic way. This is the topic of this Part, which begins with a survey of the Carnot gas engine and then considers the constant pressure cycle and its various modifications for improvement.

3. THE CARNOT ENGINE WITH GAS AS THE WORKING FLUID

A Carnot engine is a cyclic, reversible engine which exchanges heat with its environment at two, and only two, constant temperatures. Its working properties are independent of the actual working fluid involved provided the fluid is ideal. Unfortunately there is no such thing as an ideal fluid: it would be composed of molecules that have no volume and which do not interact one with the other. Fortunately, real gases have properties that imitate the ideal sufficiently well for preliminary calculations. In practice we most readily use either air or steam as the working fluid for these fluids are plentiful, cheap and chemically harmless. If the physical properties of the particular working fluid are specified the behaviour of the Carnot cycle can be predicted in quantitative form because the equation of state of the fluid is known. The various contributions to work or energy transfer are then expressed numerically in terms of the properties of the working fluid. In the present chapter we consider the features of the engine using an ideal gas as the working fluid to obtain a view of the ideal behaviour of a vapour cycle. The distinction between a gas and a vapour is historical: gases have critical temperatures below those readily attainable and so are not usually seen in the condensed form, while vapours have higher critical temperatures and are ordinarily associated with their liquids. The use of the term vapour is still convenient when the liquid phase is involved as well as the gas, as is the case for steam (see Part II).

The effects of deviations from the ideal gas behaviour are considered briefly first and the use of the ideal equation of state for calculations is substantiated. The case of steam as the working fluid is considered later, in Chapter 10.

3.1 Work Transfer and an Ideal Gas

The Carnot cycle described in Section 2.2 involves the work transfer for isothermal and adiabatic conditions and these expressions must be deduced in explicit form for the gas. It is usual to simplify the analysis by supposing the gas to have ideal properties and it is then necessary to assess the extent to which a real gas mimics the properties of the ideal.

The Validity of Using the Ideal Equation of State
It was seen in Section 1.5 that the equation of state for a mass M of gas of molecular weight m has the form, when the gas pressure is sufficiently low.

$$pV = (M/m)RT \qquad (3.1)$$

where R is the universal gas constant.

One measure of the deviation of the properties of a real gas from the ideal

28 The Carnot Engine with Gas as the Working Fluid

is given by forming the quantity $(p\tilde{V}/RT-1)$ for the gas under practical conditions: the greater this quantity differs from zero, the less 'ideal' is the behaviour. Such data for helium, argon and nitrogen are collected in Table 3.1, where it is seen that these real gases show significant deviations from the ideal behaviour only for lower temperatures and higher pressures. It will be noticed that the sign of the numerical deviation changes with temperature and pressure of a given gas, indicating rather complicated behaviour, and this will be considered further in Section 3.4. For the present it is accepted that the approximation of assuming the gas to be ideal is sufficient unless extreme accuracy of calculation is required.

Table 3.1 Experimental values for $[(p\tilde{V}/RT)-1]$ for helium, argon and nitrogen at three pressures and three temperatures. Dashes denote liquid regions.

Substance	$T_0 K$	[$p\tilde{V}/RT-1$] 1 bar	50 bars	1000 bars
helium	100	+0.0023	+0.113	—
	300	+0.0005	+0.24	+0.439
	500	+0.0003	+0.013	+0.242
argon	100	−0.0217	—	—
	300	−0.0001	−0.029	+0.675
	500	+0.0001	+0.009	+0.353
nitrogen	100	−0.010	—	—
	300	−0.002	−0.004	+0.99
	500	+0.0004	+0.021	+0.82

Reversible Adiabatic Conditions

A special relation applies between pressure p for a volume V (not necessarily the molar volume), under adiabatic conditions which are also reversible. To obtain this we remember the eqn. (1.8) describing the change of the temperature dT and specific volume dv of the gas as the result of receiving the quantity of heat q at constant volume:

$$q = C_V dT + p dv$$

Because $d(pv) = pdv + vdp$, at constant pressure this becomes

$$q = C_p dT - v dp$$

For adiabatic conditions ($q = 0$ in each case) these two expressions readily rearrange to the form

$$C_p/C_V = \gamma = -(vdp)/(pdv)$$

the quantity γ defining the ratio of the specific heat capacities at constant pressure and at constant volume. This can usually be taken to be constant so that the expression is easily integrated to give the relation between pressure and volume

$$pv^\gamma = \text{constant}, A \text{ (say)} \tag{3.2}$$

The constant A in eqn. (3.2) is found by knowing the pressure for a particular volume with γ known for the gas.

Expressions Describing Work Transfers

The (pressure–volume) work done dW if the gas is expanded from the initial volume V to the infinitesimally greater final volume $V+dV$ at constant pressure is known from the First Law of Thermodynamics to be given by

$$dW = pdV \tag{3.3}$$

For a finite expansion from the initial volume V_1 to the final volume V_2 the work done $W(2,1)$ is now finite and is obtained by summing together all the infinitesimal contributions to the work along the way. In mathematical language, $W(2,1)$ is obtained from dW by integration over the volume from V_1 to V_2. For isothermal conditions, the expansion taking place at the constant temperature T, eqn. (3.1), applies and the isothermal work done is

$$W_I(2,1) = \int_1^2 pdv = (M/m)R \int_1^2 dv/v$$

$$= (M/m)RT \log(V_2/V_1) \tag{3.4}$$

For adiabatic conditions on the other hand, no heat enters or leaves the gas during the expansion and the temperature must change. Eqn. (3.2) applies and introducing eqn. (3.3) gives instead for the adiabatic work

$$W_A(2,1) = \int_1^2 pdv = A\int_1^2 dv/v = A\int_1^2 v^{-\gamma}dv$$

Integration gives the expression

$$W_A(2,1) = C_V(T_1 - T_2) \tag{3.5}$$

This involves the initial temperature T_1 and the final temperature T_2 for a particular gas.

For the Carnot cycle (see Figure 3.1), we can now express the work done by the gas, and the work done on it, together with the heat inflow and outflow to and from the piston system, in terms of the appropriate volumes and temperatures. Remember from Section 2.4 that there are two isothermal phases and two adiabatic phases in the Carnot cycle. Repeating the cycle developed in Section 2.3, the first stage is an isothermal compression at the lower temperature T_1 (say) from point 4 to point 1. This stage involves the performance of work on the gas of amount

$$W_I(1,4) = -(M/m)RT_1 \log(V_4/V_1) \tag{3.6}$$

The next stage is an adiabatic compression from point 1 to point 2 during which the gas moves from the lower temperature T_1 to the higher temperature T_2. From (3.4), this involves the work $W_A(1,2)$ being done on the gas of amount

$$W_A(2,1) = -C_V(T_2 - T_1) \tag{3.7}$$

the negative sign denoting work done on the system. The next two stages are expansion stages. The first is an isothermal expansion from 2 to 3 during which the *gas* does work

$$W_I(3,2) = (M/m)RT_2 \log(V_3/V_2) \tag{3.8}$$

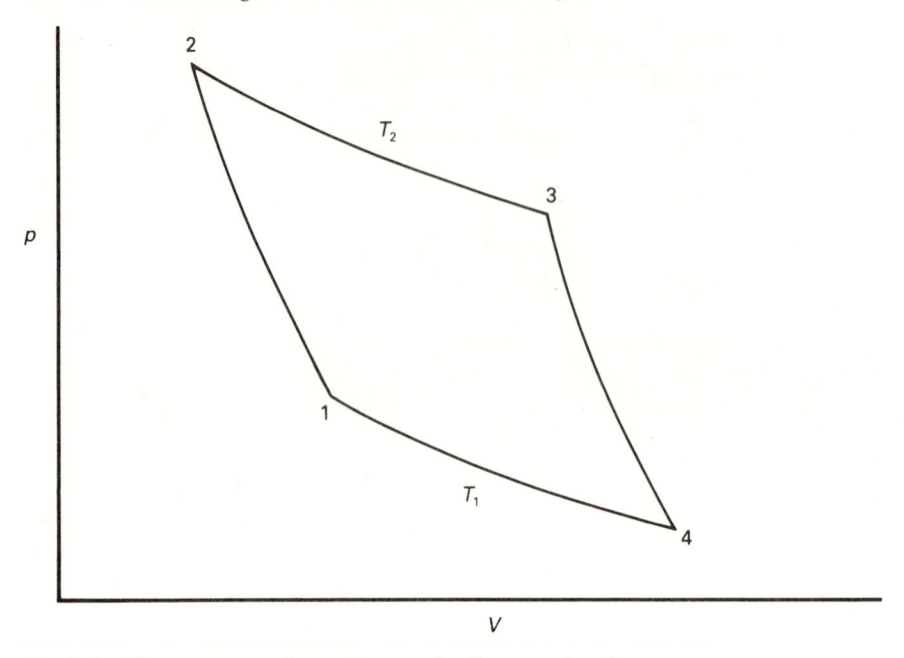

30 The Carnot Engine with Gas as the Working Fluid

Fig. 3.1 The pressure-volume diagram for the carnot cycle.

Then, finally, there is the adiabatic expansion from 3 to the initial point 1 of amount

$$W_A(3,4) = C_V(T_2 - T_1) \tag{3.9}$$

The gas began with volume V_1 and temperature T_1 and has returned to that initial state.

3.2 The Calculated Properties of the Carnot Engine

These various expressions are collected together to describe the complete Carnot cycle.

Thermal Efficiency Using an Ideal Gas

The thermal efficiency is the ratio of the net work produced in the cycle to the heat transfer into the cycle. The work done in the adiabatic phases is the sum of $W_A(3,4)$ and $W_A(1,2)$. The heat capacity C_V is constant for an ideal gas so the work depends only on the initial and final temperatures. From eqns. (3.7) and (3.9) it is seen that $W_A(3,4) = -W_A(1,2)$ so that the net adiabatic work is precisely zero. This is not so for isothermal work. From eqns. (3.6) and (3.8) this is

$$W_I(1,4) = W_I(3,2) + W_I(1,2)$$
$$= (M/m)R\{T_2 \log [V_3/V_2] - T_1 \log [V_4/V_1]\}$$

Also, $V_3/V_2 = V_4/V_1$ (see Exercises Question 3.1) so that

$$W_I = (M/m)R\{\log (V_3/V_2)\}[T_2 - T_1] \tag{3.10}$$

The heat put into the cycle is $Q = (M/m)RT_2 \log (V_3/V_2)$.

3.2 The Calculated Properties of the Carnot Engine

The thermal efficiency η of the cycle is, from the definition (2.1),

$$\eta = W_I/Q$$
$$= [T_2 - T_1]/T_2 \qquad (3.11a)$$

which is eqn. (2.1), as it must be. Eqn. (3.10) is written more simply by introducing the temperature ratio $\theta = T_2/T_1$ expressing the maximum temperature in terms of the minimum temperature: then

$$\eta = 1 - 1/\theta \qquad (3.11b)$$

It is seen that the thermal efficiency depends only on the temperature ratio and not on the pressure ratio.

Specific Work Output

The work transfer from unit mass of working fluid is given by eqn. (3.10) which can be rearranged for a perfect gas. But $\gamma = C_p/C_V$ so that

$$R = C_p - C_V = C_p[1 - (C_V/C_p)]$$
$$= C_p[1 - 1/\gamma] = C_p[(\gamma - 1)/\gamma]$$

and the expression for the specific work output W_s then takes the form

$$W_s = C_p T_1[(\gamma - 1)/m\gamma]\{(\theta - 1) \log (p_2/p_3)\} \qquad (3.12)$$

This is proportional to $(\theta - 1)$ and to the logarithm of the pressure ratio p_2/p_3. The increase of work output with increasing temperature ratio and pressure ratio is, therefore, weak.

Work Ratio

An important characteristic of an engine is the amount of usable (positive) work that is obtained from it, taking account also of the (negative) work that has to be used in compression and other causes. This is expressed by the work ratio r_w defined by

$r_w = $ [net work associated with the cycle]/usable work.

For the present case $r_w = [W_I(3,4) + W_I(2,1)]/W_I(3,4)$ or

$$r_w = [T_2 - T_1]/T_2 = 1 - (T_2/T_1) = 1 - 1/\theta$$

For this ideal case, the numerical values of the thermal efficiency and the work ratio are the same.

Specific Flow of Gas

The flow of gas necessary to run the engine is of importance since this determines its physical size. The throughput is the specific gas consumption $sgc = 1/W_I$ kg/kWs. A more conventional unit involves the kilowatt-hour (kWh) rather than the kilowatt-second (kWs). Since there are 3600 seconds in one hour, $sgc = 3600/W_I$ kg/kWh so that for the present case

$$sgc = \frac{3600\gamma}{(R/m)[T_2 - T_1] \log (p_2/p_3)} \text{ kg/kWh}$$

$$= \frac{3600 m\gamma}{C_p T_1(\gamma - 1)(\theta - 1) \log (p_2/p_3)} \text{ kg/kWh} \qquad (3.13)$$

32 The Carnot Engine with Gas as the Working Fluid

This decreases for greater differences between the reception and rejection temperatures T_2 and T_1: it also decreases as the pressure ratio increases.

Numerical Applications

The effects of the upper and lower temperatures on the efficiency of the engine are shown in Fig. 3.2 for air ($\gamma = 1.4$), for the two lower temperatures $T_1 = 2.9$ K (which is the temperature of the background cosmic radiation that pervades the universe and represents the lowest practical attainable temperature for our purposes) and for $T_1 = 19\,°\text{C} = 292$ K being the mean temperature of lakes and oceans on Earth, and so a base temperature for power plant. Notice that the calculation must *always* be made expressing temperatures in absolute (Kelvin) units. The practical application of the engine on Earth will require $T_1 > 292$ K (because that is the naturally occurring base temperature here) and the upper temperature $T_2 < 1000$ K, this being the limit set by considerations of the strengths of presently available materials from which the plant could be constructed. This gives a maximum ideal thermal efficiency of about 70.8 %.

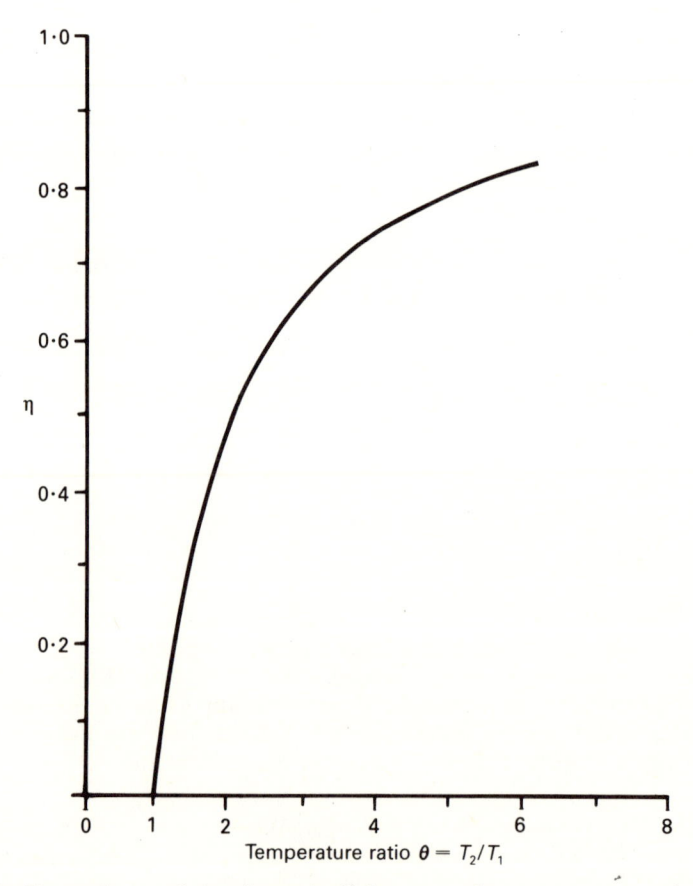

Fig. 3.2 Dependence of the thermal efficiency on the temperature ration θ for a Carnot cycle. (Air, $T_1 = 302$ K, $\gamma = 1.4$.)

3.3 Comment on the Equation of State for Real Gases

The expression (3.12) for the specific work output shows the Carnot engine to be a rather poor work source. This is shown in Fig. 3.3 where W_s is plotted as functions of the pressure and temperature ratios. W_s always remains below about 30 kJ/kg, a value about one tenth of those for the cycles considered in Chapters 4 to 7. Because the specific work output is low, the specific gas consumption is correspondingly high and values of the order of 120 kg/kJ would be expected. The Carnot engine must, therefore, itself be large. Even if an ideal gas was available, the working characteristics of the Carnot cycle would hardly be attractive for general engineering applications, even in the absence of friction.

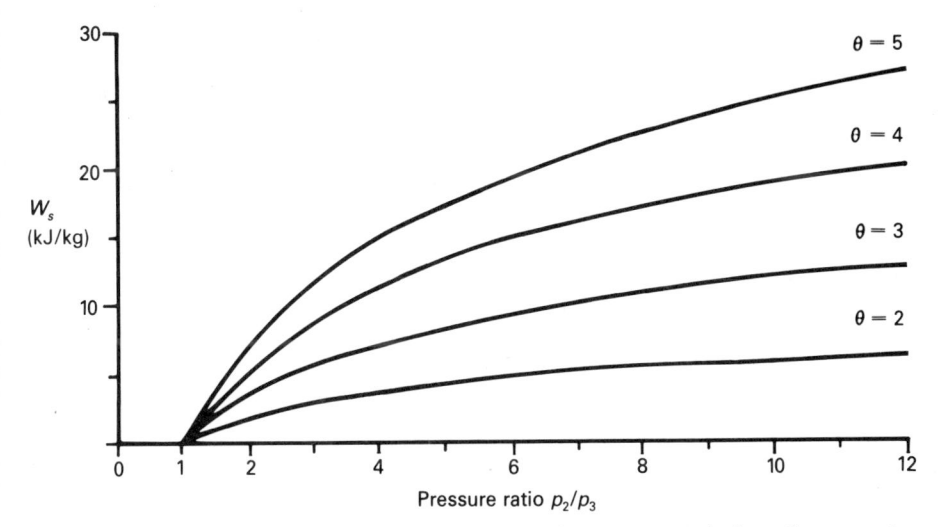

Fig. 3.3 Dependence of specific work output on the pressure ratio for a Carnot cycle (Air as the working fluid: $\gamma = 1.4$, $T_1 = 302$ K.)

3.3 Comment on the Equation of State for Real Gases

The arguments so far have involved ideal gases but they can be modified to take account of real behaviour. This is a very wide subject with an enormous literature, involving both experimental measurements and theoretical calculations. We give a very brief introduction to this work here, referring the reader who wishes to follow the detail of the work to the specialist literature.

The Virial Expansion

We consider a mole of gas. Experimentally determined $p\tilde{V}$ isotherms for a gas can be fitted empirically over a wide range of temperature and volume with the power series relation

$$p\tilde{V}/RT - 1 = B(T)/\tilde{V} + C(T)/\tilde{V}^2 + D(T)/\tilde{V}^3 + \ldots \quad (3.14)$$

This series is called the virial expansion (from the Latin *vires*, meaning forces) and gives the virial equation of state. The coefficients $B(T), C(T), D(T), \ldots$ in the expansion are called respectively the second, third and fourth virial coefficients. All vary with the temperature and are probably all positive. The virial coefficients account for the interaction forces between the molecules of the gas: B accounts for the interaction forces between pairs of molecules, C for the simultaneous interactions between triplets of molecules, D between quadruplets, and so on. Under many practical conditions it is usually sufficient to account only for the second virial coefficient B although it may be necessary to include also the third coefficient C. This is shown in Table 2 where the second and third contributions to the virial expansion are collected for nitrogen gas at three pressures. The second and third terms provide an accuracy of a few percent in the final equation of state.

Table 3.2 The Contributions of the Second and Third Terms in the Virial Expansion for Nitrogen.

Pressure	$B/\tilde{V}$	$C/\tilde{V}^2$
1 bar	−0.000 5	+0.000 003
10 bar	−0.005	+0.000 3
100 bars	−0.05	+0.03

The virial expansion can be expressed alternatively as a power series for the pressure

$$p\tilde{V}/RT - 1 = B_1(T)p + C_1(T)p^2 + \ldots$$

The virial coefficients in the volume virial expansion are related to those in the pressure virial expansion in a simple way: explicitly

$$B_1 = B/RT, \quad C_1 = (C - B^2)/(RT)^2, \text{ and so on.}$$

The virial coefficients may be measured in the laboratory or calculated using the kinetic theory of gases, but because the coefficients describe the effects of the interaction forces between the molecules of a gas, calculation requires a knowedge of these interaction forces. However such information is not known to significant accuracy for general molecules and this restricts what can be achieved by calculation. Nevertheless, the measured data can, in fact, be used to allow the properties of the interaction forces to be inferred and the study of the deviations from the ideal gas laws in a particular case can in this way be used to deduce fundamental information about the molecular structure of the gas though this aspect of the work is outside our present interests.

Explicit Expressions for Thermodynamic Functions
The thermodynamic expression for the difference between the internal energy $\tilde{U}$ per mole for the real gas and that per mole for the ideal gas $\tilde{U}_o$ is

$$\tilde{U} - \tilde{U}_o = -\int_{\tilde{V}}^{\infty} \left\{ T \frac{\partial p}{\partial T} \bigg|_{\tilde{V}} - p \right\} d\tilde{V}$$

Substituting into this expression the virial expansion (3.14) as far as the third virial coefficient gives expansion

$$\tilde{U} - \tilde{U}_o = -RT\{-(T/\tilde{V})(dB/dT) + (T/2\tilde{V}^2)(dC/dT) + \ldots\}$$

The enthalpy $\tilde{H}$ per mole is defined as $\tilde{H} = \tilde{U} + p\tilde{V}$. Again denoting the ideal

3.3 Comment on the Equation of State for Real Gases

value with the subscript o we find, using (3.14), that

$$\tilde{H} - \tilde{H}_o = \tilde{U} - \tilde{U}_o + RT\{(B/\tilde{V}) + (C/\tilde{V}^2) + \ldots\}$$

The corresponding expression for the entropy per mole is found from eqn. (4.4) of Chapter 4:

$$\tilde{S} - \tilde{S}_o = -R\{\log p + (T/\tilde{V})(\mathrm{d}B/\mathrm{d}T) + \\ [(B^2 - C)/2\tilde{V}^2] + (T/2\tilde{V}^2)(\mathrm{d}C/\mathrm{d}T) + \ldots\}$$

The specific heat capacity per mole $\tilde{C}_p$ is defined by $\tilde{C}_p = (\mathrm{d}\tilde{H}/\mathrm{d}T)_p$. Using eqn. (3.14) we find after some manipulations

$$\tilde{C}_p - \tilde{C}_{p_o} = -R\left\{\frac{T^2 \mathrm{d}^2 B}{\tilde{V} \mathrm{d}T^2}\right. \\ \left. - \frac{(B - T\mathrm{d}B/\mathrm{d}T)^2 - C + T\mathrm{d}C/\mathrm{d}T - (T^2/2)\mathrm{d}^2 C/\mathrm{d}T^2}{\tilde{V}^2} + \ldots\right\}$$

These are cumbersome formulae to use for the calculation of work and heat transfer, but may need to be used if a significant level of accuracy is required.

Empirical Equations of State

Many empirical relations have been proposed over the years to account for the measured p–$\tilde{V}$–T behaviour of gases, each being usually constructed for a specific application. Many of these are quite accurate over restricted ranges of temperature and/or pressure but lack accuracy over a wide range of these parameters. All involve one or more free parameters that must be assigned, though the wider the range of application the greater the number of parameters in general and they can therefore become very cumbersome to use in practice. Two parameters are usually all that one wishes to use, and a semi-quantitative understanding of the behaviour of a particular gas can be achieved in this way.

There is one 1-parameter and three 2-parameter equations of state that are particularly useful in this connection. Introducing the parameters a and b (which will have different values for each case), these equations are;

(i) The Clausius equation

$$p(\tilde{V} - b) = RT$$

(ii) The van der Waals equation

$$[p + a/\tilde{V}^2](\tilde{V} - b) = RT$$

(iii) The Berthelot equation

$$[p + a/T\tilde{V}^2](\tilde{V} - b) = RT$$

(iv) The Dieterici equation

$$[p\exp(a/\tilde{V}RT)](\tilde{V} - b) = RT$$

In each equation, a is a measure of the forces between the molecules and b is a measure of the volume of the molecule. The virial coefficients provided by these equations are collected in Table 3.3.

36   The Carnot Engine with Gas as the Working Fluid

Table 3.3   The Expressions for the Second and Third Virial Coefficients for Three Empirical Equations of State.

Virial Coeff.	van der Waals	Berthelot	Dieterici
$B(T)$	$b - a/RT$	$b - a/RT^2$	$b - a/RT$
$C(T)$	b^2	b^2	$b^2 - ab/RT + a^2/2(RT)^2$

If only slight corrections are necessary to the ideal equation of state Berthelot proposed the expression

$$B(T) = \frac{9RT_c}{128 p_c}\left[1 - 6\left(\frac{T_c^2}{T^2}\right)\right]$$

for the second virial coefficient, where T_c is the critical temperature and p_c is the critical pressure for the gas.

These various formulae have been included to show the theoretical complexity of moving away from the ideal gas equation. The corrections under the conditions applicable in the following chapters are generally small and their inclusion would add considerably to the complexity of the discussion without adding anything to the principles which actually concern us. We will, therefore, restrict our discussion to the ideal gas as a working fluid, realising that our discussion is deficient as to fine detail but reliable as far as the general principles and conclusions are concerned.

3.4   The Effects of Non-Ideal Components

The Carnot cycle assumes that there are no irreversibilities in the cycle but this is an ideal never achieved in practice. Although we are thinking now in terms of principles, it is important to estimate what effect irreversibilities might have on the conclusions. Let us, therefore, suppose that the expansion and compression stages of the Carnot cycle are subject to irreversible processes.

The T–s diagram for the cycle with irreversibilities will have the form shown in Fig. 3.4. The isothermal portions of the path will be unaffected (except that the overall area encased by the cycle will be reduced) but the isentropic paths will be changed to account for the associated increase of entropy now.

Isentropic Efficiencies

The two portions of the path associated with work transfer are the expansion $W_{34'}$ along 3–4' and the compression $W_{12'}$ along 1–2'. With irreversibilities, the expansion will be less complete than the ideal and the compression will be less effective. It is convenient to introduce two coefficients, one for expansion and the other for compression, to describe this incompleteness, and these are defined by the comparison between the work for the ideal and the actual systems.

Define the coefficient of isentropic expansion, η_t (often called the isentropic expansion efficiency), so that

$$\eta_t = W_{34'}/W_{34} \tag{3.14a}$$

Similarly, define the coefficient of isentropic expansion, η_c (often called the

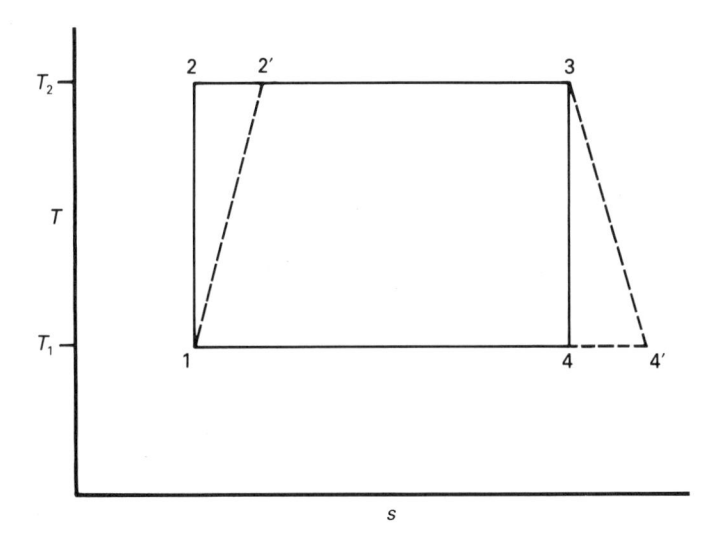

Fig. 3.4 The temperature (T)-specific entropy diagram for the Carnot cycle using gas as the working fluid for the ideal cycle (12341) and for the cycle involving irreversibilities (1 2′ 3 4′ 1).

isentropic compression efficiency), so that

$$\eta_c = W_{12}/W_{12'} \tag{3.14b}$$

These quantities are defined such that both η_t and η_c are unity for the ideal work transfer but less than unity for the actual transfer. In practice, the values of the coefficients for real compressors and turbines will typically lie in the range 0.80 to 0.97.

The Cycle Thermal Efficiency
The work and heat transfers for the system are calculated as before. For the adiabatic work transfers we find, using also the definitions (3.14a and b)

$$W_{12'} = -(1/\eta_c)W_{12} \qquad W_{34'} = \eta_t W_{34}$$

while the isothermal transfers become (since a piston is moving according to our model)

$$W_{23'} = \eta_t W_{23} \qquad W_{14'} = (1/\eta_c)W_{14}$$

The heat transfer remains Q_{12}.

The thermal efficiency now becomes, after a simple rearrangement

$$\eta = \frac{C_V(T_2 - T_1)(\eta_t - 1/\eta_c) + R(\log(V_3/V_2))(\eta_t T_2 - (1/\eta_c)T_1)}{RT_2 \log(V_3/V_2)}$$

This rather cumbersome expression can be rearranged further: we write for the ratio of the maximum and minimum temperatures $\theta = T_2/T_1$ then

$$\eta = (\eta_t - 1/\eta_c) + \frac{C_V(\theta - 1)(1 - \eta_t \eta_c)}{R \log(V_3/V_2)}$$

that is, remembering the gas is ideal

$$\eta = (1/\eta_c\theta)(\theta\eta_c\eta_t - 1) - \frac{(\theta-1)(1-\eta_t\eta_c)}{\eta_c(\gamma-1)\log(\theta p_2/p_3)},$$

It is convenient to write $a = \theta\eta_t\eta_c$ to give the final form

$$\eta = \frac{(a-1)}{\eta_c\theta} - \frac{(\theta-1)(1-a/\theta)}{\eta_c(\gamma-1)\log(\theta p_2/p_3)} \quad (3.15)$$

This is the expression for the thermal efficiency of the Carnot cycle with irreversibilities. For the efficiency to be positive the second term on the right hand side must be less than the first term. When $\eta_t = \eta_c = 1$, eqn. (3.15) reduces to the ideal form eqn. (3.10a), written $\eta(1)$. It is clear that $\eta < \eta(1)$, the Carnot cycle being affected significantly by irreversible components in the system.

The Work Output of the Cycle
The expression for the specific work output W_s is

$$W_s = C_p T_1\{(\gamma-1)\log(p_2/p_3)(\eta_t\theta - 1/\eta_c)(1/m\gamma) \\ - (\gamma-1)(\eta_t - 1/\eta_c)\} \quad (3.16)$$

This reduces to the expression (3.12), written $W_s(1)$, for the ideal cycle with $\eta_t = \eta_c = 1$. It follows that $W_s < W_s(1)$. Irreversibilities reduce the effectiveness of the Carnot cycle both through the thermal efficiency and through the ability of the cycle to provide work transfer.

3.5 Summary
1. The ideal engine has been considered (working entirely reversibly and not subject to irreversible losses in any way) using an ideal gas as a working fluid. General expressions for the thermal efficiency, the work ratio and the specific gas flow have been derived.
2. Non-adiabatic effects in the compressor or expansion (turbine) stages are accounted for by using appropriate isentropic efficiencies, which have been defined. Such effects are found to have large effects on the thermal and work transfer properties of the Carnot engine.
3. The discussion has involved an ideal gas as a working fluid but the corrections necessary to account fully for real fluids have been reviewed and been found small unless the pressure ratio is high. This means that there is interest in making analyses in terms of the ideal gas, even though in practice we must deal with real gases that are not ideal, particularly as the simple equation of state of an ideal gas reduces the complexity of the calculations substantially.

3.6 Exercises

3.1 Using the formulae (3.1) and (3.2) of the text show that the pressure and temperature of an ideal gas are related by $p^a T =$ constant where $a = (\gamma-1)/\gamma$. For the case of the isothermal and adiabatic expansions associated with the Carnot cycle show further that $V_4/V_1 = V_3/V_2$ as quoted in Section 3.1 of the text.

3.6 Exercises

3.2 A particular Carnot engine has inefficient components. Referring to Fig. 3.3, the heat accepted is $Q_h = (h_3 - h_{4'})$ while the heat rejected is $Q_c = (s_{4'} - s_1)T_1$. If the thermal efficiency of the engine is η while that for the ideal engine is $\eta(1)$ show that $\eta < \eta(1)$.

3.3 A Carnot cycle with ideal components has thermal efficiency $\eta(1)$ and specific work output $W_s(1)$ given respectively by eqns. (3.11a) and (3.12) of the main text. A Carnot cycle with inefficiencies has a thermal efficiency η and a specific work output W_s given respectively by eqns. (3.15) and (3.16). Show that

$$\frac{\eta}{\eta(1)} = \frac{(a-1)}{\eta_c(\gamma-1)} - \frac{(1-a/\theta)}{\eta_c(\gamma-1)\log(p_2/p_3)}$$

and

$$\frac{W_s}{W_s(1)} = \frac{(a-1)}{\eta_c(\theta-1)} - \frac{(1/\eta_c - \eta_t)}{C_V T_1(\gamma-1)\log(p_2/p_3)}$$

Confirm that $\eta < \eta(1)$, $W_s < W_s(1)$ and that $\eta \to \eta(1)$, $W_s \to W_s(1)$ as $\eta_t, \eta_c \to 1$.

3.4 A Carnot engine is run between the high temperature $T_2 = 700\,°C$ and the low temperature $T_1 = 30\,°C$, with air ($\gamma = 1.40$) as the working fluid. Use the expressions of Question 3.2 to determine the effects of irreversible behaviour for $(\eta_t, \eta_c) = 0.95, 0.90, 0.85, 0.80$ and 0.70, over a range of pressure ratios. What would be the effect of helium ($\gamma = 1.66$) replaced air as the working fluid? Express your results in graph form.

3.5 The actual thermal efficiency achieved in a Carnot cycle in space applications could be substantially increased over that for ordinary applications because the lower temperature T_1 can be made very low, space forming a large, very cold reservoir. Discuss this possibility remembering that real gases would have to be used in the engine. Suggest possible working fluids and estimate the efficiencies and work ratios that might be achieved using such a space power unit.

3.6 A perfect gas is made to go through the following reversible cycle with four stages:

(i) it is expanded from volume V_1 to volume V_2 ($>V_1$) at the constant temperature T_2 and at the constant pressure p_2; (ii) the pressure is now decreased at the constant volume V_2 to p_1 ($<p_2$) where the temperature is T_1 ($<T_2$); (iii) it is then compressed from volume V_2 to volume V_1 at the constant temperature T_1 and the constant pressure p_1; and (iv) the pressure is increased at the constant volume V_1 until it reaches the initial state with the pressure p_2 and temperature T_2. All the heat given out during the second stage, where the pressure is decreased at constant volume, is returned to the gas (through a regenerator) during the final stage of increasing the pressure.

Show that the net work done by the cycle is, per unit mass of gas,

$$W = R(T_2 - T_1)\log(V_2/V_1)$$

40 The Carnot Engine with Gas as the Working Fluid

and that the thermal efficiency η of the cycle (taken as the ratio of net work performed by the cycle to the heat received by the cycle) is given by $\eta = (T_2 - T_1)/T_2$. This is the ideal form of the Stirling cycle. (This cycle has been manufactured: the difficulty with the design is in constructing a regenerator with even mildly attractive heat transfer efficiency.)

3.7 A substance is passed through the following cycle:

(i) it is heated at constant pressure p_2 from the temperature T_1 to the temperature T_2 ($>T_1$); (ii) it is expanded adiabatically until its temperature is reduced to T_3; (iii) it is further cooled at a constant pressure $p_1 < p_2$ until the temperature is $T_4 < T_1, T_3$; (iv) it is compressed adiabatically until the initial conditions are re-established with temperature T_1.

The specific heat capacity of the substance is constant throughout the cycle. Show that the net specific work output of the cycle is

$$W = C_p[T_2 + T_4 - T_1 - T_3]$$

The temperatures T_2 and T_4 are <u>now</u> held fixed. Show that the condition for maximum work is $T_1 = T_3 = \sqrt{T_2 T_4}$ and that the specific work output then is

$$W_{max} = C_p[\sqrt{T_2} - \sqrt{T_4}]^2$$

3.8 Two bodies, each with the same specific heat capacity C, are at the different temperatures T_1 and T_2. They are connected together thermally by a reversible heat engine whose working brings them to the common temperature T. Show that $T = \sqrt{T_1 T_2}$ (the geometric mean of the two initial temperatures). Show further that the work done by the engine during the move to thermal equilibrium is

$$W = C[\sqrt{T_2} - \sqrt{T_1}]^2.$$

4. THE SIMPLE CONSTANT PRESSURE JOULE CYCLE

The simplest constant pressure cycle is analysed and its working characteristics are derived. As explained in the last chapter, we assume throughout that the gas is ideal and so has the equation of state (3.1). We begin with some preliminary comments on enthalpy and entropy change within an ideal gas.

4.1 Enthalpy Change and Work Transfer

It will be necessary to draw the T–s and h–s diagrams for the expansion/compression processes. Consider a closed mass of gas. The First Law of Thermodynamics can be expressed in the form

$$dQ = du + p\,dV \qquad (4.1)$$

for unit mass of gas, where du is the change of the specific internal energy consequent upon the addition of the quantity of heat dQ, and $p\,dV$ is the work done by the system at the pressure p. In terms of the specific enthalpy, $h = u + pV$, we write eqn. (4.1) instead in the form

$$dQ = dh - V\,dp \qquad (4.2)$$

According to the Second Law of Thermodynamics there is an increase ds of the specific entropy consequent upon the isothermal reversible transfer of heat at temperature T of amount $T\,ds = dQ$. We now apply eqn. (4.2) explicitly to an ideal gas, so that $V\,dp = RT(dV/V)$, and introduce the definition $C_p = dh/dT$ for the specific heat capacity at constant pressure. This gives

$$T\,ds = dh - V\,dp = C_p\,dT + RT(dV/V)$$

Division throughout by T gives

$$ds = \frac{C_p\,dT}{T} - \frac{R\,dp}{p} \qquad (4.3)$$

This is an expression for the change of the specific entropy due to changes of both temperature and pressure. The change of entropy in passing from an initial state (T_1, p_1) to the final state (T_2, p_2) is obtained by integrating eqn. (4.3) to the form

$$s_2 - s_1 = C_p \log(T_2/T_1) - R \log(p_2/p_1) \qquad (4.4)$$

This formula is important for our present purposes because it allows lines of constant pressure to be drawn on the T–s diagram. If the value of the ratio p_2/p_1 is given, the entropy difference corresponding to a given temperature ratio follows immediately from eqn. (4.4). Repeating this calculation for

42 The Simple Constant Pressure Joule Cycle

different values of T_2/T_1, with p_2/p_1 given the same value in each case, we find the entropy differences for different values of the temperature ratio, the pressure p_2 being held constant relative to some datum pressure (for instance, atmospheric pressure). A line can now be drawn on the *T–s* diagram relating temperature and entropy for the single pressure p_2 (measured in terms of p_1 as the datum): this is called a constant pressure line on the *T–s* diagram. A neighbouring constant pressure line is drawn, for the pressure ratio p_3/p_1 say, by selecting a particular entropy difference $s_3 - s_1$ and using eqn. (4.4) to calculate the corresponding values of the temperature ratio T_3/T_1. Repeating the process for different values of the pressure allows a set of constant pressure lines to be constructed.

The constant pressure lines so constructed have a very special property. Suppose we select a particular temperature on the *T–s* diagram and follow the effects of entropy change. A neighbouring constant pressure line always shows the same horizontal distance from the first for all values of the entropy (this was the way the second constant pressure line was constructed anyway): but the *vertical* distance between the two lines, describing a temperature change at constant entropy, increases with increasing entropy (see Figure 4.1). We shall see later that this feature is crucial to gas cycles, and is the reason why they are, in practice, able to allow useful work to be produced.

Let us start again from the relation

$$T ds = C_p dT + dW$$

where dW is the work done on the system. It follows that, for isentropic conditions ($ds = 0$), $dW = -C_p dT$. The negative sign indicates that this is the work the system does on the environment under these conditions. We also remember the definition of C_p and deduce that $dh = C_p dT$.

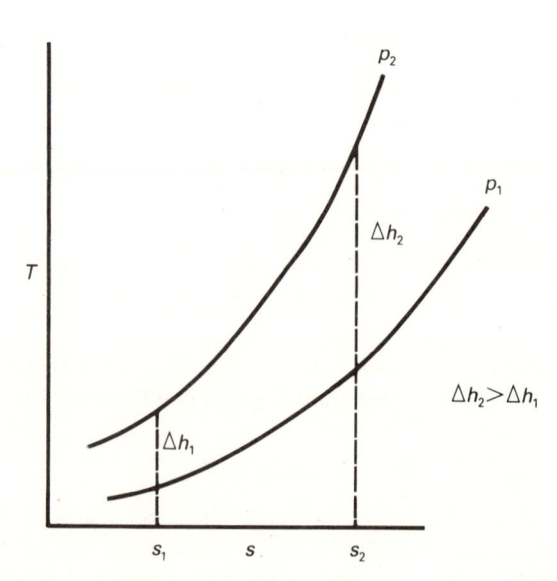

Fig. 4.1 Showing the profiles at constant pressure on the temperature-entropy diagram.

4.2 Adiabatic Conditions

For adiabatic expansion or compression of an ideal gas the two relations $pv^\gamma = $ constant (where γ is the ratio of the specific heat capacities at constant pressure C_p and constant volume C_v) and $pV = RT$ are valid. It follows immediately on combining the two relations that the temperature is related directly to the pressure by

$$T = Ap^{\frac{\gamma-1}{\gamma}}$$

where A is some constant. The constant can be eliminated if a comparison is made between the pressures and temperatures of an initial and a final state. For an adiabatic change between pressure p_1 and p_2, with corresponding temperatures T_1 and T_2, we find immediately that there is the relation

$$\frac{T_2}{T_1} = \left(\frac{p_2}{p_1}\right)^{\frac{\gamma-1}{\gamma}} = r_p^{\frac{\gamma-1}{\gamma}} \qquad (4.5)$$

where $r_p = p_2/p_1$ is the pressure ratio. It is often convenient to introduce the symbol

$$\rho_p = r_p^{\frac{\gamma-1}{\gamma}} \qquad (4.6)$$

This will be much used in the arguments to follow.

4.3 The Simple Joule Cycle

The simplest practical gas turbine cycle consists of a compression and an expansion, each at constant entropy, and a heat acceptance and a heat rejection at constant pressure. The p–V and h–s diagrams are shown in Figure 4.2. This is the simple Joule cycle (sometimes called the Brayton cycle). There are two variants of the cycle depending upon whether the turbine gas is exhausted to the atmosphere (an open cycle) or whether it is fed back to the compressor to be reheated for the turbine (a closed cycle). The practical circuits are shown in Figure 4.3. The thermodynamic arguments are the same

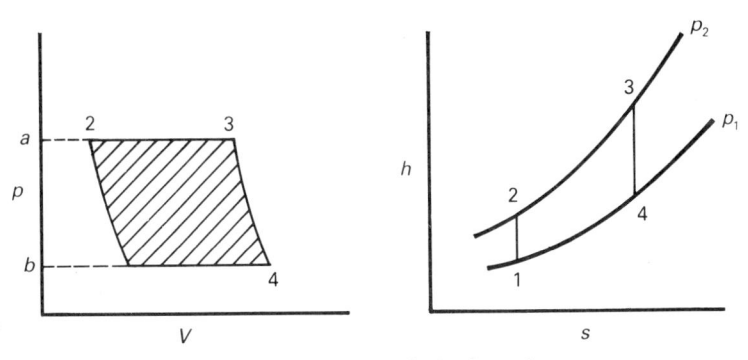

Fig. 4.2 p–V and h–S diagrams for the simple Joule cycle.

44 The Simple Constant Pressure Joule Cycle

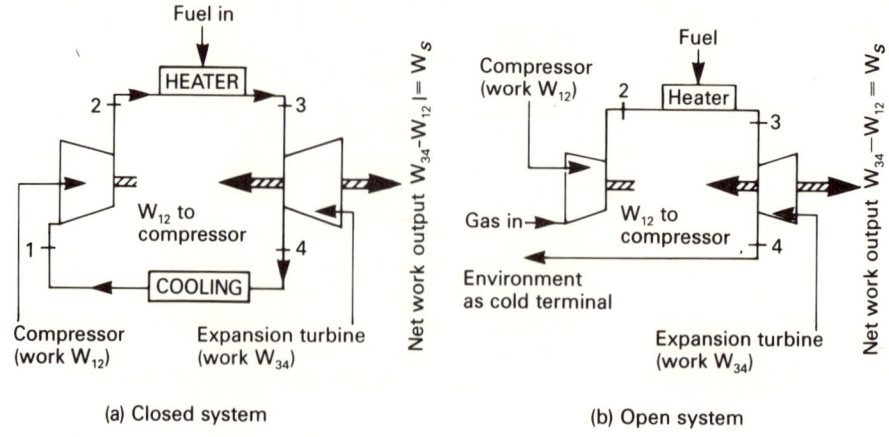

Fig. 4.3 Open and closed Joule cycles.

whether the circuit is open or closed and for convenience we will generally choose a closed circuit unless an application explicitly requires the circuit to be open.

Working Characteristics in Terms of Temperatures

Referring to Fig. 4.2, gas is compressed adiabatically by the compressor from the state 1 at pressure p_1, enthalpy h_1 and temperature T_1 to the higher pressure p_2 ($= r_p p_1$ where r_p is the pressure ratio), enthalpy h_2 and temperature T_2. The gas is then heated at the constant pressure p_2 to the temperature T_3 with enthalpy h_3, which is point 3 of the diagram. After isentropic expansion through the turbine to the initial pressure p_1, the gas has enthalpy h_4 and temperature T_4 (point 4 of the circuit diagram). The gas is then cooled, the heat being rejected, to the temperature T_1 when the circuit can start again.

The expansion work transfer due to the turbine is represented by the area a2341ba in Fig. 4.2, while the compression work transfer absorbed by the cycle is represented by the area a21ba in the Figure. The net work output of the cycle is the difference between these two work transfers, represented by the area 12341. Fuel is used in the heating stage 2–3, and this is where heat Q_{32} enters the cycle. The heat Q_{14} is rejected.

In terms of the enthalpies we can write

$$Q_{32} = h_3 - h_2 = C_p[T_3 - T_2]$$
$$Q_{14} = h_1 - h_4 = C_p[T_1 - T_4] \tag{4.7}$$

where C_p has been assumed independent of the temperature and pressure, as indeed it is for an ideal gas. This will prove a sufficient approximation for all the arguments to follow for application to real gases. Work W_{12} is done on the cycle during the compression while the cycle delivers work W_{34} during the expansion:

$$W_{12} = h_1 - h_2 = C_p[T_1 - T_2]$$
$$W_{34} = h_3 - h_4 = C_p[T_3 - T_4] \tag{4.8}$$

4.3 The Simple Joule Cycle

The total work output W by the cycle is $W = W_{12} + W_{34}$ or
$$W = C_p\{(T_3 - T_4) - (T_2 - T_1)\}$$
The thermal efficiency η of the cycle is
$$\eta = [W_{34} - W_{21}]/Q_{23}$$
and from (4.8) and (4.7) we find
$$\eta = \frac{(h_3 - h_4) - (h_2 - h_1)}{(h_3 - h_2)} = \frac{(T_3 - T_4) - (T_2 - T_1)}{(T_3 - T_2)} \tag{4.9}$$
The work ratio r_w is the total work of the cycle per unit of useful work. Using eqn. (4.8) this is
$$r_w = \frac{W_{34} + W_{12}}{W_{34}} = \frac{(h_3 - h_2) - (h_2 - h_1)}{(h_3 - h_4)}$$
$$= \frac{(T_3 - T_2) - (T_2 - T_1)}{(T_3 - T_4)} \tag{4.10}$$
The specific work output W_s is the net work of the cycle
$$W_s = W_{34} + W_{12} = (h_3 - h_4) - (h_2 - h_1)$$
$$= C_p\{(T_3 - T_4) - (T_2 - T_1)\} \tag{4.11}$$

In Terms of Pressure and Temperature Ratios
The expressions (4.9) to (4.11) are written more usefully by introducing the pressure ratio r_p of eqn. (4.5) and the ratio θ of the maximum and minimum temperatures of the cycle, $\theta = T_3/T_1$. Then
$$Q_{23} = T_3[1 - (T_2/T_3)] \tag{4.12a}$$
and
$$W = T_1[(T_3/T_1)(1 - T_4/T_1) - (T_2/T_1 - 1)] \tag{4.12b}$$
Also, remembering eqns. (4.5) and (4.6)
$$T_2/T_1 = T_3/T_4 = \rho_p \quad \text{(points on isentropic lines)} \tag{4.13}$$
and, from the chain rule
$$\frac{T_2}{T_3} = \frac{T_2}{T_1} \cdot \frac{T_1}{T_3} = \frac{\rho_p}{\theta} \tag{4.14}$$
Inserting eqns. (4.12) and (4.13) into the expressions (4.12) and then using eqn. (4.9) we obtain the alternative form for the thermal efficiency
$$\eta = \frac{\theta(1 - 1/\rho_p) - (\rho_p - 1)}{(1 - \rho_p/\theta)} = 1 - \frac{1}{\rho_p} \tag{4.15}$$
In terms of the pressure ratio we have
$$\eta = 1 - (1/r_p)^{\frac{\gamma-1}{\gamma}} \tag{4.15a}$$

46 The Simple Constant Pressure Joule Cycle

Similarly, the work ratio and specific work output become

$$r_w = 1 - \frac{\rho_p}{\theta} \qquad (4.16)$$

$$\begin{aligned}W_s &= C_p T_1 \{\theta(1 - 1/\rho_p) - (\rho_p - 1)\} \\ &= C_p T_1 (\theta - \rho_p)(1 - 1/\rho_p) \\ &= C_p T_1 (\theta - \rho_p) \eta(r_p)\end{aligned} \qquad (4.17)$$

It should be noticed that the thermal efficiency of the ideal Joule cycle depends on the pressure ratio and the nature of the gas alone and not explicitly on either the cycle temperatures or the ratio of the temperatures involved. In particular, the maximum and minimum temperatures are not explicitly involved. The work ratio does, however, depend on the nature of the gas and on both the pressure and temperature ratios, while the specific work output depends, for a given gas, on the pressure and temperature ratios and (being an absolute quantity) also on one actual temperature, which may be either the maximum or minimum temperature for the cycle. Either could have been extracted from the bracket in the formula for W_s, but it is conventional to use the lower temperature as the datum because it is usually (though by no means always) the Earth's atmosphere for a gas system and therefore is independent of the working of the cycle.

Characteristics of the Cycle

Data for the thermal efficiency of the cycle derived from (4.15) are plotted in

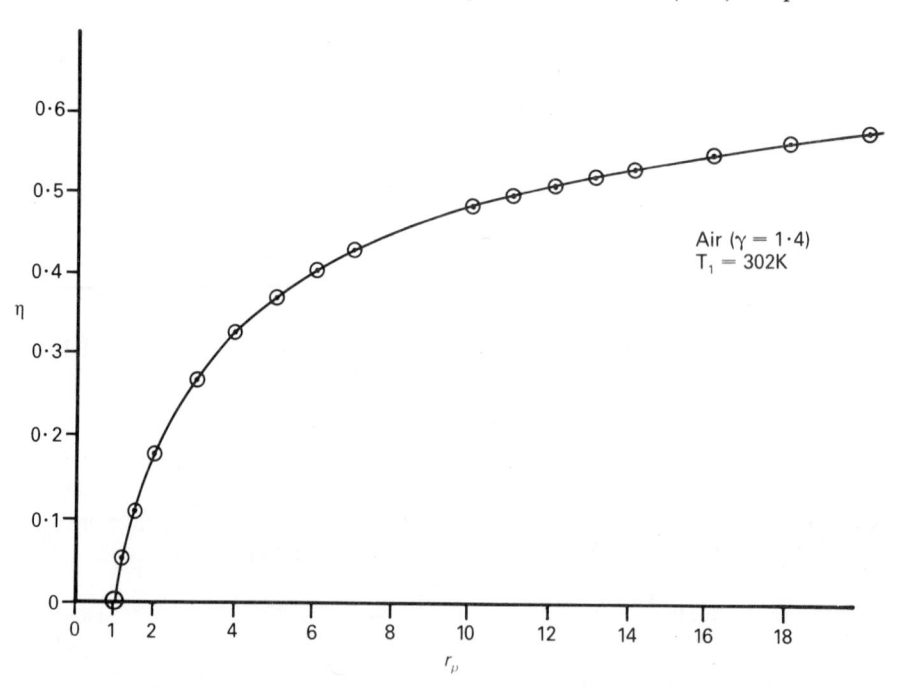

Fig. 4.4 Thermal efficiency for the simple Joule cycle as function of the pressure ratio.

Fig. 4.4. It is seen that efficiency increases as the pressure ratio increases, fast at first but then more slowly. The curve is independent of the temperature. For greater efficiency, the pressure ratio should be as high as possible. The work ratio is plotted in Fig. 4.5, the data being calculated from (4.16). The work ratio increases with the temperature ratio but, for a given temperature ratio, decreases with increasing pressure ratio. The largest values of the work ratio, therefore, require θ to be as large as possible and r_p to be near unity. Data for the specific work output, derived from (4.17), are plotted in Fig. 4.6 for the lowest temperature of 29 °C. It is seen that W_s increases with the temperature ratio and also with the pressure ratio at first. There is, however, a maximum in the W_s curve for a specific value of r_p which depends on θ. For

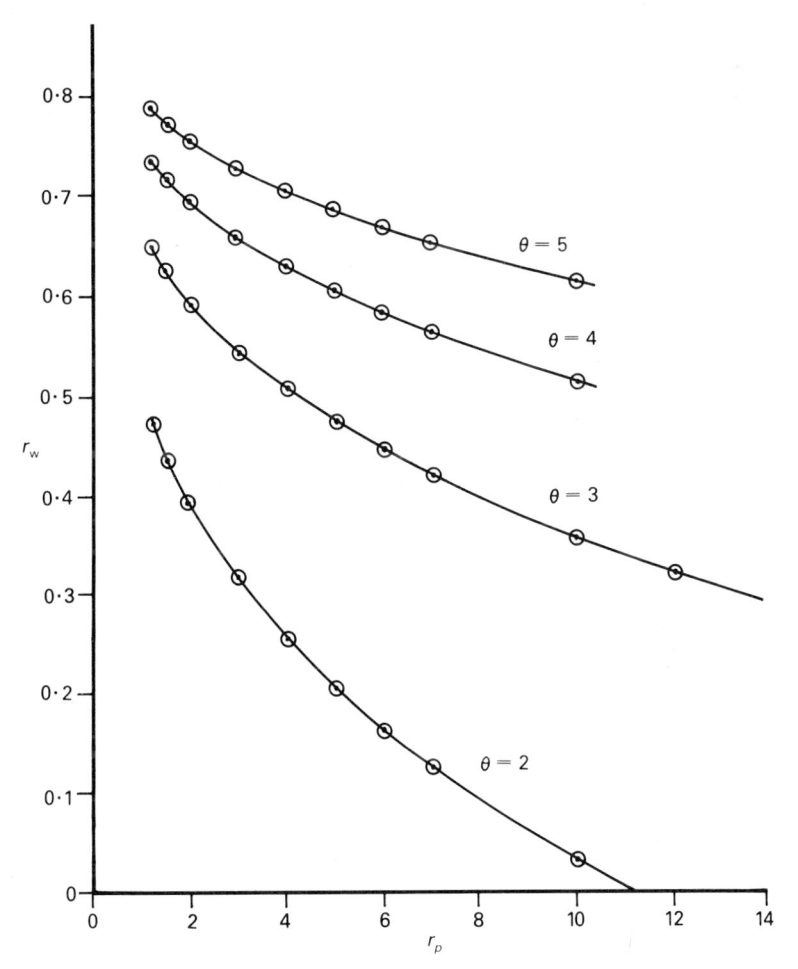

Fig. 4.5 Work ratio for the ideal Joule cycle as function of the pressure ratio for four values of the temperature ratio: Air ($\gamma = 1.4$) $T_1 = 302$ K.

48   *The Simple Constant Pressure Joule Cycle*

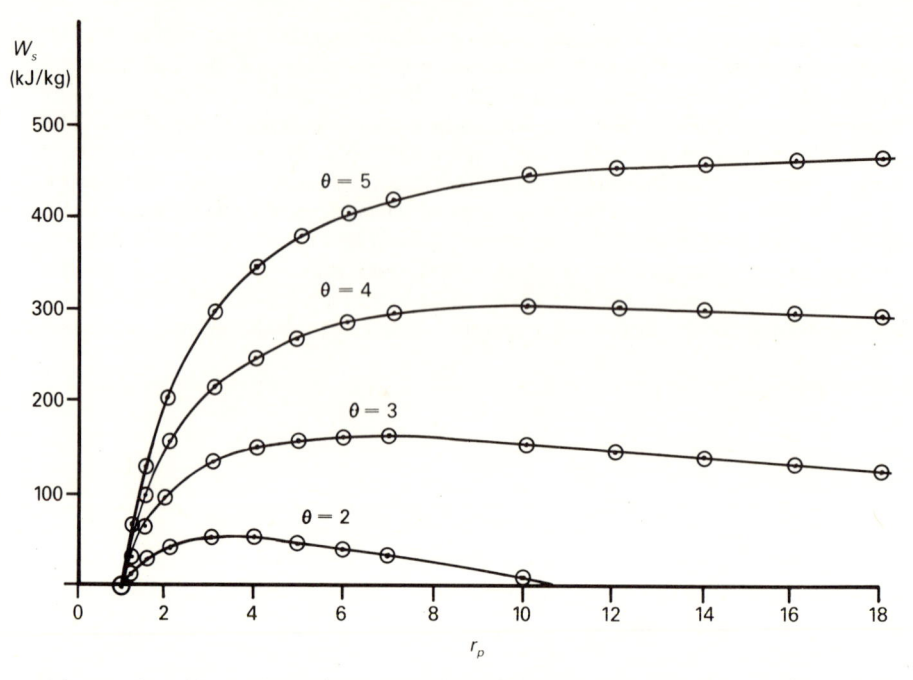

Fig. 4.6 Specific work output as function of the pressure ratio for an ideal simple Joule cycle, for four values of the temperature ratio: $T_1 = 302$ K, Air ($\gamma = 1.4$).

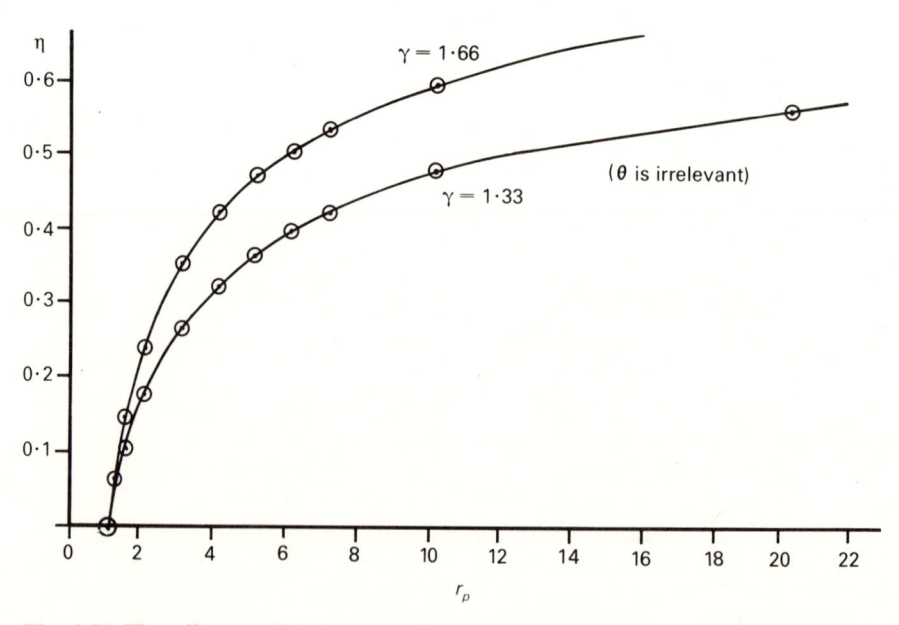

Fig. 4.7 The effect on the thermal efficiency of increasing the specific heat ratio of the gas.

values of r_p beyond this region, W_s decreases and ultimately turns negative (the cycle absorbing work from the environment). The three sets of curves together show that the best running of the cycle depends on the purpose: for the best thermal efficiency r_p should be large, but this will lead to a reduced r_w and a reduced (or even negative) W_s. If work output is the criterion, there is a limited best range for r_p and this will dictate both the values of θ and r_w that can be achieved. The value of θ is then fully determined but both r_w and W_s are increased by increasing θ. A larger value can be achieved for W_s if T_1 is increased: the metallurgical properties of the materials from which the engine is constructed place an upper limit on T_2, so raising T_1 effectively reduces the possible value for θ. The use of the cycle must involve a compromise between a variety of possible working conditions.

The characteristics of the cycle depend on the particular gas used through γ. The effect on the thermal efficiency is shown in Fig. 4.7 for air ($\gamma = 1.33$) and for a monatomic gas (for example, helium with $\gamma = 1.66$). The increase of η with increase of γ at constant r_p is clearly seen. Data for W_s for these two values of γ are plotted in Fig. 4.8 for $\theta = 4$ as typical. It is seen that increasing γ raises W_s at smaller r_p and provides a greater maximum value for W_s. W_s is reduced at large values of r_p. The practical implications of raising the value of γ for the working fluid are clearly evident from the graphs. For gases other than air a closed cycle would usually be necessary.

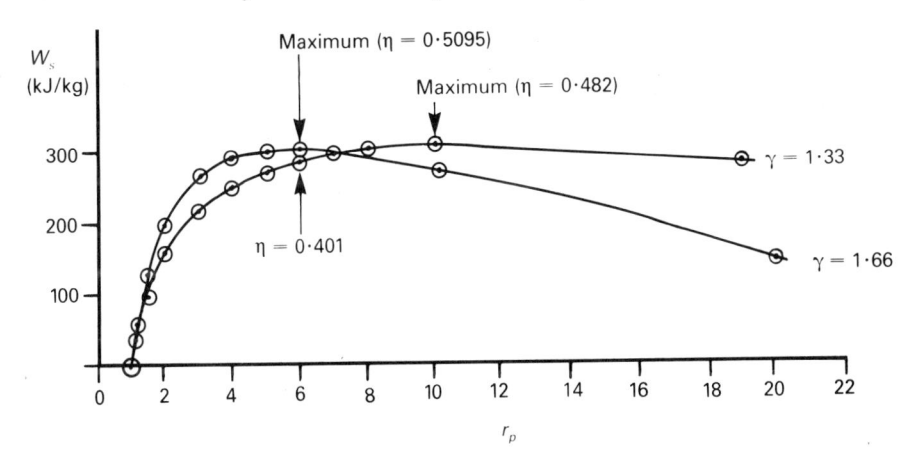

Fig. 4.8 Showing the effect of γ on the work output of the cycle.

4.4 The Condition of Maximum Specific Work Output

The data plotted in Figure 4.6 show that the work output reaches a maximum as r_p is increased, after which it falls. Using the expression (4.17) we will now show formally that, for a given temperature ratio θ, there is a range of values of the pressure ratio r_p over which the cycle has a net positive work output, but outside this range it is a net absorber of work.

The Pressure Ratio
It is seen from (4.17) that W_s is positive if

$$\theta(1 - 1/\rho_p) > \rho_p - 1$$

50 The Simple Constant Pressure Joule Cycle

that is if $\rho_p < \theta$, and if $\rho_p > 1$. These inequalities are conveniently expressed in the single form

$$1 < \rho_p < \theta \tag{4.18}$$

remembering that $\theta > 1$ if the cycle is to work at all. If $\theta < 1$ or if $\theta < \rho_p$, then W_s is negative and the system accepts work transfer from the environment. There is, then, a restricted range of pressure ratios, for a prescribed temperature ratio, for the system to function as a source for work transfer.

For the maximum work output the pressure ratio must be chosen, according to the rules of the calculus, such that

$$\frac{dW_s}{d\rho_p} = 0 \quad \text{and} \quad \frac{d^2W_s}{d\rho_p^2} < 0 \tag{4.19}$$

From eqns. (4.17) and (4.19):

$$\frac{dW_s}{d\rho_p} = C_v T_1 \{\theta/\rho_p^2 - 1\} = 0 \quad \text{if}$$

$$\theta = \rho_p^2 \tag{4.20}$$

Also, to show this represents a maximum value, we use eqn. (4.20) in conjunction with the derived second derivative, that is

$$\frac{d^2W_s}{d\rho_p^2} = -2C_p T_1 \left(\frac{\theta}{\rho_p^3}\right) = -C_p T_1 \left(\frac{1}{\rho_p}\right) < 0$$

The second derivative of W_s with respect to ρ_p is always negative (C_p, T_1 and ρ_p are positive quantities) showing W_s to have a true maximum value. For a given temperature ratio, the pressure ratio for maximum work output is

$$r_p(\text{max}) = (T_3/T_1)^{\gamma/(2\gamma-1)} \tag{4.21}$$

The pressure ratio depends on the particular gas involved (through the ratio γ), and increases with decreasing γ.

The working characteristics of the cycle for the condition of maximum work output follow by inserting eqn. (4.21) into eqns. (4.15), (4.16) and (4.17). Then we find

$$\eta = 1 - 1/\sqrt{\theta}; \quad r_w = 1 - 1/\sqrt{\theta}; \quad \text{and} \quad W_s = C_p T_1 (\sqrt{\theta} - 1)^2 \tag{4.22}$$

Comparison with eqns. (3.10) and (3.11) for the ideal cycle show that both the thermal efficiency and the work ratio are less than for the ideal.

A Numerical Comparison

We use the eqns. (4.22). For air ($\gamma = 1.4$; $C_p = 1.005$ kJ/kg K) and with $T_3 = 500\,°C = 773$ K and $T_1 = 29\,°C = 302$ K, the temperature ratio is $\theta = 2.560$, so that $\rho_p = 1.6$ (maximum specific work output) and $r_p = 5.181$. The thermal efficiency is now determined by the temperature ratio and is $\eta = 0.375$. The work ratio and specific work output are $r_w = 0.375$ and $W_s/C_p T_1 = 0.36$ so that $W_s(\text{max}) = 1.085 \times 10^2$ kJ/kg. With the inlet temperature to the turbine raised to $T_3 = 1000$ K (the lower temperature remaining the same), the temperature ratio becomes $\theta = 3.311$ and $r_p = 8.133$, and the working characteristics would be $\eta = 0.451$, $r_w = 0.451$, $W_s/C_p T_1 = 0.672$, $W_s(\text{max}) = 2.025 \times 10^2$ kJ/kg.

4.5 The Effects of Irreversibilities

The Joule cycle involves both a compressor turbine and an expansion turbine and in practice neither of these will have ideal isentropic efficiencies. The inefficiencies have important effects on the characteristics of the cycle.

Compression and Expansion Efficiencies

Expansion and compression are not now isentropic, and lead to an increase of entropy during expansion and compression. The *T–s* diagram for the cycle is shown in Fig. 4.9. The ideal cycle is represented by the figure 12341 but, with irreversibilities, it is changed to the compression 1–2', the heat transfer 2'–3, the expansion 3–4' and the heat rejection 4'–1. The areas under the regions 434'4 and 122'1 are a measure of the additional energy necessary to overcome the dissipative effects associated with non-isentropic conditions, this energy not being available for work transfer outside the cycle (useful work). This dissipated energy is expressed in terms of the isentropic coefficients for such pressure changes, and these were defined in general terms for the Carnot cycle in Section 3.4. The isentropic coefficients of expansion and compression are defined in general terms by eqns. (3.14a and b). For the present case, where the turbine outlet enthalpy is $h_{4'}$ ($>h_4$) and the compressor outlet enthalpy is $h_{2'}$ ($>h_2$), we have

$$\eta_t = \frac{(h_3 - h_{4'})}{(h_3 - h_4)} \qquad \eta_c = \frac{(h_2 - h_1)}{(h_{2'} - h_1)} \tag{4.23}$$

From these definitions it follows that $0 < \eta_t, \eta_c < 1$.

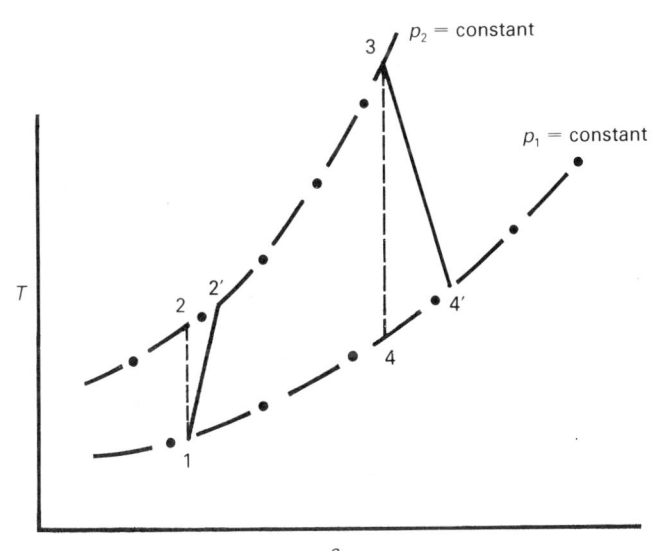

Fig. 4.9 The *T–s* diagram for the Joule cycle with less than ideally efficient expansion and compression.

The Joule Cycle with Inefficiencies

The expression for the thermal efficiency is

$$\eta = \frac{(h_3 - h_{4'}) - (h_{2'} - h_1)}{(h_3 - h_{2'})} \tag{4.24}$$

The enthalpy values for the ideal cycle can be inserted using the expressions (4.23). Then we have, introducing the temperatures as before,

Work transfer
$$= (h_3 - h_{4'}) - (h_{2'} - h_1)$$
$$= C_p[\eta_t(T_3 - T_4) - (1/\eta_c)(T_2 - T_1)]$$
$$= C_p[\eta_t T_3(1 - T_4/T_3) - T_1(1/\eta_c)(T_2/T_1 - 1)] \tag{4.25a}$$

Heat transfer
$$= h_3 - h_{2'} = (h_3 - h_1) - (h_{2'} - h_1)$$
$$= C_p[(T_3 - T_1) - (1/\eta_c)(T_2 - T_1)]$$
$$= C_p T_1[(T_3/T_1 - 1) - (1/\eta_c)(T_2/T_1 - 1)] \tag{4.25b}$$

Introducing the temperature ratio $\theta = T_3/T_1$ and using eqn. (4.13), eqn. (4.24) is reduced to the form

$$\eta = \frac{\eta_t \theta(1 - 1/\rho_p) - (1/\eta_c)(\rho_p - 1)}{(\theta - 1) - (1/\eta_c)(\rho_p - 1)} \tag{4.24a}$$

It is convenient to introduce the two new symbols

$$a = \eta_t \eta_c \theta \quad \text{and} \quad b = \eta_c(\theta - 1) + 1 \tag{4.26}$$

converting eqn. (4.25) into the form

$$\eta = \frac{(\rho_p - 1)[a/\rho_p - 1]}{(b - \rho_p)} \tag{4.24b}$$

In a similar manner the work ratio becomes

$$r_w = \frac{\eta_t(T_3 - T_4) - (1/\eta_c)(T_2 - T_1)}{\eta_t(T_3 - T_4)}$$

or, using eqns. (4.13) and (4.26)

$$r_w = 1 - \frac{\rho_p}{a} \tag{4.27}$$

Finally, the specific work output follows from eqns. (4.25a) and (4.26):

$$W_s = C_p T_1(1/\eta_c)\{a(1 - 1/\rho_p) - (\rho_p - 1)\} \tag{4.28}$$

These expressions reduce to those for the simple cycle when conditions are ideal, that is when $\eta_t = \eta_c = 1$ so that, from eqn. (4.24), $a = b = \theta$.

General Properties

With both η_t and η_c to be accounted for (4.26), (4.27) and (4.28) offer a bewildering range of possibilities. The general characteristics are collected in Fig. 4.10 for η and Fig. 4.11 for W_s. In the latter figure, the turbine and compressor efficiencies are assumed to be different and it is interesting to notice that the turbine efficiency has the more important effect on the

4.5 *The Effects of Irreversibilities* 53

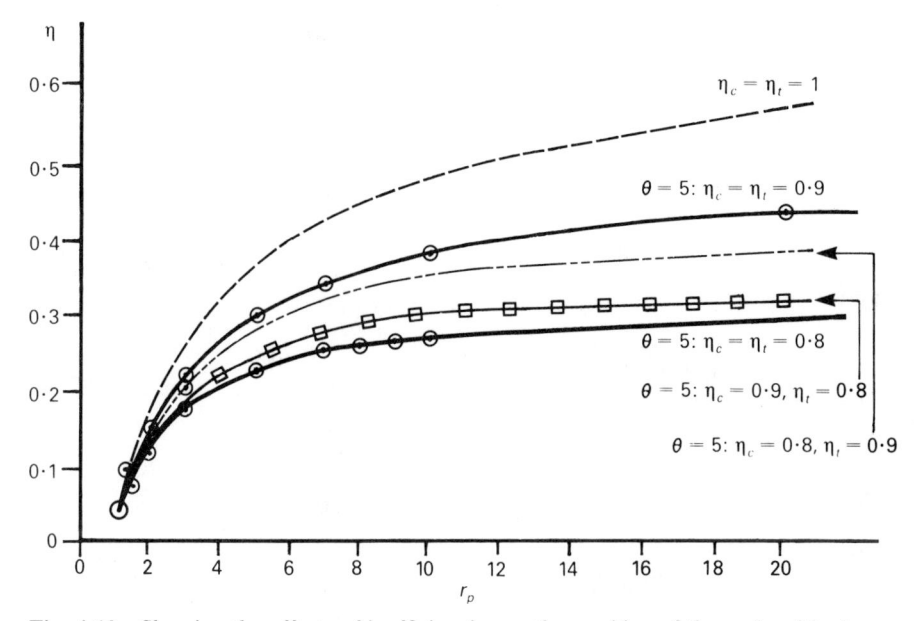

Fig. 4.10 Showing the effects of inefficiencies on the working of the cycle with air as the working fluid: $\theta = 5$.

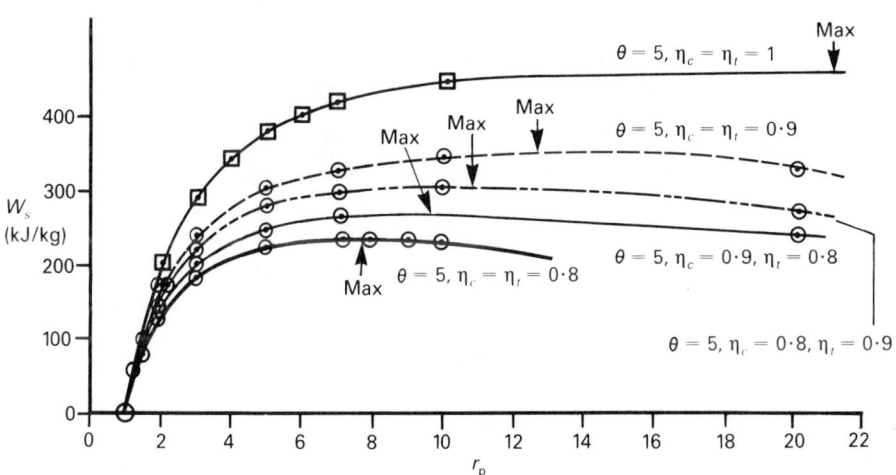

Fig. 4.11 Showing the effects of inefficiencies on the specific work output of the cycle with air as the working fluid.

characteristics of the cycle. But irreversibilities occur generally in a practical system quite apart from those associated with the turbine and compressor. For instance, there will be some pressure losses in mounting to the inlet of the turbine and, for an open system discharging to the air, there will be very substantial 'losses' in the exit side to the turbine (due to choking) which effectively reduce the pressure ratio. These arise partly from the obvious need to reduce the engine noise and, in industrial applications, to direct the outflow.

Maximum Specific Work Output

As for the simple system, so now, the various quantities vary with r_p for a fixed θ. The condition for the maximum value of W_s follows as before by differentiating W_s with respect to ρ_p to account for changes of the pressure ratio. The result is that W_s will have a maximum value $W_s(\text{max})$ if $\rho_p^2 = a$. Because $a<1$, the pressure ratio for maximum work is reduced by the irreversibilities. Also

$$W_s(\text{max}) = (C_p T_1/\eta_c)[a^{1/2} - 1]^2$$

and

$$r_p = [\eta_t \eta_c T_3/T_1]^{\frac{\gamma-1}{\gamma}} = a^{\frac{\gamma-1}{\gamma}} \qquad (4.29)$$

This is to be compared with eqn. (4.21) for the ideal case.

As a numerical example, take the temperature ratio of the previous case for the ideal cycle considered in Section 4.6. For air, $T_3 = 773$ K and $T_1 = 302$ K so that $\theta = 2.560$. If the isentropic efficiency of the compressor η_c is 0.85 and the isentropic efficiency of the turbine η_t is 0.87 (not untypical values) then from (4.24) $a = 1.893$ and $b = 2.326$. This gives $\rho_p = 1.376$ and so a pressure ratio $r_p = 3.056$, a drastic reduction on the previous value of 5.18. The thermal efficiency of the cycle is $\eta = 0.149$ (where previously it was 0.370). The work ratio is $r_w = 0.273$ (it was 0.375 before) and $W_s/C_p T_1 = 0.166$, giving the maximum value of the specific work output $W_s(\text{max}) = 5.047 \times 10$ kJ/kg (against the previous ideal value of 1.093×10^2 kJ/kg). The irreversibilities have reduced the maximum work output by a factor of 2.47.

If the turbine inlet temperature is raised to $T_3 = 1000$ K the situation is improved to some extent, but irreversibilities are still very effective at reducing the performance. Now, the temperature ratio is $\theta = 3.311$, $a = 2.448$ and $b = 2.964$. The pressure ratio is $r_p = 4.79$. This leads to the thermal efficiency $\eta = 0.228$, $r_w = 0.361$ and $W_s = 1.13 \times 10^2$ kJ/kg. These values compare with the values without irreversibilities $\eta = 0.451$, $r_w = 0.451$ and $W_s = 2.04 \times 10^2$ kJ/kg, and show clearly the adverse effects of irreversible features on the maximum work capability of the system.

4.6 Summary

1. The simple Joule (or Brayton) cycle is considered which has heat reception at a high and constant pressure and heat rejection at a lower, constant pressure. Such an arrangement is called a constant pressure cycle.
2. The expansion and compression are at first supposed to be isentropic processes. The thermal efficiency, work ratio and specific work output are considered for this simple constant pressure cycle (Section 4.3). The thermal efficiency depends, for a given gas, on the pressure ratio but not on the temperature or temperature ratio.
3. The work output increases to a maximum value with increasing pressure ratio, after which it becomes smaller. The conditions necessary for maximum work output are derived in Section 4.4.
4. The effects of irreversibilities in the expansion and compression phases are

considered in Section 4.5. The thermal efficiency now joins the other characteristics in depending on the temperature ratio. These effects reduce the working characteristics of the cycle in a way investigated in that section.

4.7 Exercises

4.1 Use the formulae of Section 4.3 to construct the graphs 4.4, 4.5 and 4.6 for air. It is required to achieve j a specific work output for the cycle of 350 kJ/kg and a thermal efficiency of 0.5. What is the pressure ratio required to do this? What, then, is the work ratio for the cycle?

4.2 Use the formulae of Section 4.5 to construct the graphs 4.10 and 4.11 for $\theta = 2$, 3, 4 and 6, with air as the working fluid. What conditions are necessary to achieve a thermal efficiency of 0.5 if the isentropic efficiencies of the turbine and compressor are each 0.89? What, then, is the specific work output for the cycle? How does this output compare with that for the ideal cycle at the same pressure ratio?

4.3 Which of the inefficiencies, in the turbine or in the compressor, have the greater effect on the performance characteristics of the Joule cycle, across the range of temperature ratios? Account for your conclusions in physical terms.

4.4 The characteristics of the simple Joule cycle have been deduced in the text by calculating separately the net work transfer and the heat input. It was however seen in Chapter 2 that the thermal efficiency of the cycle can also be expressed as the ratio of the net heat transfer (heat energy accepted minus heat energy rejected) and the heat accepted. The specific work output is then the difference between the heat accepted and the heat rejected. Using Fig. 4.9, derive the formulae (4.24b), (4.27) and (4.28) again using the appropriate energy transfers without referring specifically to the work transfers. (This alternative method of representation provides a check on the formulae derived by the method of the main text.)

4.5 For a Joule cycle involving inefficiencies in the compression and expansion stages (see Fig. 4.9), the heat input is reduced by the inefficient compression (by the amount 2–2' of the Figure) while the heat rejected is increased by the inefficient expansion (by the amount 4–4' of the Figure). Explain these effects in terms of the behaviour of the gas at these two stages. Describe in these terms the adverse effects of irreversibilities on the thermal efficiency and specific work output.

4.6 The direct use of the gases of combustion as the working fluid in a gas turbine system has the advantage of avoiding a heat exchange unit for heating the working fluid. For combustion gases $\gamma = 1.33$ is usually an acceptable approximation. Set down the overall advantages and disadvantages of using this arrangement in practice as an open cycle unit in comparison with using a closed cycle with helium as the working fluid.

5. REHEAT AND INTERCOOLING

It is the ability of the cycle to do work that is of special interest so it is desirable to improve this capability. This can be achieved by breaking the expansion into more than one stage and adding heat transfer between the stages in a process of reheating the gas, the addition being called reheat. A comparable procedure can be applied at the cold end of the system, which has one or more stages of cooling, and is called intercooling. We shall consider these modifications in this chapter. It should be said at the beginning that reheat is much more often applied in practice than intercooling, for reasons that will become clear as we develop our arguments.

5.1 The Principle of Reheat

The central feature for reheat is the fact that the enthalpy difference between two constant pressure lines increases with the entropy. If the expansion between some initial state 3, say (see Figure 5.1), on the T–s diagram is stopped at an intermediate pressure p_j (point 5), and the gas reheated to the initial temperature (point 6) before continuing the expansion to point 7, the work transfer is

$$(h_3 - h_5) + (h_6 - h_7) = (h_3 - h_4) + [(h_6 - h_7) - (h_5 - h_4)]$$
$$> (h_3 - h_4)$$

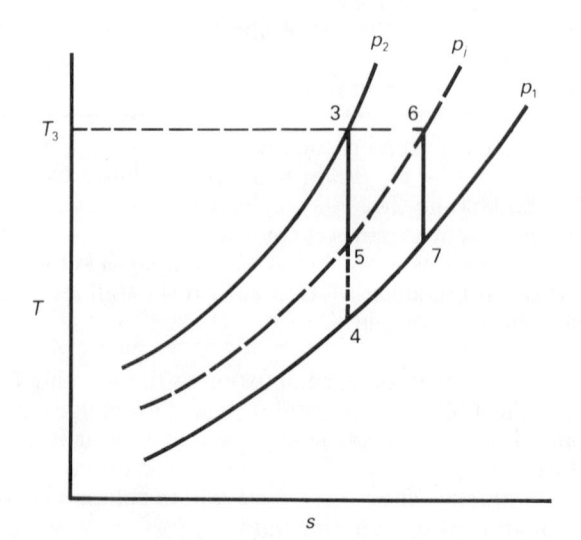

Fig. 5.1 Showing the effect of an intermediate pressure on expansion work transfer.

This is, then, greater than the work transfer involved in the expansion of the gas from state 3 to state 4 directly. Because the low temperature end of the cycle is not affected, the net work output for the system is increased by an amount that will depend on the choice of p_j.

This is the principle of reheat. Clearly it will be necessary to choose the value of the intermediate pressure carefully to achieve the maximum increase in the specific work output, and it could be advantageous in principle to add more than one reheat stage to the system. Again, because heat transfer to the working fluid is involved in the reheat, the thermal efficiency will be affected: we might guess, from thermodynamic considerations, that the thermal efficiency will in fact be decreased to some extent.

5.2 The Cycle with Reheat

Suppose reheat is added to a simple Joule cycle. The circuit is shown in Figure 5.2 and the full T–s diagram in Figure 5.3. The upper and lower pressures are respectively p_2 and p_1 and $p_2 = r_p p_1$. The intermediate pressure is p_j, whose best value is yet to be selected.

Heat transfer to the working fluid takes place between points 2 and 3 and the first expansion at constant entropy (between pressures $p2$ and pj, takes place between points 3 and 5. There is then a further heat transfer to the gas between points 5 and 6, followed by the final isentropic expansion between points 6 and 4. The gas is then cooled at constant pressure between points 4 and 1, after which it is compressed isentropically to return to the starting point 1 on the T–s diagram.

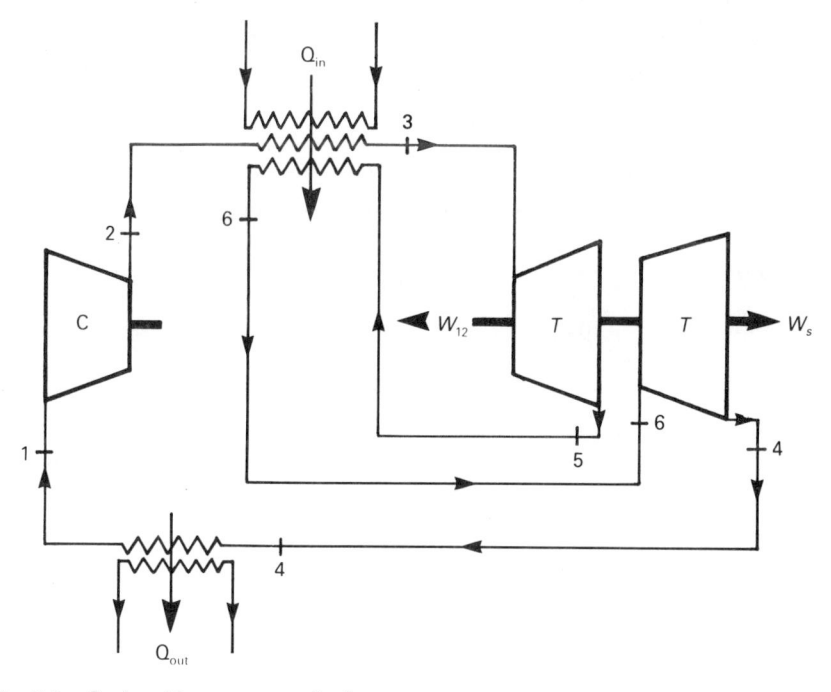

Fig. 5.2 Cycle with one stage of reheat.

58 Reheat and Intercooling

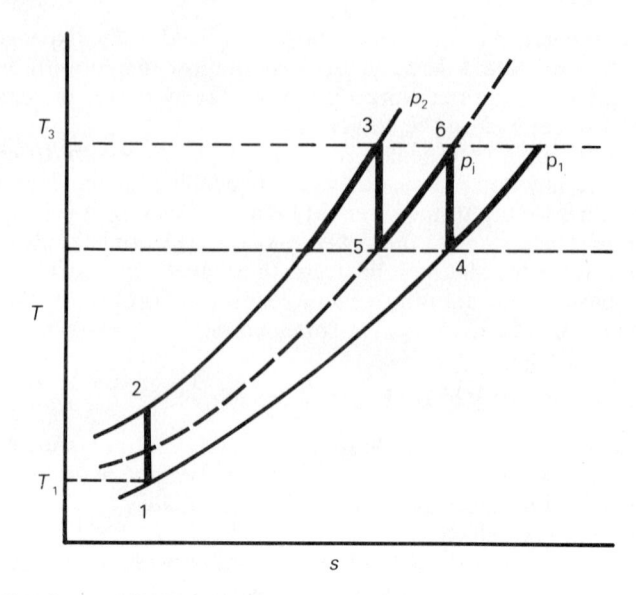

Fig. 5.3 *T–s* diagram for a single stage of complete reheat.

Specific Work Output is Maximised

The work and heat transfers are readily written down:

work transfer = $C_p\{(T_3 - T_5) + (T_6 - T_4) - (T_2 - T_1)\}$

heat transfer = $C_p\{(T_3 - T_2) + (T_6 - T_5)\}$

The work transfer is written alternatively

$$W_s = C_p\{T_3(1 - T_5/T_3) + T_6(1 - T_4/T_6) - T_1(T_2/T_1 - 1)\} \tag{5.1}$$

We now introduce the relationships between pressure and temperature ratios

$$T_3/T_5 = (p_2/p_j)^{\frac{\gamma-1}{\gamma}} \qquad T_6/T_4 = (p_j/p_1)^{\frac{\gamma-1}{\gamma}} \tag{5.2}$$

so that eqn. (5.1) becomes

$$W_s = C_p T_1 \{\theta[1 - (p_j/p_2)^{\frac{\gamma-1}{\gamma}}] + \theta[1 - (p_1/p_j)^{\frac{\gamma-1}{\gamma}}] - \rho_p + 1\} \tag{5.3}$$

We will choose the pressure p_j such that the specific work output W_s has the maximum value for the overall pressure ratio p_2/p_1 by maximising the expansion work transfer: the compression work transfer remains unaffected. This condition will, in fact, be found to be $p_j = \sqrt{p_1 p_2}$.

The condition for W_s to have a maximum value is again the requirements (4.19), but now subject to variations of p_j. In particular, we require that $(dW_s/dp_j) = 0$. Then

$$\frac{dW_s}{dp_j} = C_p T_1 \theta \left(\frac{\gamma-1}{\gamma}\right) \left\{ -p_j^{-(\frac{\gamma-1}{\gamma})-1} p_2^{-(\frac{\gamma-1}{\gamma})} + p_1^{(\frac{\gamma-1}{\gamma})} p_j^{-(\frac{\gamma-1}{\gamma})-1} \right\}$$

5.2 The Cycle with Reheat

But

$$\left(\frac{\gamma-1}{\gamma}\right) - 1 = -\frac{1}{\gamma} \quad \text{and} \quad -\left(\frac{\gamma-1}{\gamma}\right) - 1 = -\left(\frac{2\gamma-1}{\gamma}\right)$$

so that

$$\frac{dW_s}{dp_j} = C_p T_1 \theta \left(\frac{\gamma-1}{\gamma}\right) \left\{ -p_j^{-\frac{1}{\gamma}} p_2^{-\left(\frac{\gamma-1}{\gamma}\right)} + p_1^{\left(\frac{\gamma-1}{\gamma}\right)} p_j^{-\left(\frac{2\gamma-1}{\gamma}\right)} \right\}$$

This total expression vanishes if that in the bracket vanishes, that is if

$$p_j^{-\frac{1}{\gamma}} p_2^{-\left(\frac{\gamma-1}{\gamma}\right)} = p_1^{\left(\frac{\gamma-1}{\gamma}\right)} p_j^{-\left(\frac{2\gamma-1}{\gamma}\right)}$$

or

$$p_j^{2\left(\frac{\gamma-1}{\gamma}\right)} = (p_1 p_2)^{\left(\frac{\gamma-1}{\gamma}\right)}$$

This reduces to the condition

$$p_j^2 = p_1 p_2 \quad \text{or} \quad p_j = \sqrt{p_1 p_2} \tag{5.4}$$

Apparently p_j must be the geometrical mean between the upper and lower pressures for the cycle if W_s is to have a maximum value.

From eqn. (5.2), this result gives also

$$(p_2/p_j)^{\left(\frac{\gamma-1}{\gamma}\right)} = (\rho_p)^{1/2} \quad (p_j/p_1)^{\left(\frac{\gamma-1}{\gamma}\right)} = (\rho_{pi})^{1/2}$$

giving $(p_2/p_1)^{\left(\frac{\gamma-1}{\gamma}\right)} = \rho_p$ as it should. Also

$$T_3/T_5 = T_6/T_4 = \sqrt{\rho_p} \tag{5.5}$$

Thermal Efficiency and Work Transfer

With these results we can write

$$\text{work transfer} = W_s = C_p T_1 \{ 2\theta(1 - 1/\sqrt{\rho_p}) - (\rho_p - 1) \}$$
$$\text{heat transfer} = C_p T_3 \{ (1 - T_2/T_3) + (1 - T_5/T_6) \} \tag{5.6}$$

But

$$T_5 = T_4, \quad T_6 = T_3, \quad T_5/T_6 = (T_5/T_3)(T_3/T_6) = 1/\sqrt{\rho_p}$$

so that

$$\text{heat transfer} = Q = C_p T_3 \{ (1 - \rho_p/\theta) + (1 - 1/\sqrt{\rho_p}) \} \tag{5.7}$$

On dividing eqn. (5.6) by eqn. (5.7) we obtain, after a little rearrangement, the expression for the thermal efficiency η of the cycle

$$\eta = \frac{2\theta(1 - 1/\sqrt{\rho_p}) - (\rho_p - 1)}{(\theta - \rho_p) + \theta(1 - 1/\sqrt{\rho_p})} \tag{5.8}$$

The work ratio r_w and specific work output W_s are obtained from eqn. (5.6) as

$$r_w = 1 - \frac{\sqrt{\rho_p}}{2}(\sqrt{\rho_p} + 1) \tag{5.9}$$

and

60 Reheat and Intercooling

$$W_s = C_p T_1 \{2\theta(1 - 1/\sqrt{\rho_p}) - \rho_p + 1\} \tag{5.10}$$

The data calculated from these expressions are collected in Figs. 5.4, 5.5 and 5.6, as functions of the pressure ratio r_p for several values of the temperature ratio θ. The comparison between Figs. 4.4 and 5.4 for the ideal case shows, as we suspected earlier, that the thermal efficiency is reduced by reheat. Comparison between Figs. 4.5 and 5.5 shows, on the other hand, that reheat improves the work ratio: comparison between Figs. 4.6 and 5.6 shows that the specific work output is improved for all values of the pressure ratio but the improvement is greater for higher pressure ratios.

5.3 Reheat with Irreversibilities

For practical application it is necessary to account for non-isentropic behaviour in the turbines and in the compressor. The most general case is when the two turbines have different isentropic efficiencies. We therefore suppose that the isentropic efficiency of the turbine spanning the pressure range p_2 and

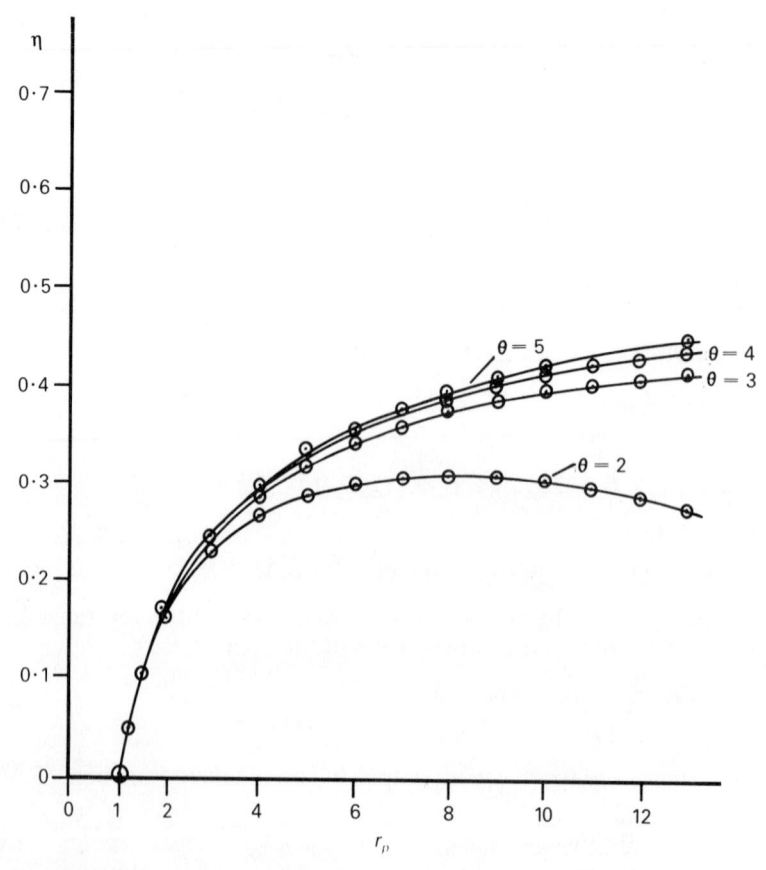

Fig. 5.4 One reheat stage (complete) ideal case; Air: $T_1 = 302$ K.

5.3 Reheat with Irreversibilities

p_j is η_{t2}, and that the isentropic efficiency spanning the pressure range p_j and p_1 is η_{t1}. The isentropic efficiency of the compressor is η_c. Written explicitly

$$\eta_{t2} = \frac{T_3 - T_{5'}}{T_3 - T_5} \quad \eta_{t1} = \frac{T_6 - T_{4'}}{T_6 - T_4} \quad \eta_c = \frac{T_2 - T_1}{T_{2'} - T_1} \tag{5.11}$$

The associated T–s diagram for the circuit now is given in Fig. 5.7.

The specific work output is, by the same methods as before, given by

$$\begin{aligned}W_s &= C_p\{T_3 - T_{5'}) + (T_6 - T_{4'}) - (T_{2'} - T_1)\} \\ &= C_p\{\eta_{t1}(T_3 - T_5) + \eta_{t2}(T_6 - T_4) - (1/\eta_c)(T_2 - T_1) \\ &= C_p T_1\{\theta[\eta_{t2}(1 - 1/\rho_{pj}) + \eta_{t1}(1 - 1/\rho_{pi})] - (1/\eta_c)(\rho_p - 1)\}\end{aligned} \tag{5.12}$$

where we have used eqn. (5.2).

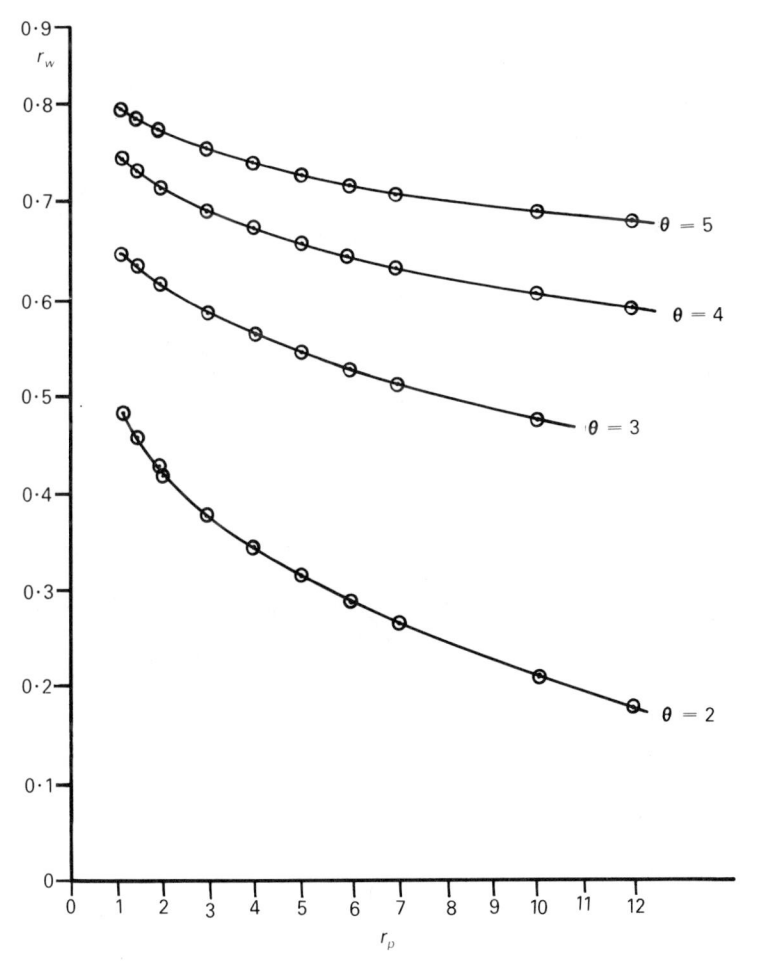

Fig. 5.5 Reheat: Work ratio; $\eta_t = \eta_c = 1$; $T_1 = 302$ K; Air ($\gamma = 1.4$).

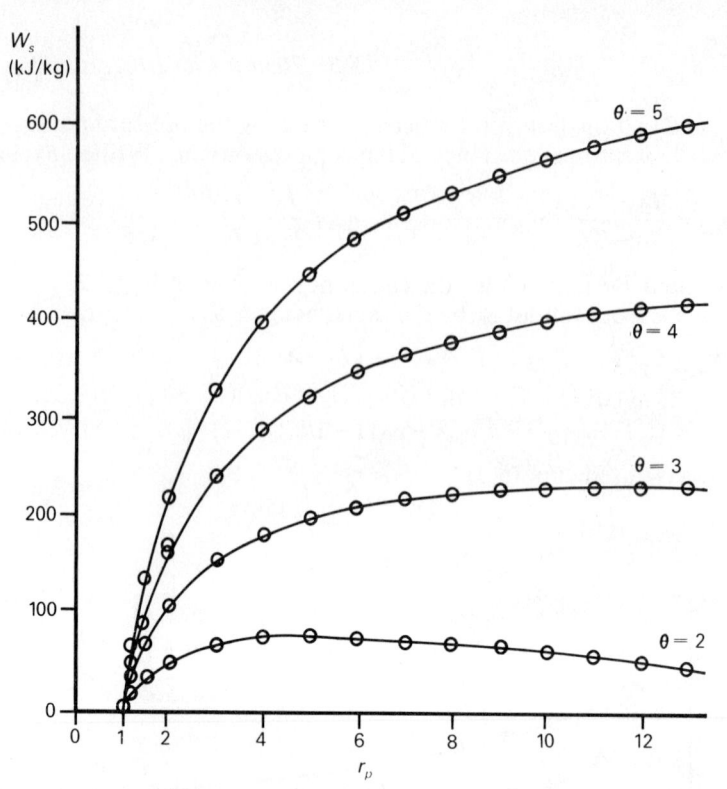

Fig. 5.6 One reheat stage (complete): ideal case. Air: $T_1 = 302$ K.

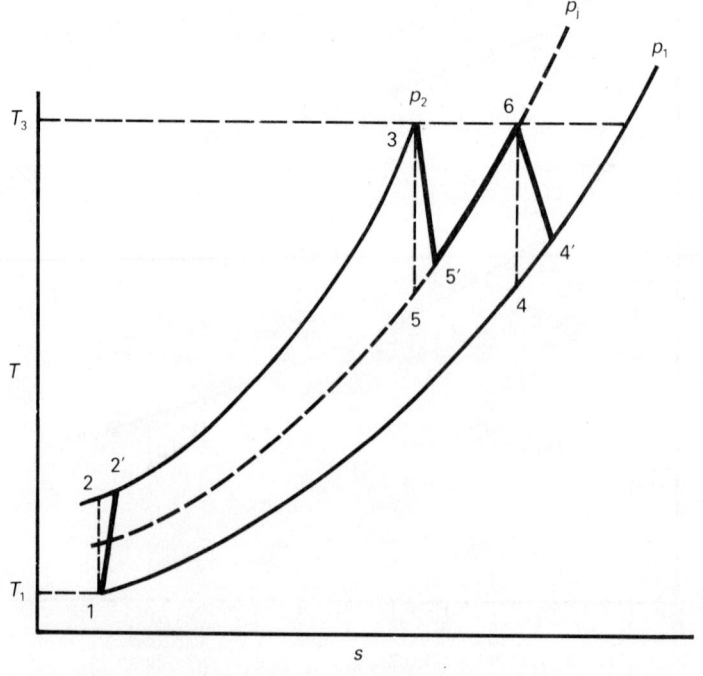

Fig. 5.7 The T–s diagram for one stage of complete reheat with inefficiencies.

The Intermediate Pressure
The intermediate pressure is to be selected so that W_s has a maximum value. Using precisely the same criterion and method as before (in Section 5.2) it is found that the cycle has maximum work output if

$$p_j = \sqrt{(\eta_1/\eta_2)}\sqrt{p_1 p_2} \tag{5.13}$$

This will be very different from the original condition (5.4) in practice and becomes the same if the two turbine stages have the same isentropic efficiency. We can, therefore, still use the condition (5.4) for the selection of the intermediate pressure in all cases. We will make no distinction in the following between η_{t1} and η_{t2}, writing both as η_t. Eqn. (5.12) now becomes

$$W_s = (C_p T_1/\eta_c)\{2a(1 - 1/\sqrt{\rho_p}) - \rho_p + 1\} \tag{5.14}$$

The Working Characteristics
The heat transfer Q into the cycle with reheat is

$$\begin{aligned} Q &= C_p\{(T_3 - T_{2'}) - (T_6 - T_{5'})\} \\ &= C_p\{(T_3 - T_1) - (T_{2'} - T_1) + (T_6 - T_3) - (T_{5'} - T_3)\} \\ Q &= C_p T_1\{(\theta - 1) - (1/\eta_c)(\rho_p - 1) - \theta\eta_t(1/\sqrt{\rho_p})^{1/2} - 1)\} \end{aligned} \tag{5.15}$$

where we have used eqn. (5.5).

An expression for the thermal efficiency η follows by dividing eqn. (5.14) by eqn. (5.15)

$$\eta = \frac{2a(1 - 1/\sqrt{\rho_p}) - \rho_p + 1}{b - \rho_p + a(1 - 1/\sqrt{\rho_p})} \tag{5.16}$$

Here, a and b are defined by eqn. (4.26).

The work ratio follows from eqn. (6.12)

$$r_w = 1 - \frac{\sqrt{\rho_p}(\sqrt{\rho_p} + 1)}{2a} \tag{5.17}$$

and the specific work output is given by eqn. (5.14). The effects of non-isentropic behaviour are seen in Figs. 5.8 and 5.9.

5.4 Intercooling

The division of the expansion stage at the high temperature side of the cycle to increase the expansion work transfer has its counterpart in the division of compression at the low temperature side to reduce the compression work transfer. The full compression occurs now through an intermediate pressure and the overall enthalpy change on compression is reduced from the change due to a single stage. The net work transfer is consequently increased because the expansion work transfer is unaffected. The use of an intermediate pressure regime during compression is called intercooling and if each compression stage starts from the same low temperature the intercooling is said to be complete. The layout and T–s diagram for the cycle are shown respectively in Figs. 5.10 and 5.11. It will be realised that the heat transfer into the system is increased by intercooling so the thermal efficiency is reduced.

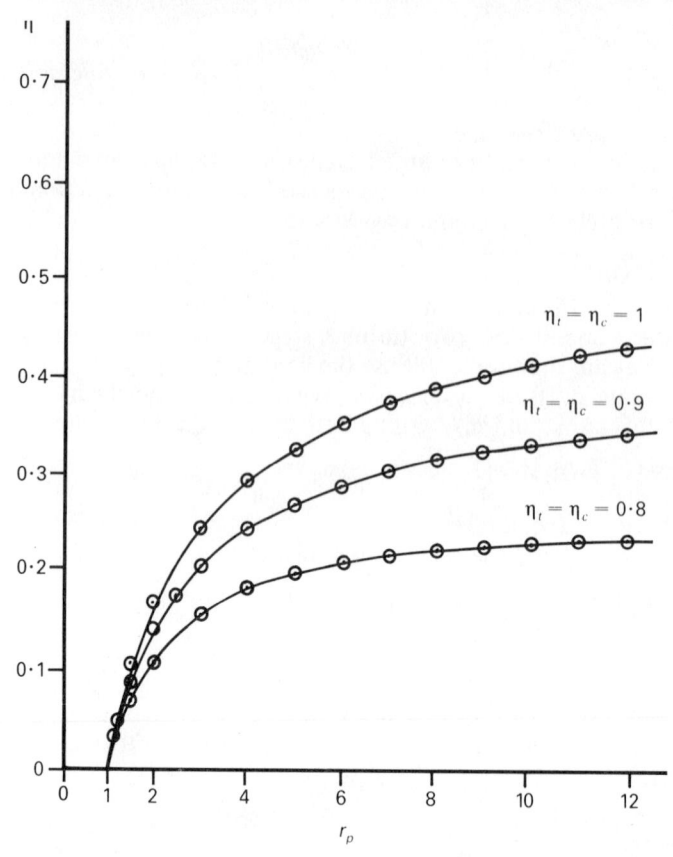

Fig. 5.8 Thermal Efficiency: One stage complete reheat. $\theta = 4$: $T_1 = 302$ K: Air.

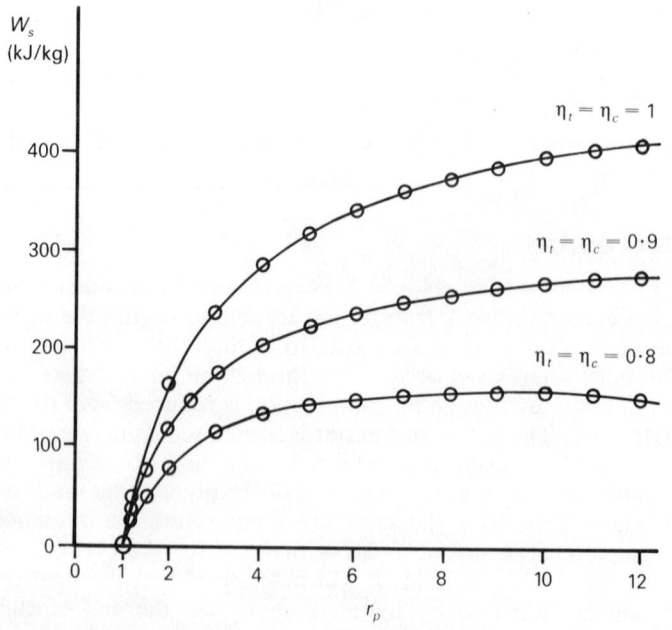

Fig. 5.9 Specific Work Output: One stage complete reheat; $\theta = 4$: $T_1 = 302$ K: Air.

5.4 Intercooling   65

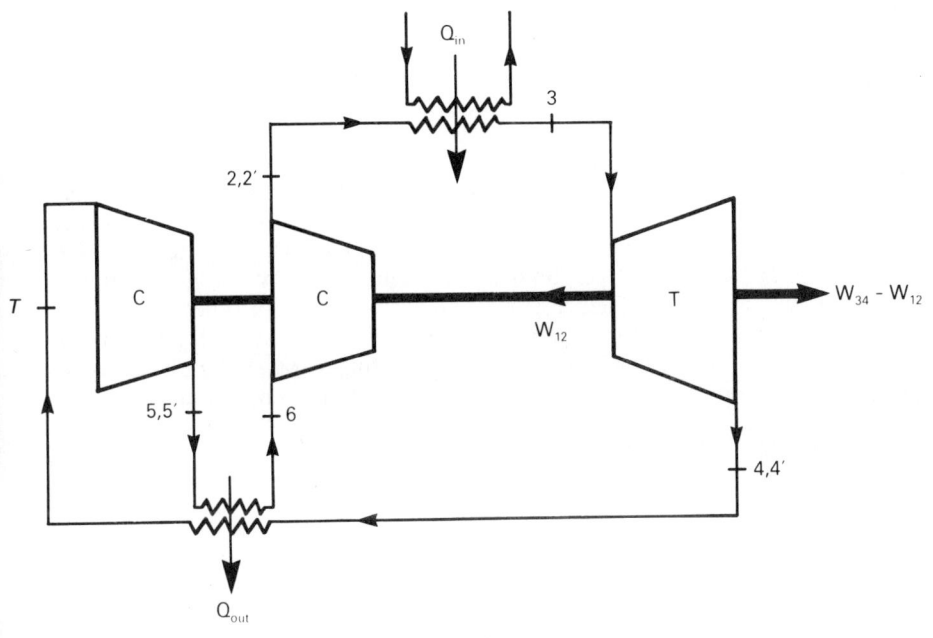

Fig. 5.10 Cycle with one stage of intercooling.

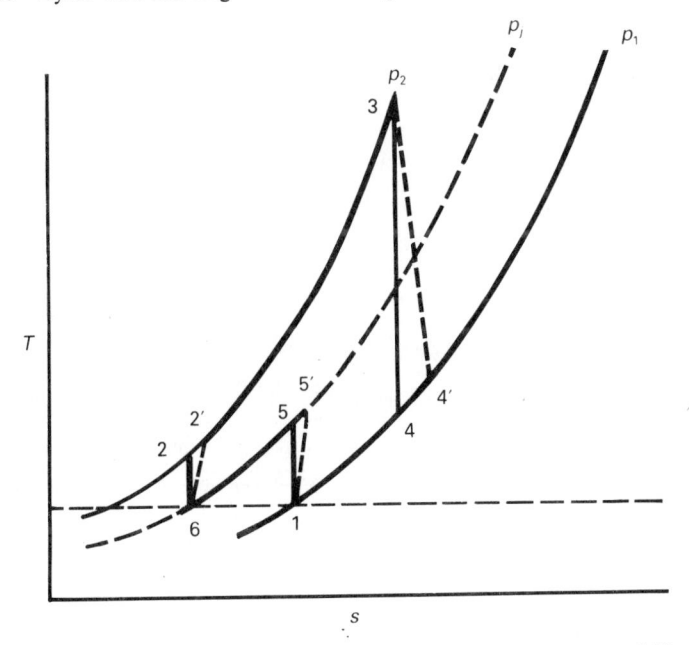

Fig. 5.11 *T–s* diagram for one stage of intercooling including irreversibility.

Selecting the Intermediate Pressure
Consider a single intermediate pressure stage in intercooling. The selection of the pressure will be made to give the minimum compression work transfer. The total work transfer by the cycle will be increased because the compress-

ion work is reduced but the expansion work remains unaffected. The arguments of Section 5.2 for selecting the intermediate pressure for reheat apply again and the result (5.4) (that the intermediate pressure is the geometric mean of the upper and lower cycle pressures) applies again.

The Working Characteristics

The gas is compressed, from the low temperature T_1, through the two stages 1–5' and 6–2' and heated to temperature T_3, point (see Fig. 5.11). It is then expanded through 3–4' and cooled to T_1 (point 1) to complete the cycle. The arguments applied to the cycle analysis in the previous discussions are applied again now. The resulting expressions differ from those for reheat in that the heat transfer into the system (between points 2' and 3) is slightly affected by intercooling but there is no supplementary heating associated with the use of the intermediate pressure. The expression for the thermal efficiency is:

$$\eta = \frac{a(1 - 1/\rho_p) - 2(\sqrt{\rho_p} - 1)}{b - \sqrt{\rho_p}} \tag{5.18}$$

The work ratio is

$$r_w = 1 - \frac{2\sqrt{\rho_p}}{a(\sqrt{\rho_p} + 1)} \tag{5.19}$$

while the specific work output is

$$W_s = (C_p T_1 / \eta_c)\{a(1 - 1/\rho_p) - 2(\sqrt{\rho_p} - 1)\} \tag{5.20}$$

These expressions differ from eqns. (5.16), (5.17) and (5.14) for reheat alone.

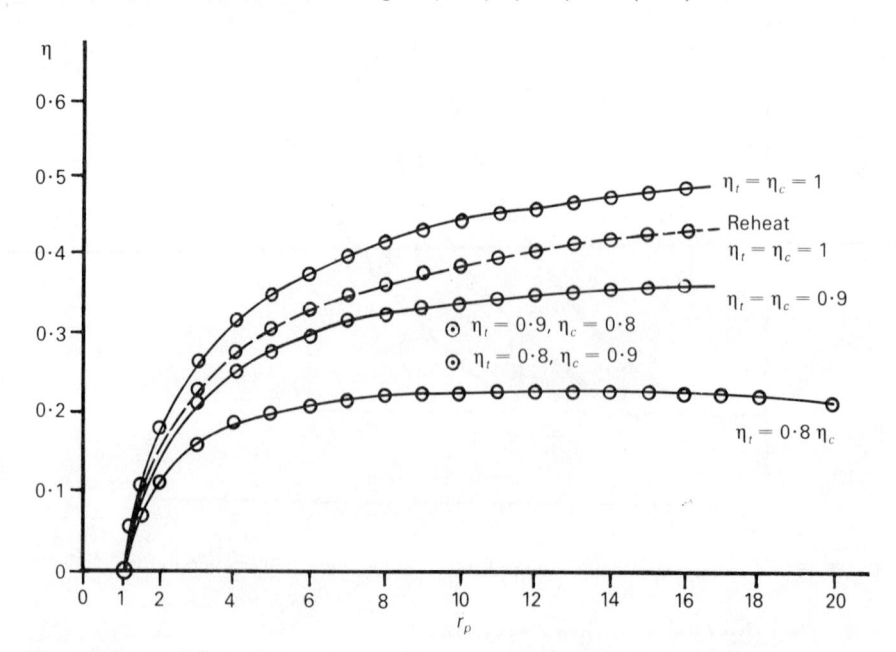

Fig. 5.12 Thermal Efficiency: Intercooling (complete); $\theta = 4$, Air ($\gamma = 1.4$), $T_1 = 302$ K.

5.5 Reheat and Intercooling 67

The data provided by these formulae are collected in Figs. 5.12, 5.13 and 5.14, for the single case of θ = 4 as typical. The corresponding data for one stage of fully efficient reheat are included in each case for comparison. It is seen from Fig. 5.12 that intercooling provides a significant increase over reheat in the thermal efficiency of the cycle, which is consistent with there being no supplementary heating in intercooling to correspond with that for reheat. Also, inefficiencies in the expansion and compression stages affect the thermal efficiency quite markedly, the turbine inefficiency being the more important.

5.5 Reheat and Intercooling

Both reheat and intercooling can be applied simultaneously to the cycle, the T–s diagram then having the form shown in Fig. 5.15. For one stage of

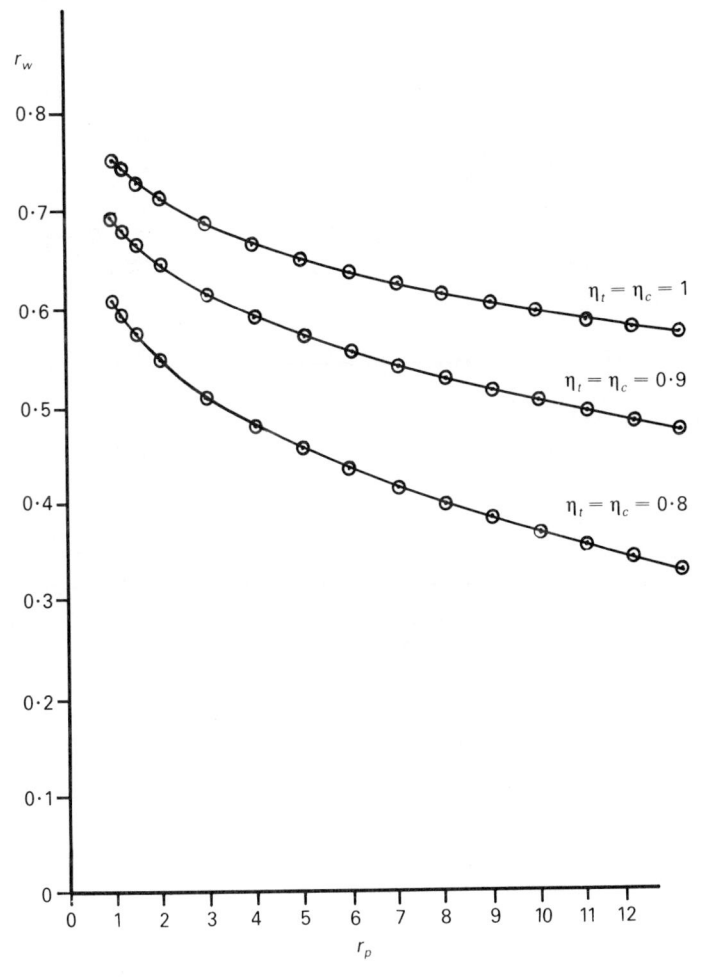

Fig. 5.13 Work ratio: Intercooling; θ = 4, T_1 = 302 K (Air).

68 Reheat and Intercooling

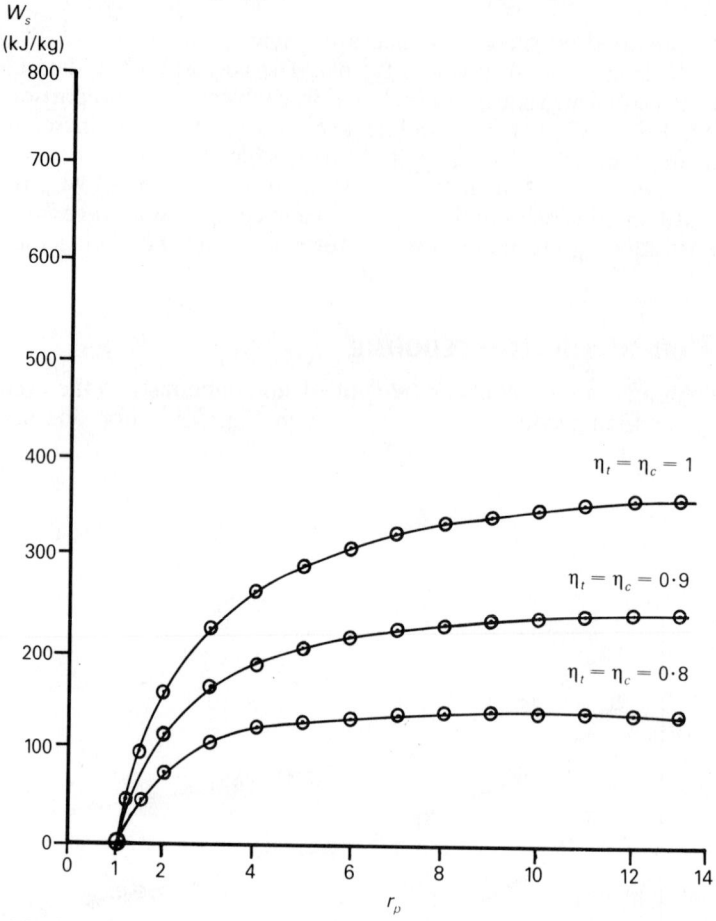

Fig. 5.14 Specific work output intercooling; $\theta = 4$, $T_1 = 302$ K (Air).

complete reheat and one stage of complete intercooling the thermal efficiency is given by

$$\eta = \frac{2\{a(1-1/\sqrt{\rho_p})-(\sqrt{\rho_p}-1)\}}{b-\sqrt{\rho_p}+a(1-1/\sqrt{\rho_p})} \tag{5.21}$$

The work ratio is given by

$$r_w = 1 - \frac{\sqrt{\rho_p}}{a} \tag{5.22}$$

and the specific work output by

$$W_s = C_p T_1 (1/\eta_c)\{2[a(1-1/\sqrt{\rho_p})-(\sqrt{\rho_p}-1)]\} \tag{5.23}$$

The data derived from these expressions are collected in Figs. 5.16, 5.17 and 5.18.

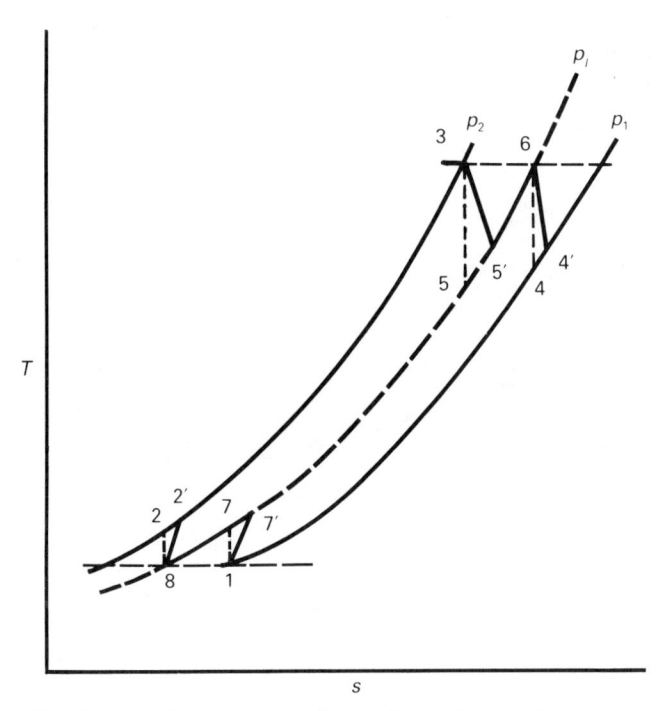

Fig. 5.15 *T–s* diagram for one stage of complete reheat and one stage of complete intercooling.

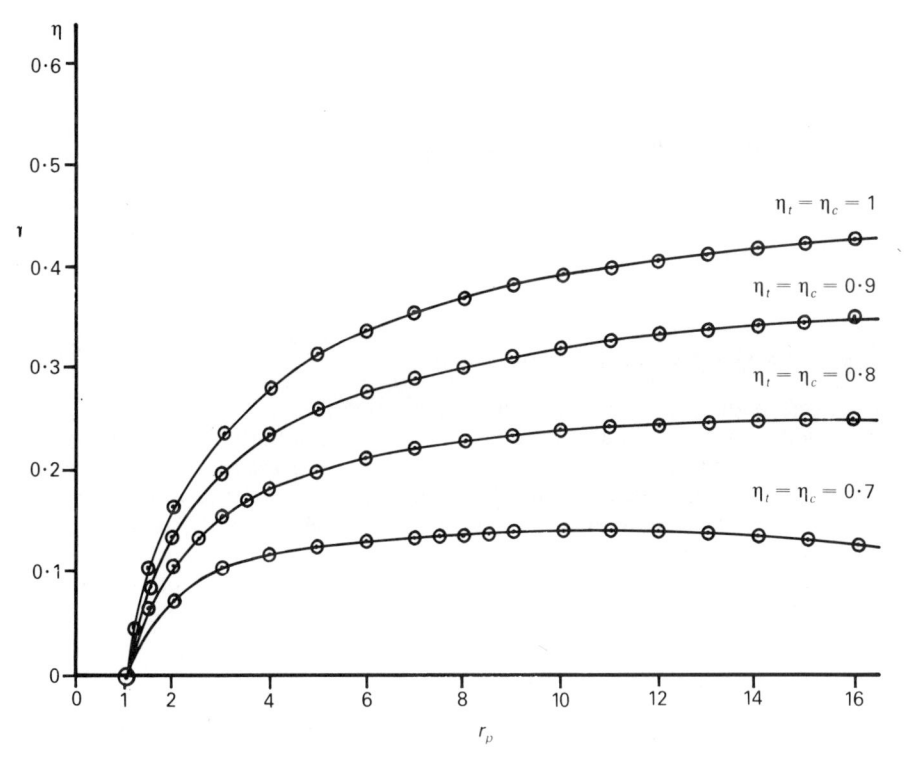

Fig. 5.16 Thermal efficiency: reheat and intercooling; $\theta = 4$, $T_1 = 302$ K (Air).

70 Reheat and Intercooling

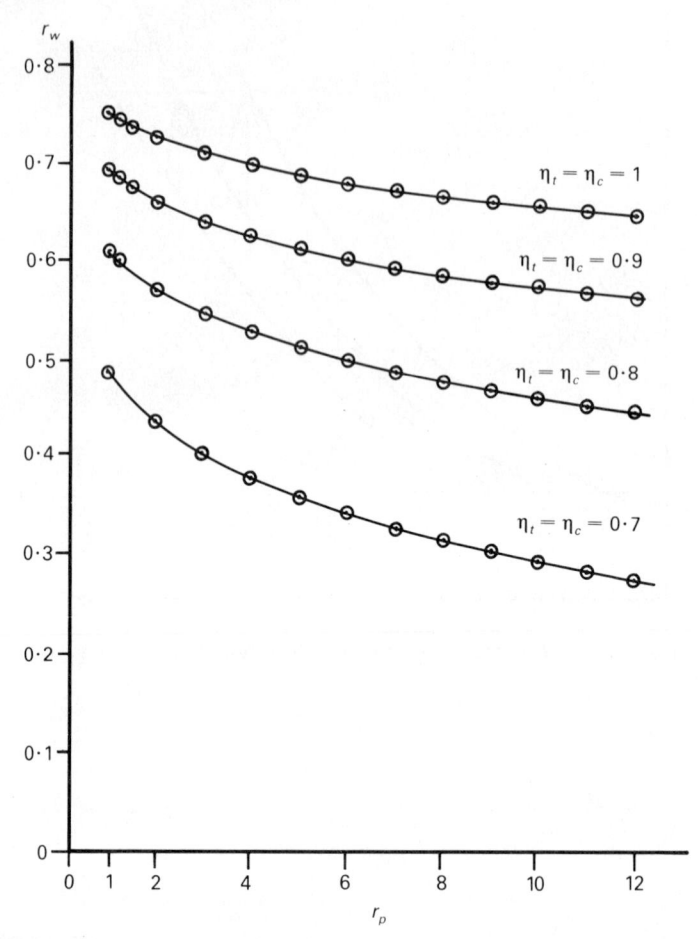

Fig. 5.17 Work ratio: $\theta = 4$ Air.

5.6 Multi-Stage Cycles

Only one stage of reheat or intercooling has been treated here but more than one stage can be applied in practice. This enhances the effects of the single stage, improving the specific work output and, to a lesser extent, the work ratio. The pressure range $p_2 - p_1$ now is spanned by more than one intermediate pressure. The arguments of Section 5.2.1 are applicable again to achieve maximum specific work output.

As an example, consider two stages of reheat without intercooling, the intermediate pressures being p_j and p_k (see Fig. 5.19). Applying eqn. (5.4) to consecutive pairs of pressure lines we have

$$p_j = (p_2 p_k)^{1/2} \qquad p_k = (p_j p_1)^{1/2}$$

and, in terms of p_2 and p_1, this gives

$$p_j = (p_2^2 p_1)^{1/3} \qquad p_k = (p_2 p_1^2)^{1/3} \tag{5.24}$$

5.6 Multi-Stage Cycles

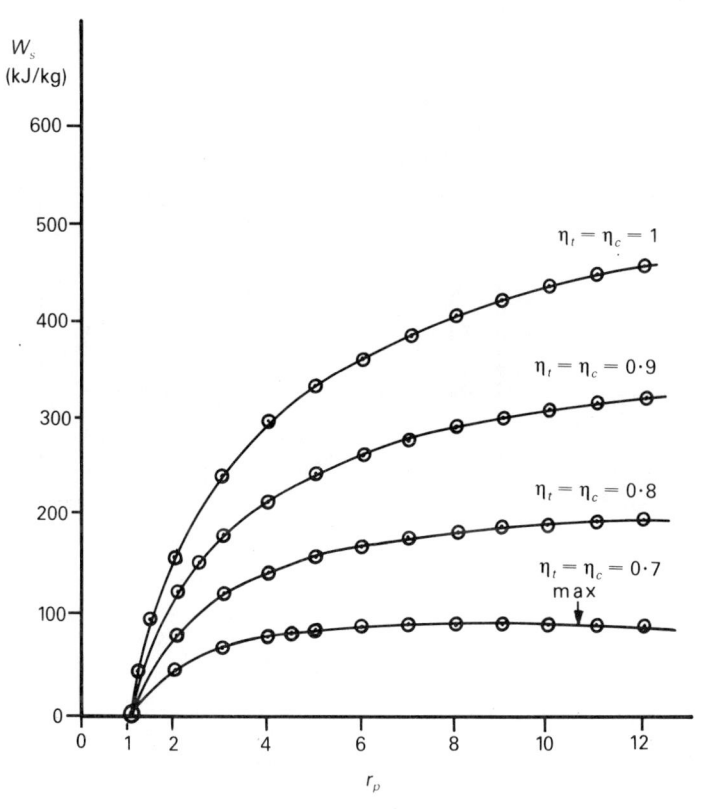

Fig. 5.18 Showing the dependence of the specific work output Ws on the pressure ratio, γ_ϕ, with reheat and intercooling. Air, $\theta = 4$, $T_1 = 302$ K.

The corresponding temperature ratios are given by

$$T_2/T_j = T_j/T_1 = \rho_p^{1/3}$$

Expressions for the thermal efficiency, work ratio and specific work output are derived in the usual way: explicitly

$$\eta = \frac{3a(1 - (1/\rho_p^{1/3})) - \rho_p + 1}{b - \rho_p + 2a(1 - 1/\rho_p^{1/3})} \tag{5.25}$$

$$r_w = 1 - \frac{\rho_p^{1/3}(\rho_p - 1)}{3a(\rho_p^{1/3} - 1)} \tag{5.26}$$

and

$$W_s = C_p T_1(1/\eta_c)\{3a(1 - 1/\rho_p^{1/3}) - \rho_p + 1\} \tag{5.27}$$

Further stages of reheat can be added in the same way. Intercooling can also be supplemented by further compression stages (see also Section 7.3).

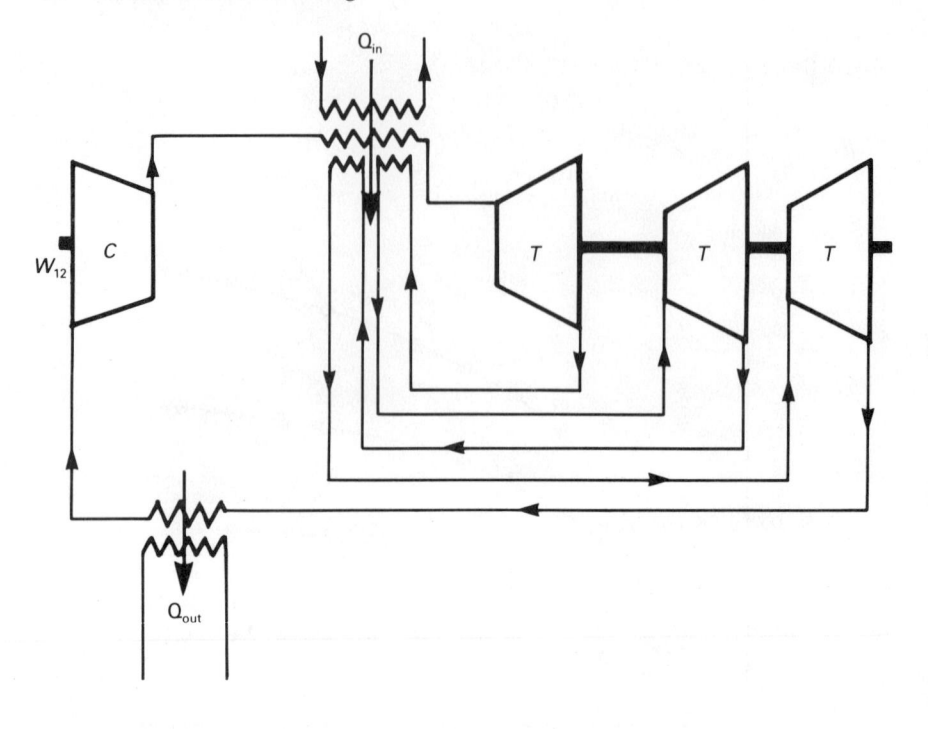

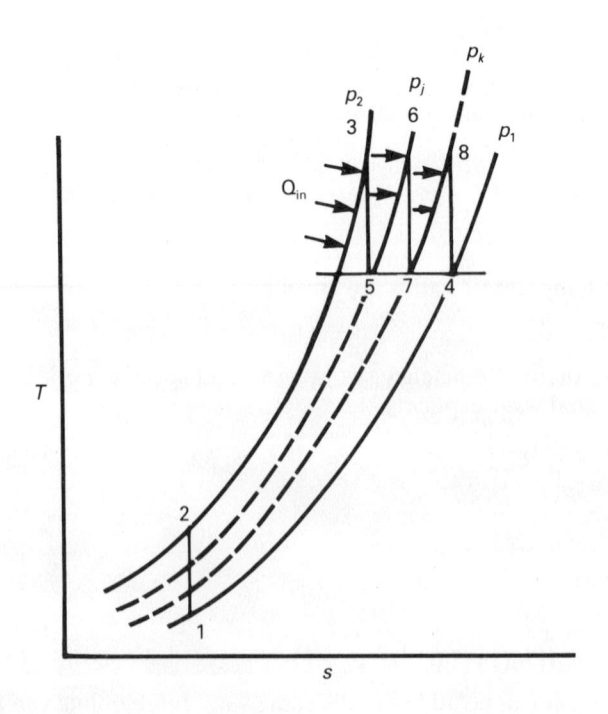

Fig. 5.19 Two stages of reheat.

5.7 Isothermal Compression

An indefinitely large number of intercooling stages would effectively provide an isothermal compression stage to the cycle. This limiting case can be considered by supposing the cycle to have the T–s diagram shown in Fig. 5.20, where non-ideal expansion is included and there is no reheat. The stage 1–2 is now isothermal and the compression work transfer is $W_{12} = RT_1 \log r_p$, where R is the gas constant. Remembering that, for an ideal gas, $R = C_p - C_v$ (where C_p and C_v are the specific heat capacities of the gas at constant pressure and constant volume respectively), it is found that the thermal efficiency of the cycle is given by

$$\eta = [\eta_t \theta(\rho_p - 1) - \rho_p(1 - 1/\gamma) \log r_p]/[\rho_p(\theta - 1)] \tag{5.28}$$

The work ratio is expressed by

$$r_w = 1 - \left(\frac{\gamma-1}{\gamma}\right)\frac{\rho_p}{\eta_t(\rho_p - 1)} \log r_p \tag{5.29}$$

and the specific work output by

$$W_s = C_p T_1 \eta_t \{(\theta/\rho_p)(\rho_p - 1) - (1/\eta_t)(1 - 1/\gamma) \log r_p \tag{5.30}$$

The dependence of these quantities on the pressure ratio r_p is plotted in Figs. 5.21 and 5.22.

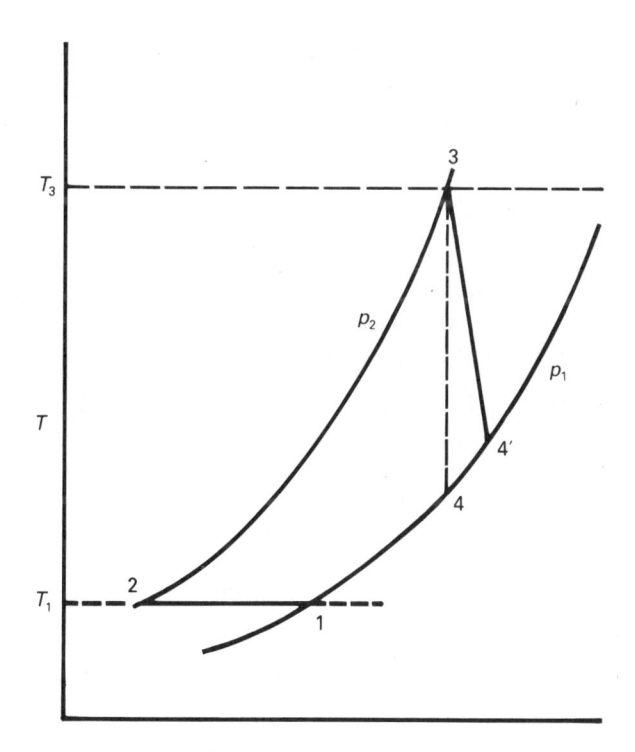

Fig. 5.20 T–s diagram for isothermal compression.

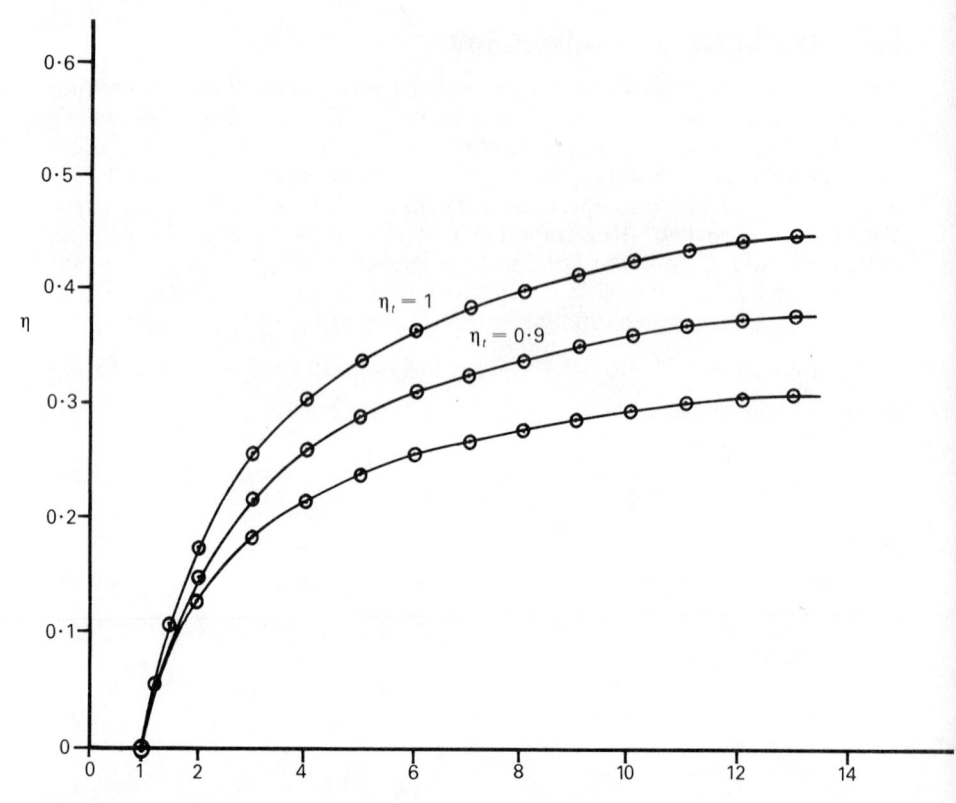

Fig. 5.21 Showing the dependence for air of the thermal efficiency, η on the pressure ratio r_p, $\theta = 4$, $T_1 = 302$ K.

5.8 Summary

1. The specific work output of the Joule cycle can be increased by replacing the single isentropic expansion (or compression) by a succession of smaller expansions (or compressions) involving a series of intermediate pressures.
2. The intermediate pressure stages are chosen to maximise the specific work output. This provides the geometric mean of the higher and lower pressure stages as the appropriate choice for the intermediate stage.
3. For expansion, the expansion work transfer is made a maximum by expanding the working gas through a series of intermediate pressures, bringing the temperature of the gas back to the initial temperature after each expansion. Such a process is called complete reheat. The reheat is incomplete if the temperature after expansion is not returned to the initial value.
4. Reheat involves the expenditure of more heat transfer into the system so, although the specific work output is increased, the thermal efficiency is lower than for the simple cycle.
5. The behaviour of the cycle with one stage of complete reheat is

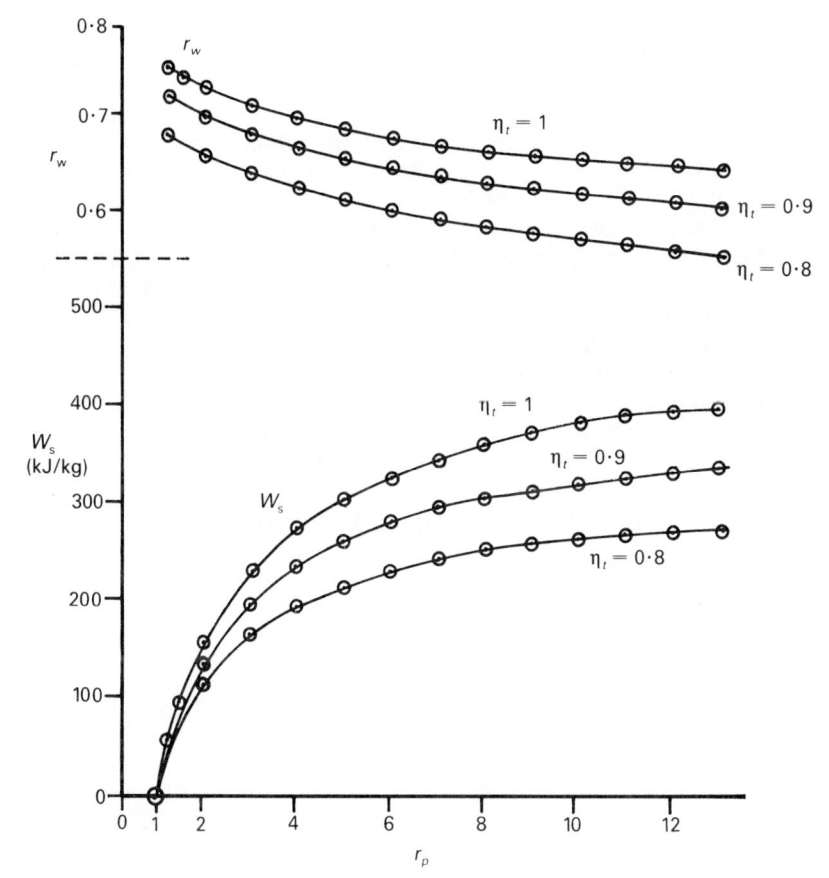

Fig. 5.22 Showing the dependence of the specific work output W_s, and the work ratio r_p on the pressure ratio, $\theta = 4$, $T_1 = 302$ K.

investigated with full isentropic change and in the presence of inefficiencies.
6. The introduction of intermediate pressure stages for the heat rejection process reduces the compression work transfer. The selection of the pressure stages for minimum work of compression follows the same rule as for maximum work of expansion. The introduction of intermediate pressure stages in the compression stage is called intercooling: intercooling is complete if each expansion stage is brought back to the common lowest temperature for the cycle, otherwise it is incomplete.
7. Intercooling does not involve an increase of heat transfer into the system and so provides an increase of specific work output without a corresponding deterioration of the thermal efficiency.
8. The transfer characteristics for a cycle with one stage of intercooling are investigated both with and without inefficiencies of the expansion/compression stages.

76 Reheat and Intercooling

9. The transfer characteristics of the cycle with both reheat and intercooling are investigated.
10. Isothermal compression is the limiting case of intercooling when the number of stages becomes indefinitely large. This condition is investigated in detail.
11. The results of all the deliberations in this chapter are summarised numerically in Figs. 5.23 (for the thermal efficiency) and 5.24 (for the work output): we have chosen the case $\theta = 4$ as representative. It is seen that the specific work output of the simple Joule cycle is increased by introducing intermediate pressure stages. Intercooling increases W_s by a few per cent, reheat by rather more than 20 % and combined reheat and intercooling by about 30 % or more for r_p about 10. These values are increased if θ is increased. This is a worthwhile augmentation of work output. The thermal efficiency of the simple cycle is, on the other hand, reduced by intercooling, rather more so by reheat and most by a combination of reheat and intercooling. The decrease then is rather in excess of 20 %. The increase of W_s at the cost of a decrease of η is a general result for all temperature and pressure ratios.
12. The effects of non-isentropic behaviour are marked and it is generally

J   - Simple Joule cycle
I   - Joule cycle with intercooling alone
R   - Joule cycle with reheat alone
I + R - Joule cycle with reheat and undercooling

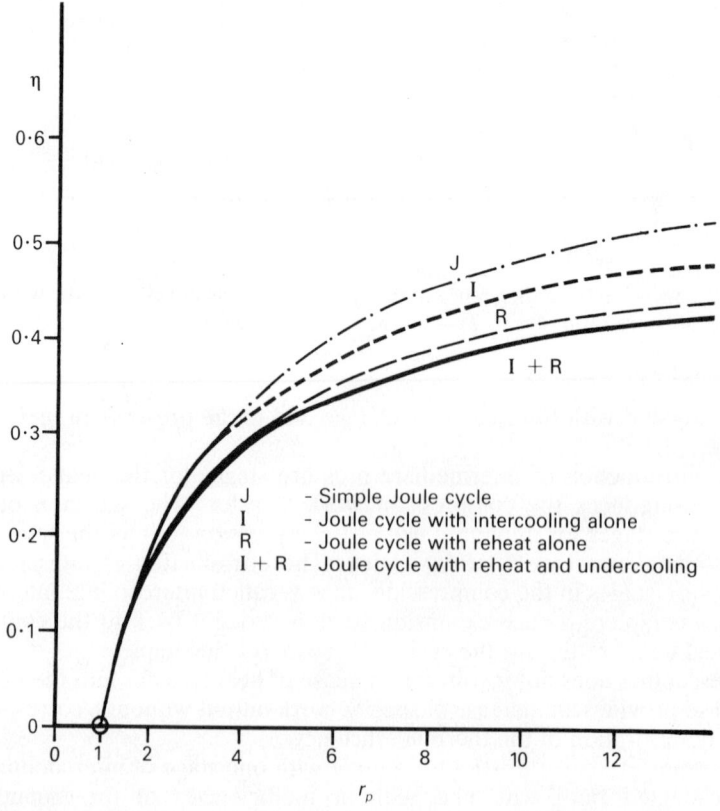

Fig. 5.23 Thermal Efficiency: Summary; $\theta = 4$: Air: $T_1 = 302$ K.

Fig. 5.24 Summary; $\theta = 4$: Air: $T_1 = 302$ K.

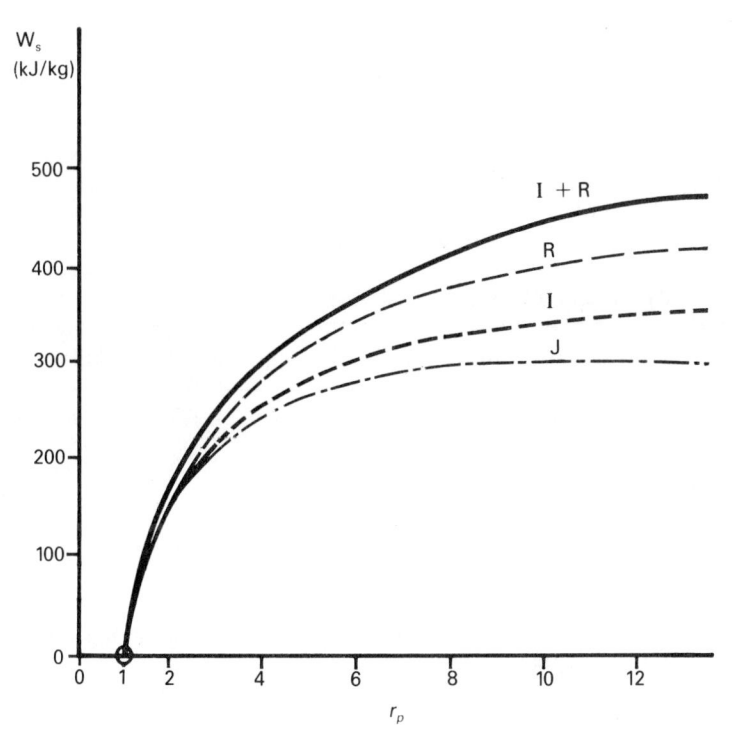

The Specific Work Output for various cases of the Joule cycle for air:
J   - Simple Joule cycle
I   - Joule cycle with intercooling alone
R   - Joule cycle with reheat alone
I + R - Joule cycle with reheat and intercooling

advantageous to include reheat (and perhaps intercooling) provided the isentropic efficiencies for expansion and compression are in excess of 0.85.

13. Reheat is generally more effective than intercooling in that it provides a substantial increase of work output without excessive loss of thermal efficiency and yet holds to a minimum the cost and weight of pipework ancillary to the main cycle. This is particularly important for mobile power plants.

5.9 Exercises

[It is presumed that a microcomputer is available for use: if this is not so, derive the formulae in the Questions and take one numerical situation (the same in each case) as a method of comparison between the different cycle configurations.]

5.1 A Joule cycle has one stage of complete reheat, the work transfer characteristics being as described in Section 5.3. By taking successively $\theta = 3$

78 Reheat and Intercooling

and $\theta = 5$ in eqns. (5.18), (5.19) and (5.20), investigate how deviations from the full isentropic efficiencies of expansion and compression affect the working behaviour of the cycle. Plot your results graphically.

5.2 The cycle has one stage of intercooling added, the work transfer characteristics being as in Section 5.5. Construct companion curves to Figs. (5.16), (5.17) and (5.18) for $\theta = 3$ and $\theta = 5$. What conclusions can be drawn about the effects of irreversibilities on the working of the cycle?

5.3 The arguments of Section 5.2 are incomplete in that it was not demonstrated that $d^2W_s/dp_j^2 < 0$ for the choice for p_j expressed by equation (5.4). Make that demonstration now by showing, using (5.3a) and (5.4), that

$$\frac{d^2W_s}{dp_j^2} = -C_p T_1 \{((2\gamma-1)/\gamma) p_1^{(\frac{3\gamma-1}{\gamma})} p_2^{(\frac{3\gamma-1}{2\gamma})} - (1/\gamma) p_1^{-(\frac{\gamma+1}{2\gamma})} p_2^{-(\frac{3\gamma-1}{\gamma})}\}$$

concluding that the expression within the bracket is always positive within the working characteristics of the cycle. This is interpreted as showing that the choice $p_j = \sqrt{p_1 p_2}$ does, indeed, provide a maximum value for the specific work output of the cycle.

5.4 A Joule cycle has one stage of complete reheat added, the single intermediate pressure being chosen to be $p_j = \alpha\sqrt{p_1 p_2}$ where α is a number in the range $\sqrt{1/r_p} < \alpha < \sqrt{r_p}$. Writing $n = (\gamma-1)/\gamma$, show that the expression for the specific work output of the cycle is now

$$W_s(\alpha) = C_p T_1 \{\theta(1 - \alpha^n/\sqrt{\rho_p}) + \theta(1 - 1/\alpha^n \rho_p) - \rho_p + 1\}$$

For chosen values of r_p and θ, establish that $W_s(\alpha) < W_s(\alpha = 1)$ in the range available to α. This shows that the specific work output for the conditions considered in Section 5.2 is indeed the maximum value.

Show further that the corresponding expression for the thermal efficiency is

$$\eta = \frac{2\theta - \theta/\sqrt{\rho_p}(\alpha^n + 1/\alpha^n) - \rho_p + 1}{2\theta - \rho_p - \alpha^n(\theta/\sqrt{\rho_p})}$$

Deduce the effect of moving away from the condition of maximum specific work output (that is, of varying α) on the thermal efficiency of the cycle. Explain your results in terms of the appropriate T–s diagram.

5.5 A gas turbine system has n stages of complete reheat but no intercooling. The gas can be supposed ideal, and the isentropic efficiencies for expansion and compression are respectively η_t and η_c. Show that the temperature ratio between consecutive intermediate pressure stages is given by

$$T_j/T_{j-1} = \rho_p^{1/(n+1)} = d_n$$

for the jth and $(j-1)$th pressure stages. If there are no intermediate pressure stages, $n = 0$ and $\rho_p = d_0$.

Derive expressions for the thermal efficiency, work ratio and specific work output for the cycle in the form

5.9 Exercises

$$\eta = \frac{a(n+1)(1-1/d_n) - d_0 + 1}{b - d_0 + na(1-1/d_n)}$$

$$r_w = 1 - \frac{d_n(d_0 - 1)}{(n+1)a(d_n - 1)}$$

$$W_s = (C_p T_1/\eta_c)\{a(n+1)(1-1/d_n) - d_0 + 1\}$$

Show that these expressions reduce to those of Section 5.3 for the case of a single reheat stage ($n = 1$).

Taking $\theta = 4$ as typical, calculate the values of η, r_w and W_s for the cases of 1, 2, 3, 5, 7 and 10 stages of complete reheat expressing your results in graph form. What conclusions can be drawn from your data about the *practical* use of more than one stage of reheat?

5.6 The gas turbine system of Question 5.1. is given n stages of intercooling and the reheat stages are removed. Show that the thermal efficiency, work ratio and specific work output of the new cycle are given by

$$\eta = \frac{a(1 - 1/d_0) - (n+1)(d_n - 1)}{b - nd_n}$$

$$r_w = 1 - \frac{(n+1)d_o(d_n - 1)}{a(d_0 - 1)}$$

$$W_s = (C_p T_1/\eta_c)\{a(1 - 1/d_0) - (n+1)(d_n - 1)\}$$

Establish that these expressions reduce to those of Section 5.4 for the case of a single intercooling stage ($n = 1$).

Again taking $\theta = 4$ as typical, calculate the values of these quantities for the cases of 1, 2, 3, 5, 7 and 10 stages of complete intercooling. Express the effect of intercooling on these quantities in graph form.

5.7 The cycle now has reheat restored to yield n stages of complete reheat and m stages of complete intercooling. Show that the transfer characteristics of the cycle are described by the expressions

$$\eta = \frac{a(n+1)(1 - 1/d_n) - (m+1)(1 - d_m)}{b - d_n + n(1 - 1/d_n)}$$

$$r_w = 1 - \frac{(m+1)(1 - d_m)}{a(n+1)(1 - 1/d_n)}$$

$$W_s = (C_p T_1/\eta_c)\{a(n+1)(1 - 1/d_n) - (m+1)(1 - d_m)\}$$

Show that these expressions reduce to those of Setting 5.5 for the case of a single stage of reheat and a single stage of intercooling.

Taking the special case $\theta = 4$ as typical, calculate the values of these quantities for various combinations of reheat and intercooling with $n, m = 1, 2, 3, 5, 7$ and 10, expressing your conclusions in graph form.

80 Reheat and Intercooling

5.8 Compare and contrast the results of Questions 5.4, 5.5, 5.6 and 5.7 from the point of view of the *practical* application of reheat and intercooling, for (a) stationary and (b) mobile gas turbine plant. How would your conclusions be affected if (a) the plant was to be used as the sole source of work transfer and (b) if the plant was to be used only briefly for occasionally boosting the peak load of a separate base load power plant?

5.9 A gas turbine cycle has isothermal cooling and n stages of complete reheat. Show that the thermal efficiency, work ratio and specific work output are described respectively by the expressions

$$\eta = \frac{\eta_t \theta(n+1)(1-1/d_n) - (1-1/\gamma)\log r_p}{n\theta(1-1/d_n) + \theta - 1}$$

$$r_w = 1 - \frac{1}{\eta_t} \cdot \frac{\gamma - 1}{\gamma} \cdot \frac{\log r_p}{(n+1)\theta(1-1/d_n)}$$

$$W_s = (C_p T_1 \eta_t \{(n+1)\theta(1-1/d_n) - (1/\eta_t)(1-1/\gamma)\log r_p\}$$

Plot numerical values for these expressions as functions of the pressure ratio for 1, 2, 3, 5, 7 and 10 reheat stages in graph form. Compare the calculated features with those obtained from Question 5.3 for the different numbers of stages of intercooling. Discuss the practical applicability of isothermal cooling and its desirability in practice. What might be its value in a space vehicle?

5.10 Derive the formulae expressing the various combinations of reheat and intercooling considered in this chapter by calculating in each case the heat accepted Q_{in} and the heat rejected Q_{out} by the cycle, using the expressions

$$\eta = [Q_{in} - Q_{out}]/Q_{in} \quad \text{and} \quad W_s = [Q_{in} - Q_{out}]$$

This provides an essentially independent check on the derivation of the various expressions.

6. JOULE CYCLE WITH HEAT EXCHANGE

The ideal simple Joule cycle has useful work characteristics even though they fall short of the Carnot ideal. Like the Carnot cycle, the Joule cycle is, however, very adversely affected by irreversible behaviour and both the thermal efficiency and the specific work output suffer badly. Ways of improving the work ratio and the specific work output were treated in the last chapter. In this chapter we focus attention on methods of improving the thermal efficiency and this is important for the economical working of the plant. The method used is to incorporate an internal heat exchange system in which part, or even a major part, of the heating of the gas before expansion is achieved using heat transfer from the expanded (exhaust) gas which is of no further use for work transfer by further expansion.

6.1 The Concept of Internal Heat Exchange

The gas that leaves the turbine still has an appreciable enthalpy content even though it is too low to provide significant work transfer in an expansion turbine. This heat, in the simple cycle, passes to the cooler or to the atmosphere and is lost to the plant. It could, however, be used to pre-heat the

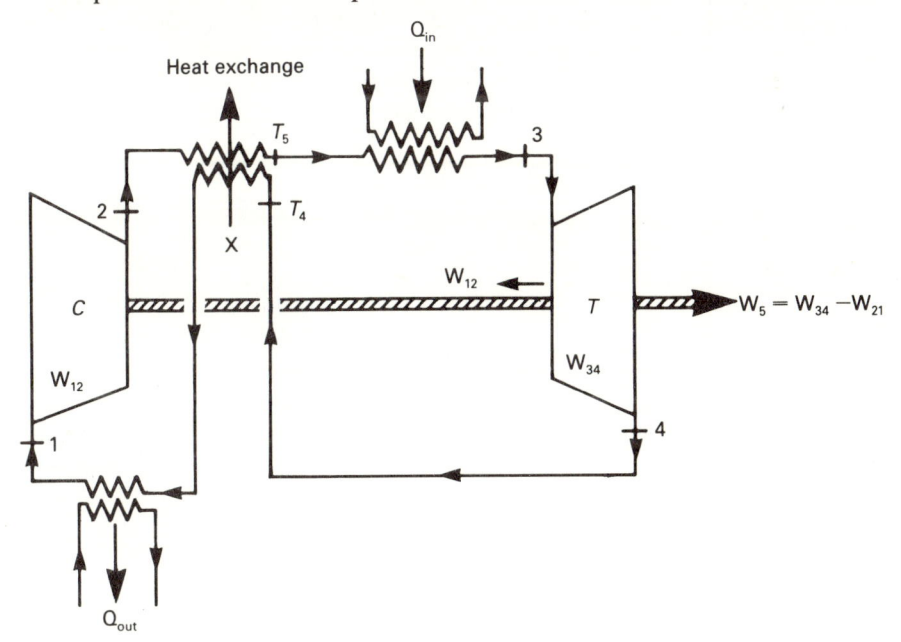

Fig. 6.1 The simple Joule cycle with heat exchange added.

82 Joule Cycle with Heat Exchange

gas entering the heater before it is finally cooled and in this way regenerative heating is employed in the cycle. This is the aim of the heat exchanger.

The circuit is shown in Figure 6.1 and the associated T–s diagram in Figure 6.2. The gas is compressed between points 1 and 2 (the two diagrams follow the same notation) and is heated to the state 3 at the constant pressure p_2. Expansion at constant entropy in the turbine is denoted by the line joining points 3 and 4. The gas is then taken at the constant pressure p_1 (with $p_2 = p_1 r_p$) to the low temperature point T_1 (point 1) but on the way is brought into contact with the gas being heated (point 5). Ideally, the incoming gas is raised to the same temperature as the outgoing gas at point 4, that is $T_4 = T_5$ and the line 4–5 in Figure 6.2 is horizontal. With this arrangement the heat input is between points 5 and 3 and the heat between points 2 and 5 is provided by the already expanded gas from the turbine. Less heat transfer is needed to raise the gas to the prescribed temperature so the thermal efficiency of the cycle must be raised. The expansion and compression enthalpies are unaffected by the heat exchange process leaving the work ratio and the specific work output unchanged. The heat exchange process, therefore, improves the thermal efficiency of the cycle without affecting the work transfer.

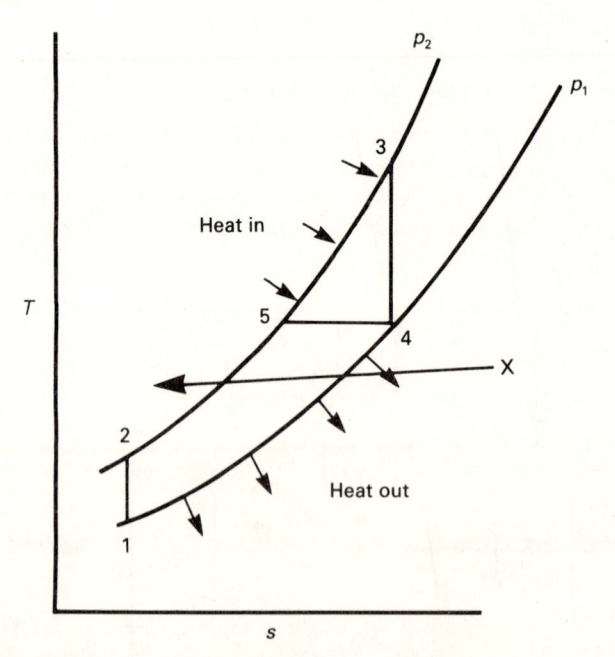

Fig. 6.2 The T–s diagram for the simple Joule cycle with heat exchange (X denotes heat exchange).

6.2 Characteristics of the Cycle

Expressions for the working characteristics of the system are readily obtained by calculating the work and heat transfers. From Fig. 6.2 it follows that

work transfer = $(h_3 - h_4) - (h_2 - h_1) = C_p\{(T_3 - T_4) - (T_2 - T_1)\}$
heat transfer = $(h_5 - h_3) = C_p(T_5 - T_3)$.
Also $T_4 = T_5$.

The Thermal Efficiency
This is expressed in the form

$$\eta = \frac{(T_3 - T_4) - (T_2 - T_1)}{(T_3 - T_5)}$$

$$= \frac{T_3(1 - T_4/T_3) - T_1(T_2/T_1 - 1)]}{(T_3 - T_2) - (T_5 - T_2)}$$

$$\eta = \frac{T_3(1 - T_4/T_3) - T_1(T_2/T_1 - 1)}{T_3(1 - T_2/T_3) - T_4(1 - T_2/T_4)} \quad (6.1)$$

where in the denominator of the second form we have written

$$T_3 - T_5 = T_3 - T_5 + T_2 - T_2$$

Noting that, from eqns. (4.13) and (4.14)

$$T_4/T_3 = 1/\rho_p \quad T_2/T_1 = \rho_p \quad T_3/T_2 = (T_3/T_1)(T_1/T_2) = \theta/\rho_p$$
$$T_4/T_2 = (T_4/T_3)(T_3/T_2) = \theta/\rho_p^2$$

we have successively

$$\eta = \frac{(1 - 1/\rho_p) - (1/\theta)(\rho_p - 1)}{(1 - \rho_p/\theta) - (1/\rho_p)(1 - \rho_p^2/\theta)}$$

$$= \frac{(1 - 1/\rho_p) - (1/\theta)(\rho_p - 1)}{(1 - 1/\rho_p)}$$

$$\eta = 1 - \rho_p/\theta \quad (6.2)$$

This is an entirely different form from eqn. (4.15) for the system without a heat exchanger. Because $\rho_p < \theta$, it is seen immediately that the thermal efficiency of the system with heat exchange is greater than that without. Because the presence of a heat exchanger does not affect the work transfer in any way the work ratio and the specific work output are as for the simple case. The effect of the heat exchanger is, then, to increase the thermal efficiency without affecting the work transfer characteristics of the system.

Data for the thermal efficiency are plotted in Fig. 6.3.

Efficiency at Maximum Work Output
The maximum work output occurs, as before (see Section 4.4), when $\rho_p^2 = \theta$. The thermal efficiency with the heat exchanger then reduces to the form of the system without heat exchange. There is, then, *under this condition* no advantage in including a heat exchanger in the circuit since, at the point of maximum work output, the heat exchanger makes no contribution to the system.

84 Joule Cycle with Heat Exchange

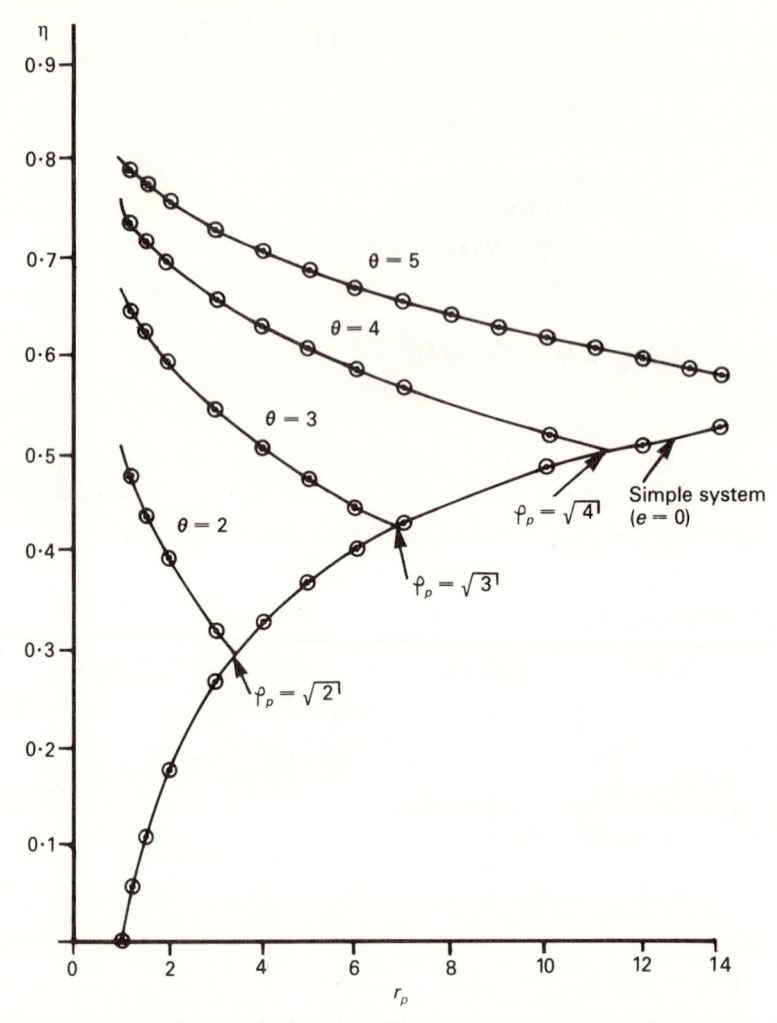

Fig. 6.3 Thermal efficiency with ideal heat exchange ($e = 1$). For air $\gamma = 1.40$; $\eta_t = \eta_c = 1$; $T_1 = 302$ K.

The reason for this becomes clear when we write the condition in terms of the temperatures in the circuit. We have

$$\rho_p \rho_p = (T_3/T_4)(T_2/T_1) = T_3/T_1$$

for maximum work output. This leads immediately to the condition $T_2 = T_4$, the exit temperature from the compressor being equal to the exit temperature from the turbine. At this stage the cycle clearly offers no net heat transfer. If $\rho_p > \theta$, then $T_2 > T_4$ and the heat exchanger would transfer heat to the already expanded gas: work transfer is then into the cycle from the environment. We see that the heat exchanger is important for pressure ratios less than that for maximum work output, but it is positively disadvantageous for larger pressure ratios, actually reducing the thermal efficiency of the original system.

6.3 Less than Perfect Heat Exchange

In practical systems the heat exchange will not be perfect and the point 4 in Fig. 6.4 for the return circuit will not raise the incoming fluid to the point 5. Rather it will raise it to the lower temperature $T_{5'}$. The efficiency factor e for the heat exchange is chosen to describe how close point 5' is to the ideal point 5. Explicitly

$$e = \frac{(h_{5'} - h_2)}{(h_5 - h_2)} = \frac{T_{5'} - T_2}{T_5 - T_2} \tag{6.3}$$

This means we write $T_{5'} - T_2 = e(T_5 - T_2)$.

The work done is, as before, entirely unaffected by the heat exchanger but the heat transfer is now

$$\begin{aligned} T_3 - T_{5'} &= (T_3 - T_2) - (T_{5'} - T_2) \\ &= (T_3 - T_2) - e(T_5 - T_2) \end{aligned}$$

Also, $T_5 = T_4$. The expression for the thermal efficiency is

$$\eta = \frac{T_3\{(1 - T_4/T_3) - (T_1/T_3)(\rho_p - 1)\}}{T_3(1 - T_2/T_3) - eT_4(1 - T_2/T_4)}$$

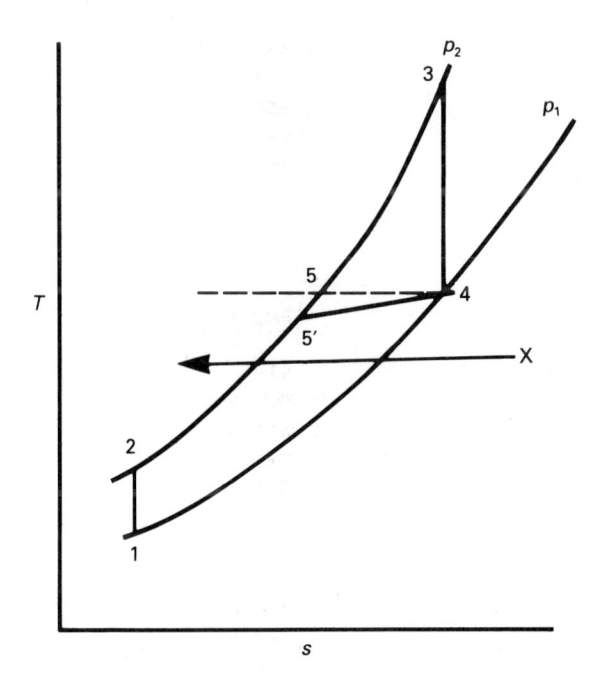

Fig. 6.4 Showing the effect of imperfect heat exchange. The compressed gas is heated to the temperature $T_{5'} < T_5$ and enthalpy $h_3 - h_{5'}$, $h_3 - h_5$ must be supplied by the heat input.

86 Joule Cycle with Heat Exchange

Introducing the symbols for the several ratios as before we have

$$\eta = \frac{(1-1/\rho_p) - (1/\theta)(\rho_p - 1)}{(1-\rho_p/\theta) - (e/\rho_p)(1-\rho_p^2/\theta)} \tag{6.4}$$

Data for the thermal efficiency of the system with less than ideal heat exchange are collected in Fig. 6.5. It is seen that the system is sensitive to the efficiency of the heat exchange process, but this sensitivity is obvious at the lower values of the pressure ratio if e is not far from unity. Eqn. (6.4) reduces to that of Section 6.2, eqn. (6.2) if the heat transfer is ideal, that is $e = 1$. It reduces to that of the simple cycle (Section 4.3, eqn. (4.15)) if there is no heat transfer, that is if $e = 0$.

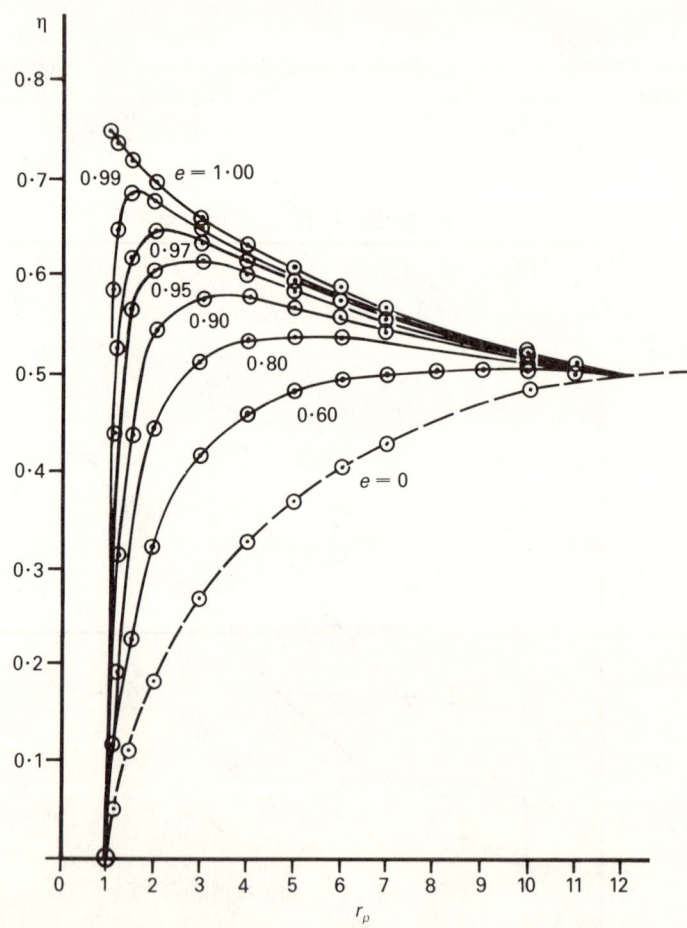

Fig. 6.5 Thermal efficiency as function of the pressure ratio for different values of the effectiveness for heat exchange, e. $\theta = 4$; $T_1 = 302$ K. For air: $\gamma = 1$; $\eta_t = \eta_c = 1$.

6.4 A Less than Perfect System

Although we have taken account of the inefficiencies in the heat exchanger unit, we have not yet accounted for the corresponding inefficiencies in the turbine and condenser sets as well. This we do now, deriving expressions to describe the total system with less than perfect components. The work transfer will now be affected as well as the heat transfer so the thermal efficiency, work ratio and specific work output will all be affected by the irreversibilities of the system. The T–s diagram is shown in Fig. 6.6. The work transfer is unaffected by the heat exchange stage and so the work transfer is as in Section 4.5, eqn. (4.28): this is work transfer

$$\begin{aligned} &= C_p\{(T_3 - T_{4'}) - (T_{2'} - T_1)\} \\ &= C_p\{\eta_t(T_3 - T_4) - (1/\eta_c)(T_2 - T_1)\} \\ &= C_p T_1\{\eta_t \theta(1 - 1/\rho_p) - (1/\eta_c)(\rho_p - 1)\} \end{aligned} \qquad (6.5)$$

The heat transfer has one new feature in that the lower temperature for heat reception is now $T_{2'}$ and not T_2. The effectiveness e of the heat exchanger, involving the lower heat entry temperature, is now to be written

$$e = \frac{T_{5'} - T_{2'}}{T_5 - T_{2'}} \qquad (6.6)$$

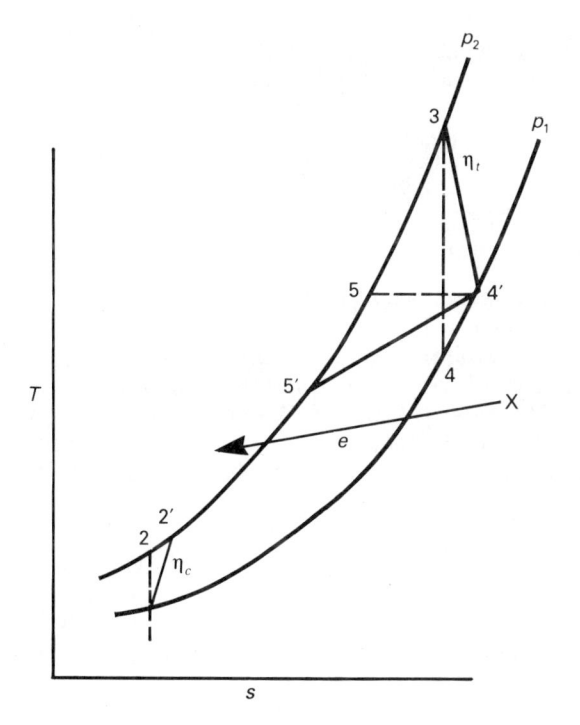

Fig. 6.6 The Joule cycle with heat exchange for an imperfect system with losses in the compression and expansion and in the heat exchange stages (X denotes heat exchange).

Joule Cycle with Heat Exchange

This gives, using eqn. (6.6),

$$\text{heat transfer} = C_p(T_3 - T_{5'}) = C_p\{(T_3 - T_{2'}) - (T_{5'} - T_{2'})\}$$
$$= C_p\{(T_3 - T_{2'}) - e(T_5 - T_{2'})\}$$
$$= C_p\{(T_3 - T_1) - (T_{2'} - T_1) - e(T_{4'} - T_{2'})\}$$

or

$$Q_{23} = C_p\{T_1(T_3/T_1 - 1) - (1/\eta_c)T_1(T_2/T_1 - 1) - e(T_{4'} - T_{2'})\} \quad (6.7)$$

But

$$(T_{4'} - T_{2'}) = (T_{4'} - T_3) - (T_{2'} - T_3)$$
$$= \eta_t T_3(T_4/T_3 - 1) - (T_{2'} - T_1) + (T_3 - T_1) \quad (6.8)$$
$$= T_1\{\eta_t\theta(T_4/T_3 - 1) - (1/\eta_c)(T_2/T_1 - 1) + (T_3/T_1 - 1)\}$$

Inserting eqn. (6.8) into eqn. (6.7) gives finally for the heat transfer,

$$Q_{23} = T_1\{(\theta - 1) - (1/\eta_c)(\rho_p - 1) - \\ - e[\eta_t\theta(1/\rho_p - 1) - (1/\eta_c)(\rho_p - 1) + (\theta - 1)]\} \quad (6.9)$$

On dividing eqn. (6.5) by eqn. (6.9) we obtain the expression for the thermal efficiency

$$\eta = \frac{a(1 - 1/\rho_p) - (\rho_p - 1)}{(b - \rho_p) - e[a(1/\rho_p - 1) - (\rho_p - 1) + (b - 1)]}$$

where a and b are defined by eqn. (4.26). This expression is written more simply if we introduce the quantity c defined by

$$c = 1 - 1/\rho_p \quad (6.10)$$

Then

$$\eta = \frac{ac - \rho_p + 1}{(b - \rho_p)(1 - e) - eac} \quad (6.11)$$

Data showing the dependence of η on the various degrees of irreversibility in the system are shown in Fig. 6.7, for $\theta = 4$ as typical. There is a virtually endless range of possibilities but eight cases are set down as representative. Curves A and B refer to perfect compression and expansion: curve A has $e = 0$ and curve B has $e = 1$. Curves I, II and III refer to $\eta_t = \eta_c = 0.9$: curve I has $e = 0$, curve II has $e = 1$ and curve III has $e = 0.9$. Curves i, ii and iii refer to $\eta_t = \eta_c = 0.8$: curve i has $e = 0$, curve ii has $e = 1$ and curve iii has $e = 0.9$. The drastic effects of irreversibilities on the thermal efficiency of the gas turbine system are very evident from the figure.

The eqn. (6.11) reduces to eqn. (6.2) of Section 6.2 for perfect heat exchange ($e = 1$) with fully isentropic expansion and compression ($\eta_t = \eta_c = 1$), and to the eqn. (6.4) of Section 6.3 when the heat exchange is less than perfect. Finally, it reduces to the eqn. (4.25) of Section 4.5 when $e = 0$ (no heat exchange).

The work ratio and specific work output follow in a similar way: in fact they are the same as in Section 4.3, eqns. (4.27) and (4.28). We write these expressions again here for completeness, introducing the quantity c defined in eqn. (6.10):

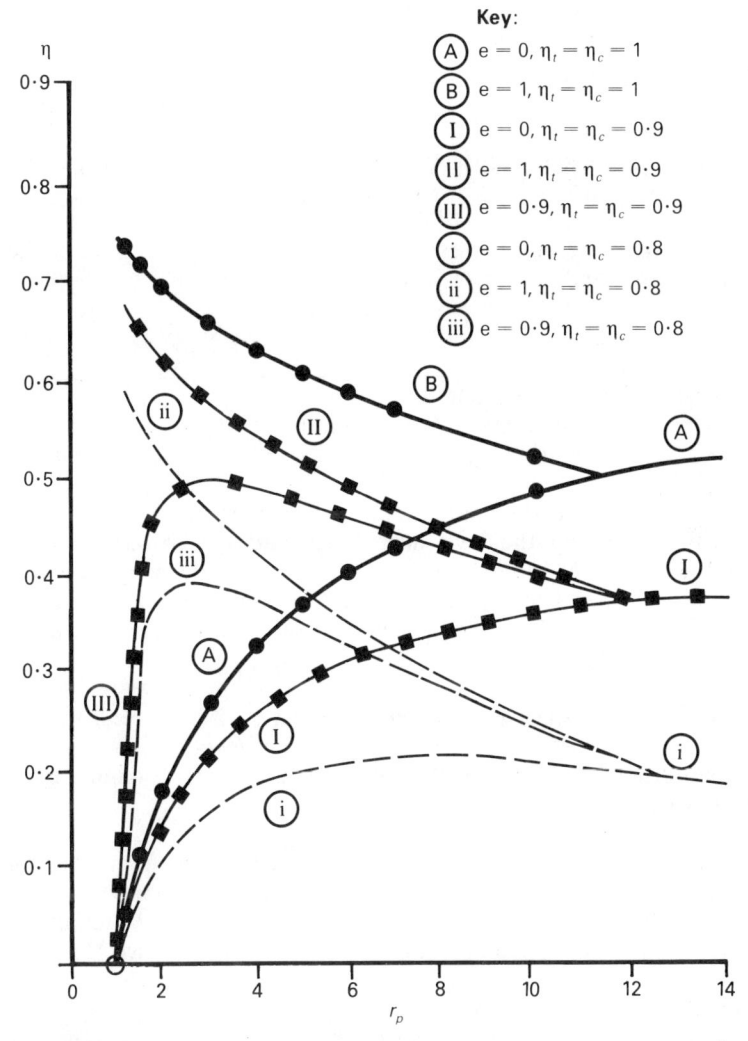

Key:
- Ⓐ $e = 0$, $\eta_t = \eta_c = 1$
- Ⓑ $e = 1$, $\eta_t = \eta_c = 1$
- Ⓘ $e = 0$, $\eta_t = \eta_c = 0.9$
- Ⓘⓘ $e = 1$, $\eta_t = \eta_c = 0.9$
- Ⓘⓘⓘ $e = 0.9$, $\eta_t = \eta_c = 0.9$
- ⓘ $e = 0$, $\eta_t = \eta_c = 0.8$
- ⓘⓘ $e = 1$, $\eta_t = \eta_c = 0.8$
- ⓘⓘⓘ $e = 0.9$, $\eta_t = \eta_c = 0.8$

Fig. 6.7 The dependence of the thermal efficiency η on the pressure ratio r_p for various values of e, η_t and η_c.

$$r_w = 1 - \rho_p/a$$
$$W_s = (C_p T_1/\eta_c)\{ac - \rho_p + 1\} \tag{6.12}$$

Reference can be made to Fig. 4.10 for the dependence of W_s on r_p.

6.5 Summary

1. The thermal efficiency of the simple Joule cycle is improved by regenerative heating where enthalpy of the expanded gas is transferred to the cold compressed gas before the latter is heated. This pre-heating reduces the heat transfer into the cycle for a given temperature ratio.

90 Joule Cycle with Heat Exchange

2. The work ratio and the specific work output of the cycle remain unaffected by the introduction of heat exchange.
3. The thermal efficiency now depends on the temperature ratio (unlike the simple Joule cycle: see Fig. 6.3) and the dependence on the pressure ratio is now radically different.
4. The thermal efficiency of the cycle with heat exchange is very dependent on the effectiveness of the heat exchange process (see Fig. 6.4).
5. The working characteristics of the less than perfect cycle are deduced (see Fig. 6.6) where neither the expansion/compression processes nor the heat exchange are ideal.

6.6 Exercises

6.1 Construct a simple computer program to reproduce Fig. 6.6 for yourself, but now including also the situations for which $\eta_t = 0.95$, $\eta_c = 0.95$ and $e = 0.95$. (Save your program—you will need it in Question 6.3.)

6.2 In the main text, the expressions for the thermal efficiency have been derived by calculating the heat acceptance and the work output. Again deduce these several expressions alternatively by deriving in each case expressions for the heat input Q_{in} and heat rejection Q_{out} and forming $\eta = (Q_{in} - Q_{out})/Q_{in}$. This allows a check on the correctness of the derived expressions.

6.3 The numerical analyses in the main text have assumed air is the working gas ($\gamma = 1.4$). Reconstruct Fig. 6.6 using helium ($\gamma = 1.66$) as the working gas. What differences arise in the two cases? Why is helium not more generally used as the working fluid?

6.4 A gas turbine system is to be constructed for maximum work output, thermal efficiency not being the prime consideration. The temperature ratio is to be 4.5, the lower temperature being atmospheric. Whether a heat exchange system is included or not is according to your choice. The pressure ratio also is open to your choice. What value would you choose and why? Sketch the cycle and set down the corresponding T–s diagram. What then would be the working characteristics of the cycle? How important is a heat exchanger in a gas turbine system?

7. THE COMPLETE SYSTEM

The simple Joule cycle has been investigated in Chapter 4 and the various modifications that can improve its performance have been introduced separately in the succeeding chapters. The separate modifications involve either changes in the heat transfer process or in the work transfer process but not in both. Consequently, each modification has a beneficial effect on some cycle characteristics (for instance, heat exchange improves the thermal efficiency while reheat improves the specific work output) but leaves others essentially unaffected. All the characteristics are improved if all the modifications are incorporated at the same time to form what might be called the complete system. The simultaneous modification of heat and work transfer to form the complete cycle is the subject of the present chapter, and is essentially the combination of the analyses of Chapters 5 and 6. We can say immediately that this represents the ultimate modification to the simple gas turbine system but may not be necessary, or indeed appropriate, in a particular practical application. The practical aspects of the modifications will be considered in the final section.

7.1 Total and Complete Systems

We begin by defining what we mean by the complete system, but must first consider the total system. The total system is a gas turbine system, either closed or open, which contains heat exchange, complete reheat and complete intercooling. Generally there will be only one stage, or perhaps two stages, of reheat and intercooling. This can be a practical system and its predicted heat and work transfer characteristics have a real relevance. The complete system is the total cycle where the number of reheat and intercooling stages are each indefinitely large. This must be the ideal and, as we shall see in Section 7.4, the heat and work characteristics then, in fact, approach those on the ideal Carnot cycle.

7.2 Heat Exchange with Reheat

We consider a cycle with heat exchange and one stage of complete reheat but without intercooling.

Ideal Heat Exchange and Isentropic Change
The circuit plan is shown in Fig. 7.1 and the associated T–s diagram in Fig. 7.2.
 The work transfer is

$$W = (C_p T_1/\eta_c)\{2\theta(1 - 1/\sqrt{\rho_p}) - (\rho_p - 1)\} \tag{7.1}$$

while the heat transfer is

$$Q = C_p\{2T_3(1 - 1/\sqrt{\rho_p})\} \tag{7.2}$$

92 The Complete System

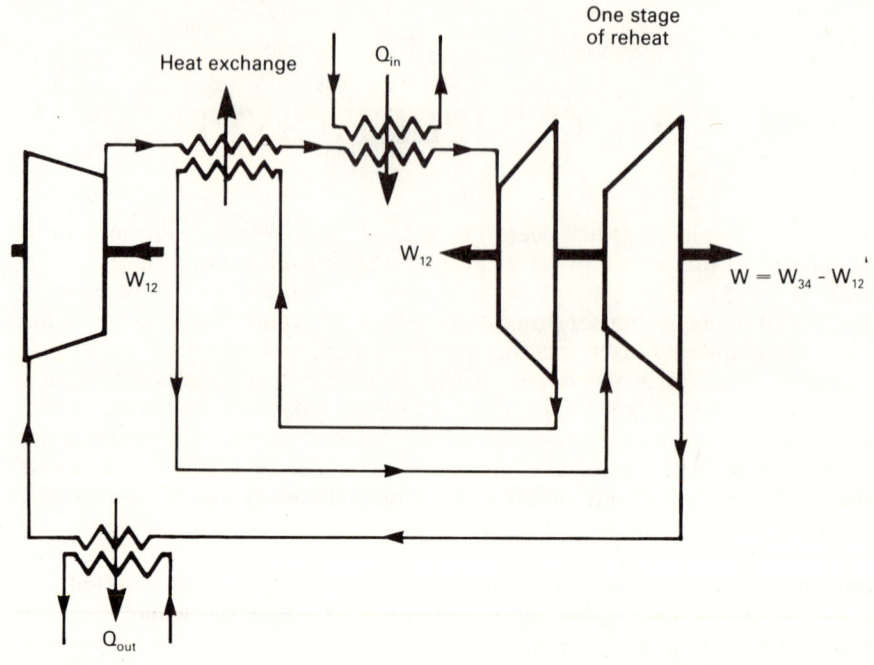

Fig. 7.1 The circuit diagram for a closed gas cycle with heat exchange and one stage of complete reheat.

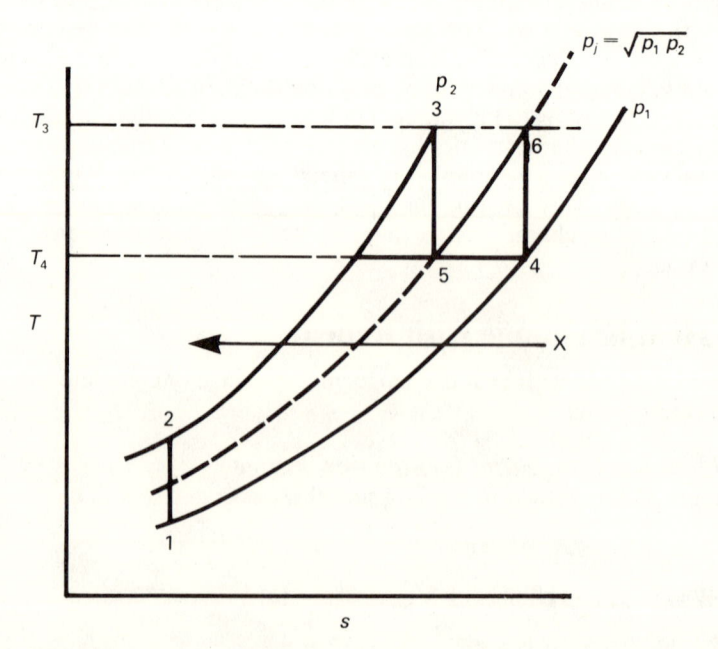

Fig. 7.2 The temperature-specific entropy diagram for heat exchange with reheat.

7.2 Heat Exchange with Reheat

Dividing eqn. (7.1) by eqn. (7.2) provides the expression for the thermal efficiency

$$\eta = 1 - \frac{\sqrt{\rho_p}(\sqrt{\rho_p}+1)}{2\theta} \tag{7.3}$$

The work ratio and specific work output are, using eqn. (7.1)

$$r_w = 1 - \frac{\sqrt{\rho_p}(\sqrt{\rho_p}+1)}{2\theta} \tag{7.4}$$

and

$$W_s = (C_p T_1 / \eta_c)\{2\theta(1 - 1/\sqrt{\rho_p}) - \rho_p + 1\} \tag{7.5}$$

The data derived from these expressions are plotted in Figs. 7.3 and 7.4.

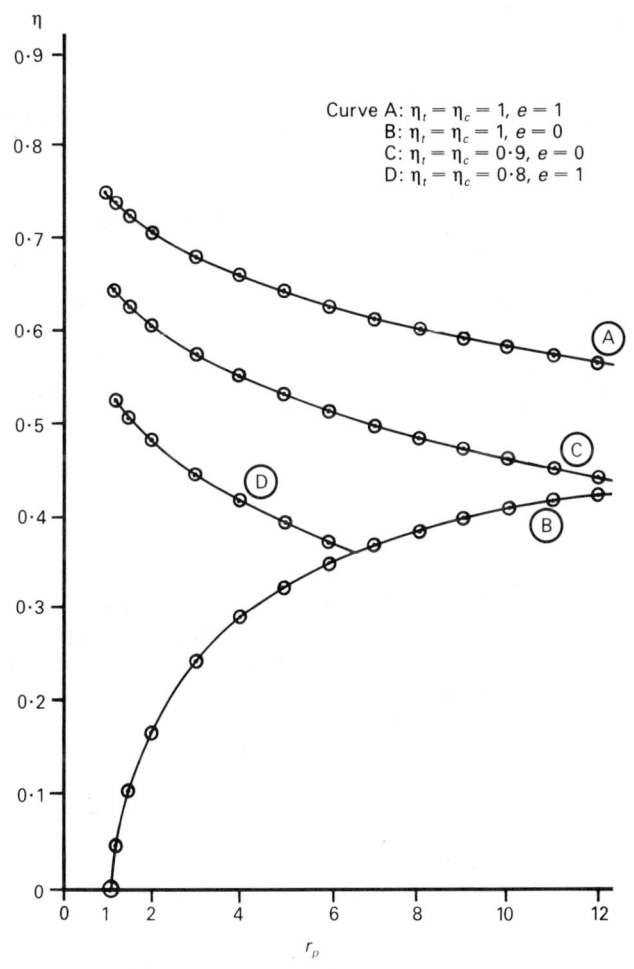

Fig. 7.3 Thermal efficiency: heat exchange and reheat (one stage). $\theta = 4$, Air, $T_1 = 302$ K.

94 *The Complete System*

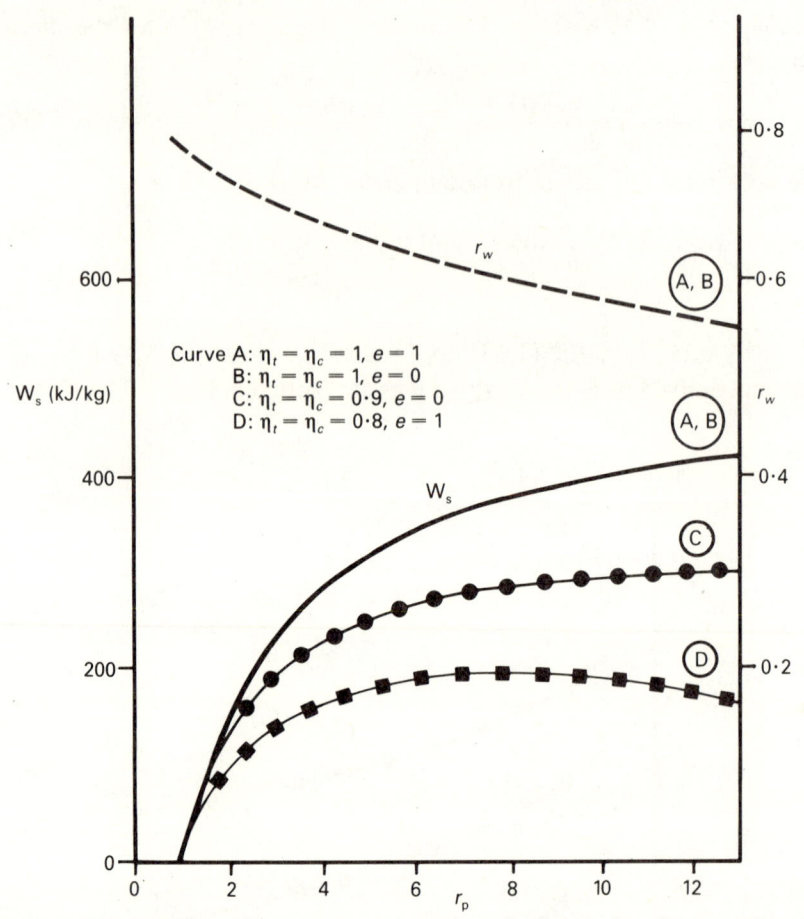

Fig. 7.4 The specific work output and work ratio for a closed gas cycle with irreversibilities. (Air, $\theta = 4$, $T_1 = 302$ K). Heat exchange and one reheat stage.

Curve A: $\eta_t = \eta_c = 1$, $e = 1$
B: $\eta_t = \eta_c = 1$, $e = 0$
C: $\eta_t = \eta_c = 0.9$, $e = 0$
D: $\eta_t = \eta_c = 0.8$, $e = 1$

Inefficient Heat and Work Transfer

The expressions (7.2) to (7.4) must be modified to take account of the non-isentropic pressure changes and the less than perfect heat transfer. The heat exchange spans the full pressure range. The *T–s* diagram is shown in Fig. 7.5 and we use that notation here. For this more realistic system we have

$$\begin{aligned}\text{work transfer} \quad W &= C_p\{(T_3 - T_{5'}) + (T_6 - T_{4'}) - (T_{2'} - T_1)\} \\ &= C_p T_1\{2\eta_t \theta(1 - 1/\sqrt{\rho_p}) - (1/\eta_c)(\rho_p - 1)\} \\ &= (C_p T_1/\eta_c)\{2a(1 - 1/\sqrt{\rho_p}) - \rho_p + 1\} \end{aligned} \qquad (7.6)$$

For the heat transfer we have

$$\begin{aligned} Q &= C_p\{(T_3 - T_{7'}) + (T_6 - T_{5'})\} \\ &= C_p\{(T_3 - T_{2'}) + (T_{4'} - T_{2'}) + (T_6 - T_3) - (T_{5'} - T_3)\}.\end{aligned}$$

Introducing the various pressure ratios in place of the temperature ratios in

7.2 Heat Exchange with Reheat

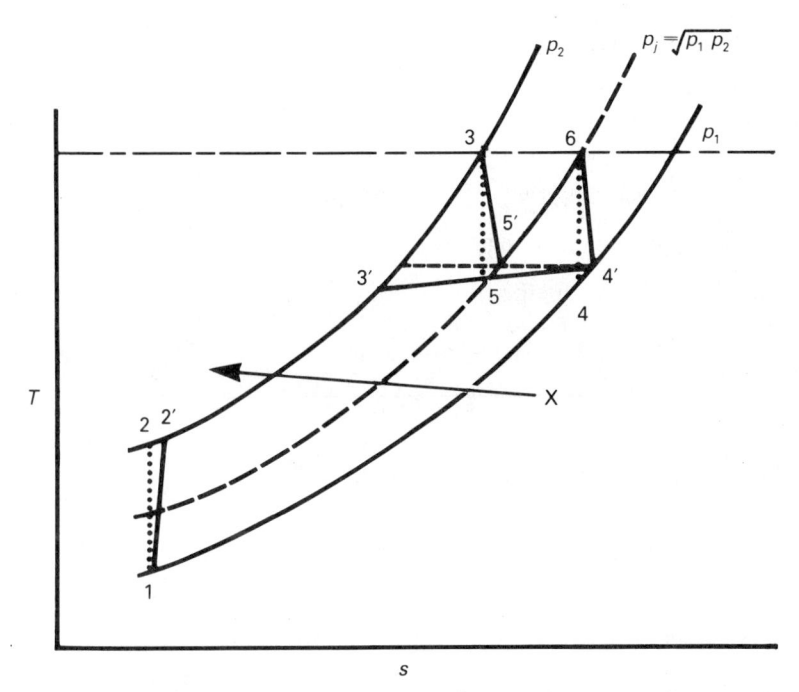

Fig. 7.5 The temperature-specific entropy diagram for a closed gas cycle with imperfect heat exchange, compression and one stage of reheat.

usual way it follows after some manipulation that

$$Q = (C_p T_1/\eta_c)\{b - \rho_p + a(1 - 1/\sqrt{\rho_p}) - ef\} \tag{7.7}$$

where
$$f = a\left(\frac{1}{\sqrt{\rho_p}} - 1\right) + 1 - \rho_p$$

and e is defined by eqn. 6.3. The thermal efficiency follows by dividing eqn. (7.6) by eqn. (7.7) to give

$$\eta = \frac{2a(1 - 1/\sqrt{\rho_p}) - \rho_p + 1}{\left\{b - \rho_p + a(1 - 1/\sqrt{\rho_p}) - e\left[a\left(\frac{1}{\sqrt{\rho_p}} - 1\right) + 1 - \rho_p\right]\right\}} \tag{7.8}$$

where, as before, we have introduced a and b according to eqn. (4.26), that is

$$a = \eta_t \eta_c \theta \quad \text{and} \quad b = \eta_c(\theta - 1) + 1$$

The work ratio and specific work output are obtained from eqn. (7.6) in the form

$$r_w = 1 - \frac{\sqrt{\rho_p}(\sqrt{\rho_p} + 1)}{2a} \tag{7.9}$$

and

$$W_s = (C_p T_1/\eta_c)\{2a(1 - 1/\sqrt{\rho_p}) - \rho_p + 1\} \tag{7.10}$$

Data from these expressions are plotted in Figs. 7.3 and 7.4.

7.3 Heat Exchange with Intercooling

The circuit diagram is shown in Fig. 7.6 and the *T–s* diagram in Fig. 7.7 for a system with less than perfect volume change and heat transfer. The analysis follows the previous pattern precisely when it is found that

work transfer $= (C_p T_1/\eta_c)\{a(1-1/\rho_p) - 2(\sqrt{\rho_p}-1)\}$

and heat transfer $= (C_p T_1/\eta_c)\{b - \sqrt{\rho_p} - e[a(\frac{1}{\rho_p}-1) - \sqrt{\rho_p}+1]\}$

This gives for the thermal efficiency

$$\eta = \frac{a(1-1/\rho_p) - 2(\sqrt{\rho_p}-1)}{b - \sqrt{\rho_p} - e[\,a(\frac{1}{\rho_p}-1) - \sqrt{\rho_p}+1]} \qquad (7.11)$$

The work ratio and specific work output are

$$r_w = 1 - \frac{2\rho_p}{a(\sqrt{\rho_p}+1)}, \qquad (7.12)$$

$$W_s = (C_p T_1/\eta_c)\{a(1-1/\rho_p) - 2(\sqrt{\rho_p}-1)\} \qquad (7.13)$$

The data derived from these equations are shown in Figs. 7.8 and 7.9.

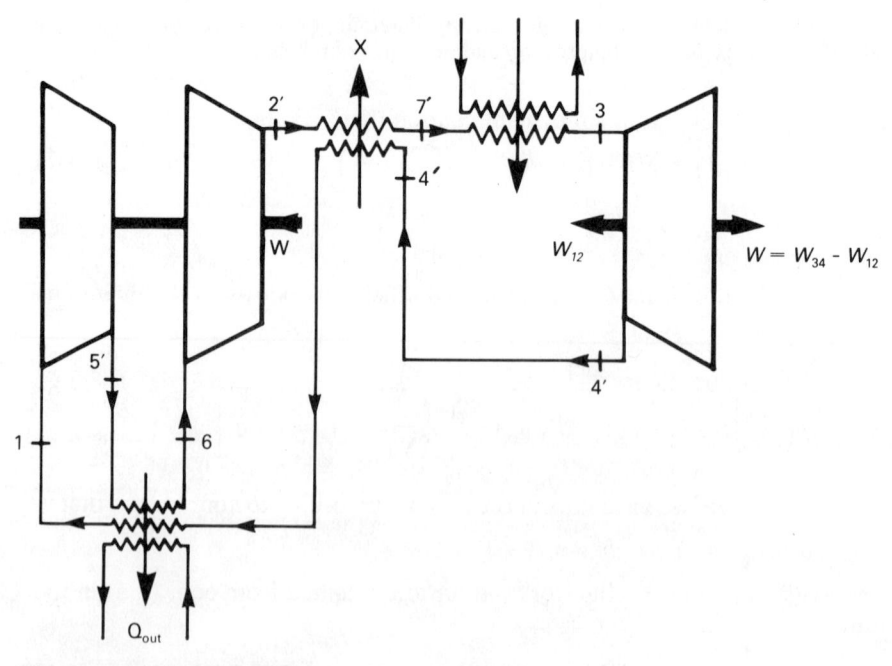

Fig. 7.6 The circuit diagram for a closed gas cycle with heat exchange and one stage of intercooling.

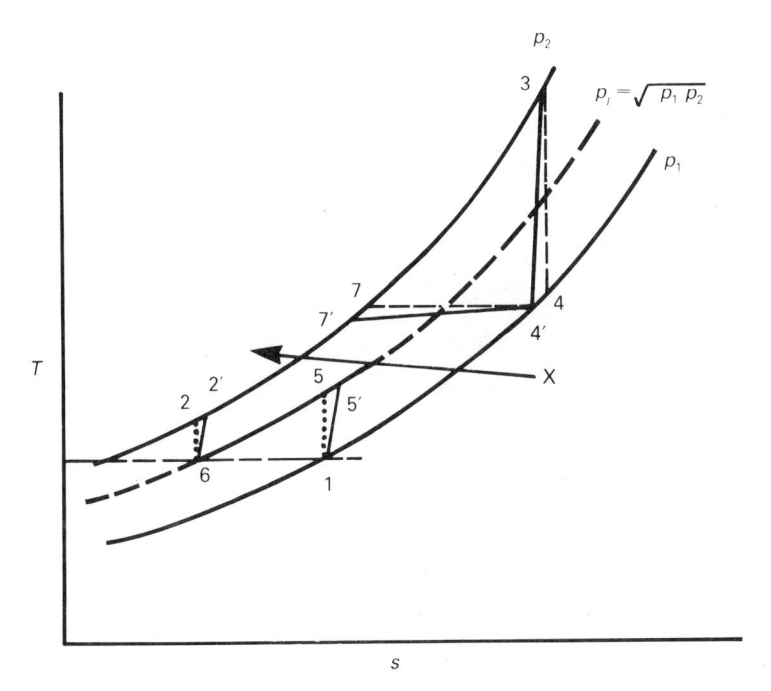

Fig. 7.7 The temperature-specific entropy diagram for a gas cycle with heat exchange and one stage of intercooling.

7.4 The Complete System

All the improvements to the simple gas turbine are used if heat exchange is added to the reheat/intercool cycle considered in the last section. There will be one or more stages of complete reheat and complete intercooling, together with a heat exchange system which will show some inefficiency, though this may be small.

The complete system can, in principle at least, be expanded to contain an indefinitely large number of reheat and intercooling stages. In this case the network of intermediate pressures becomes indefinitely fine and both the heat reception and heat rejection transfers for the cycle take place increasingly isothermally as the number of intermediate pressure stages increases. In the limit of an indefinite number of stages, the heat transfer is precisely isothermal and conditions are the same as for the Carnot cycle and this must represent an ideal limit for our studies. We will now see how this result is derived.

n Complete Reheats and Intercoolings with Heat Exchange

The work and heat transfer are readily obtained using the arguments of Section 7.3 for n complete stages of reheat and n complete stages of intercooling with heat exchange linking the incoming gas to the exhaust gas flow. With $p_2 = r_p p_1$ marking the upper temperature we write, as before, $d_n = (\rho_p)^{1/(n+1)}$ and obtain

98 The Complete System

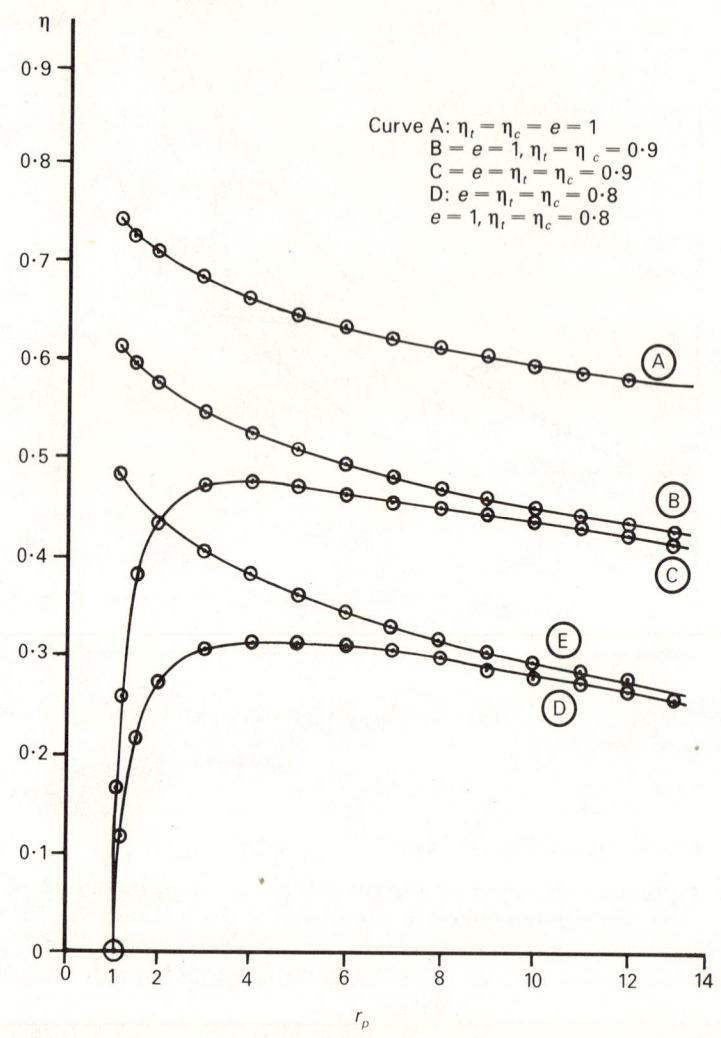

Fig. 7.8 Heat exchange + one stage of intercooling. ($\theta = 4$, Air, $T_1 = 302$ K)°.

Curve A: $\eta_t = \eta_c = e = 1$
B = $e = 1$, $\eta_t = \eta_c = 0.9$
C = $e = \eta_t = \eta_c = 0.9$
D: $e = \eta_t = \eta_c = 0.8$
$e = 1$, $\eta_t = \eta_c = 0.8$

work transfer $W = (n+1)(C_p T_1/\eta_c)\{a(1-1/d_n) - d_n + 1\}$
heat transfer $Q = (C_p T_1/\eta_c)\{b - d_n + an(1 - 1/d_n) + ef\}$
where $f = \eta_c[\eta_t(d_n - 1) - \theta/d_n + 1]$

and e is defined by eqn. (6.7). The thermal efficiency for the cycle is then $\eta = W/Q$ or

$$\eta = \frac{(n+1)[a(1-1/d_n) - d_n + 1]}{b - d_n + an(1 - 1/d_n) + ef} \qquad (7.14)$$

The work ratio r_w is

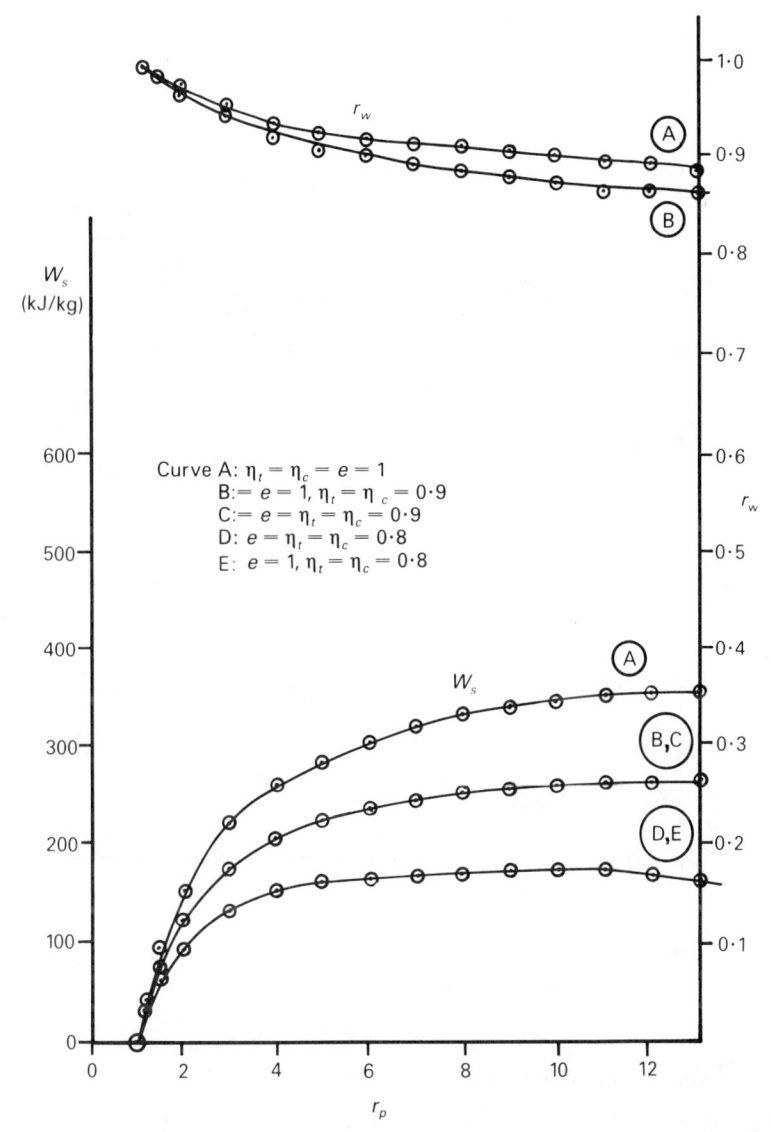

Curve A: $\eta_t = \eta_c = e = 1$
B: $e = 1, \eta_t = \eta_c = 0.9$
C: $e = \eta_t = \eta_c = 0.9$
D: $e = \eta_t = \eta_c = 0.8$
E: $e = 1, \eta_t = \eta_c = 0.8$

Fig. 7.9 Specific work output and work ratio of a closed gas cycle with heat exchange and one stage of intercooling. (Air, $\theta = 4$, $T_1 = 302$ K.)

$$r_w = 1 - \frac{d_n}{a} \tag{7.15}$$

while the specific work output W_s is

$$W_s = (n+1)(C_p T_1/\eta_c)\{a(1 - 1/d_n) - d_n + 1\} \tag{7.16}$$

These expressions are plotted in Figs. 7.10 and 7.11 for $\theta = 4$ and $n = 2, 5$ and 10, for the ideal cases with $\eta_t = \eta_c = 1$ and with $e = 1$. The effects of irreversibilities are shown in Figs. 7.12 and 7.13 for $\theta = 4$ and $n = 1$ and 5.

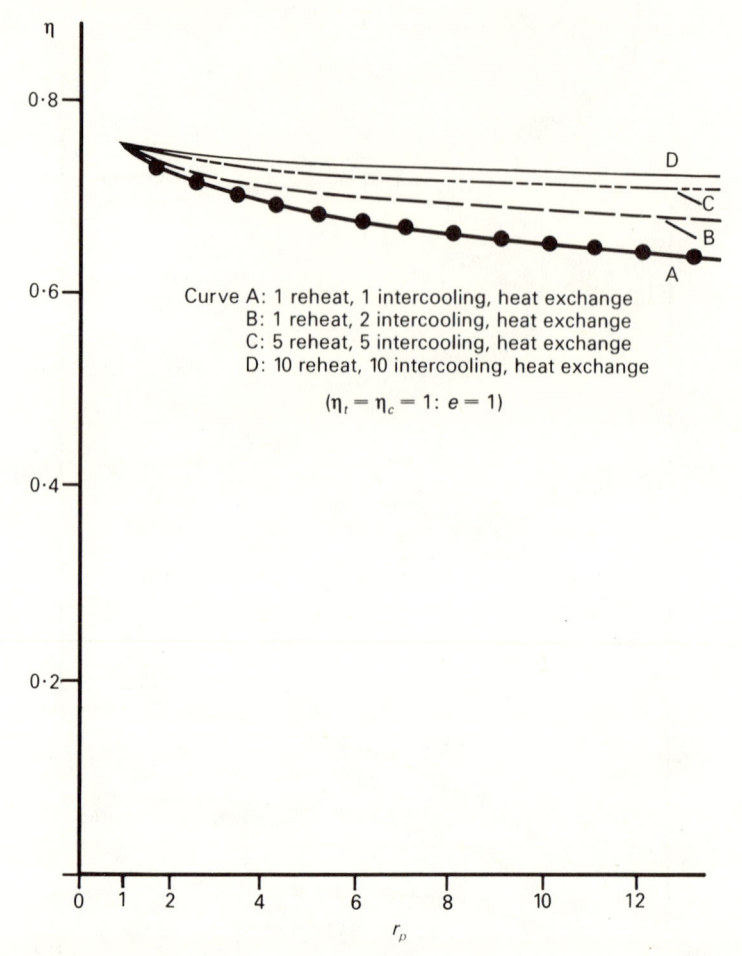

Fig. 7.10 The thermal efficiency of a closed gas cycle with reheat, intercooling and heat exchange. (Air, $\theta = 4$, $T_1 = 302$ K.)

Curve A: 1 reheat, 1 intercooling, heat exchange
B: 1 reheat, 2 intercooling, heat exchange
C: 5 reheat, 5 intercooling, heat exchange
D: 10 reheat, 10 intercooling, heat exchange

$(\eta_t = \eta_c = 1; e = 1)$

7.5 The Total System

Now there is an indefinitely large number of reheat and intercooling stages with reheat. It is not necessary that the volume changes and heat transfer should have ideal efficiencies.

The heat reception and heat rejection for the cycle become essentially isothermal in the limit $n \to \infty$. In this limit $d_n \to 1$, because there is an indefinitely large number of intermediate pressure stages. The pressure ratio for each stage is very closely unity, and becomes unity in the limit.

The values of the expressions (7.15) to (7.17) in the limit $d_n = 1$ are not readily derivable since they each involve the division of zero by zero and this is an indeterminate number. The limit is most easily explored by supposing n to be very large although not indefinitely large. Then we can set

$$d_n = 1 + \delta, \tag{7.17}$$

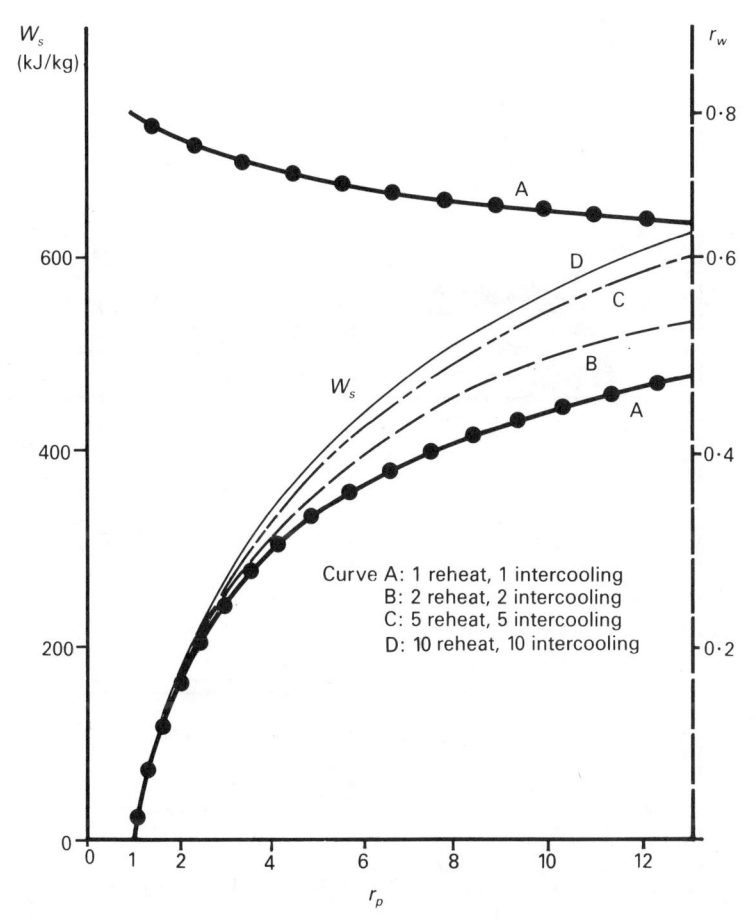

Fig. 7.11 The specific work output and work ratio for a close gas cycle with reheat, intercooling and heat exchange. (Air, $\theta = 4$, $T_1 = 302$ K.)

Curve A: 1 reheat, 1 intercooling
B: 2 reheat, 2 intercooling
C: 5 reheat, 5 intercooling
D: 10 reheat, 10 intercooling

where δ is a number which can be made indefinitely small: the approach of δ to zero corresponds in the cycle to the limit of an indefinitely large number of intermediate stages. Because δ is vanishingly small we can write $1/d_n = (1+\delta)^{-1} = 1-\delta$. Inserting eqn. (7.17) into the expression (7.14) we find after some manipulation and cancellations that

$$\eta = \frac{(n+1)\delta(a-1)}{(an-1)+(b-1)(1-e)} \tag{7.18}$$

The expressions (7.15) and (7.16) for the work ratio and specific work output become respectively

$$r_w = 1 - \frac{1+\delta}{a} \tag{7.19}$$

$$W_s = (n+1)(C_p T_1/\eta_c)[\delta(a-1)] \tag{7.20}$$

102 The Complete System

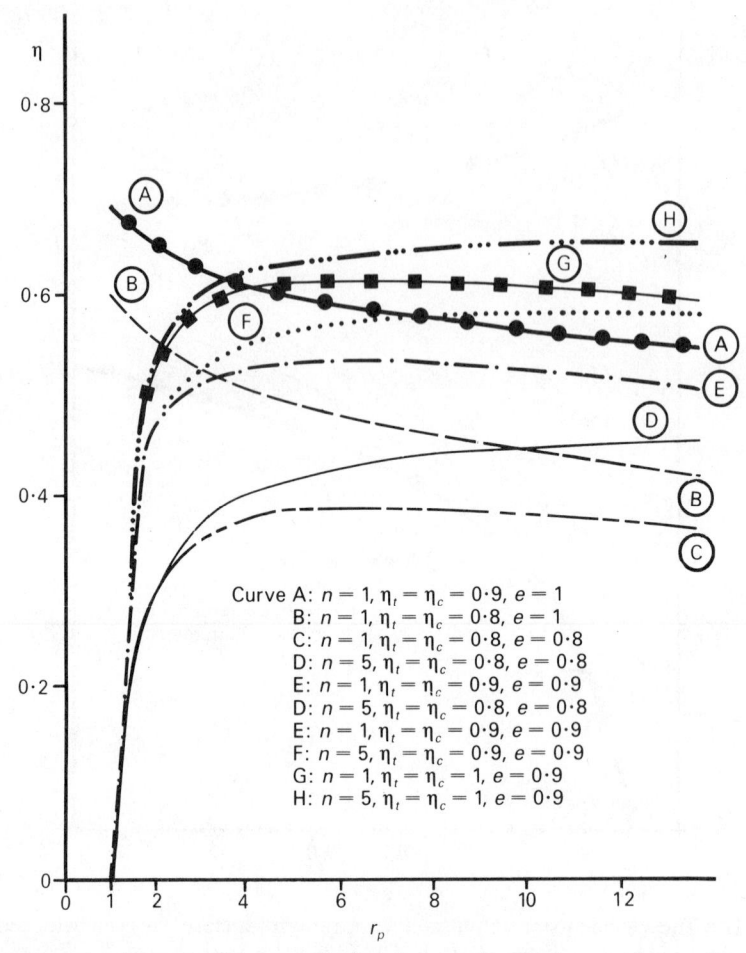

Fig. 7.12 The thermal efficiency of a closed gas cycle with heat exchange, h stages of reheat and n stages of intercooling. (Air, $\theta = 4$, $T_1 = 302$ K.)

Curve A: $n = 1$, $\eta_t = \eta_c = 0\cdot 9$, $e = 1$
B: $n = 1$, $\eta_t = \eta_c = 0\cdot 8$, $e = 1$
C: $n = 1$, $\eta_t = \eta_c = 0\cdot 8$, $e = 0\cdot 8$
D: $n = 5$, $\eta_t = \eta_c = 0\cdot 8$, $e = 0\cdot 8$
E: $n = 1$, $\eta_t = \eta_c = 0\cdot 9$, $e = 0\cdot 9$
D: $n = 5$, $\eta_t = \eta_c = 0\cdot 8$, $e = 0\cdot 8$
E: $n = 1$, $\eta_t = \eta_c = 0\cdot 9$, $e = 0\cdot 9$
F: $n = 5$, $\eta_t = \eta_c = 0\cdot 9$, $e = 0\cdot 9$
G: $n = 1$, $\eta_t = \eta_c = 1$, $e = 0\cdot 9$
H: $n = 5$, $\eta_t = \eta_c = 1$, $e = 0\cdot 9$

We move towards the ideal by assuming the heat transfer to be perfect, that is $e = 1$. Then eqn. (7.18) becomes

$$\eta = \frac{(1 + 1/n)(a - 1)}{(a - 1/n)} \tag{7.21}$$

In the limit $n \to \infty$ the number of intermediate pressure stages becomes indefinitely large, which is described by setting $\delta = 0$. Then eqns. (7.21) and (7.20) reduce to the forms

$$\eta \to 1 - 1/a \quad \text{and} \quad W_s \to 0 \quad (\delta \to 0) \tag{7.22}$$

If the expansion and compression of the gas can also be supposed perfect ($\eta_t = \eta_c = 1$) the cycle becomes ideal in each component: then eqns. (7.22) reduce finally to

7.5 The Total System

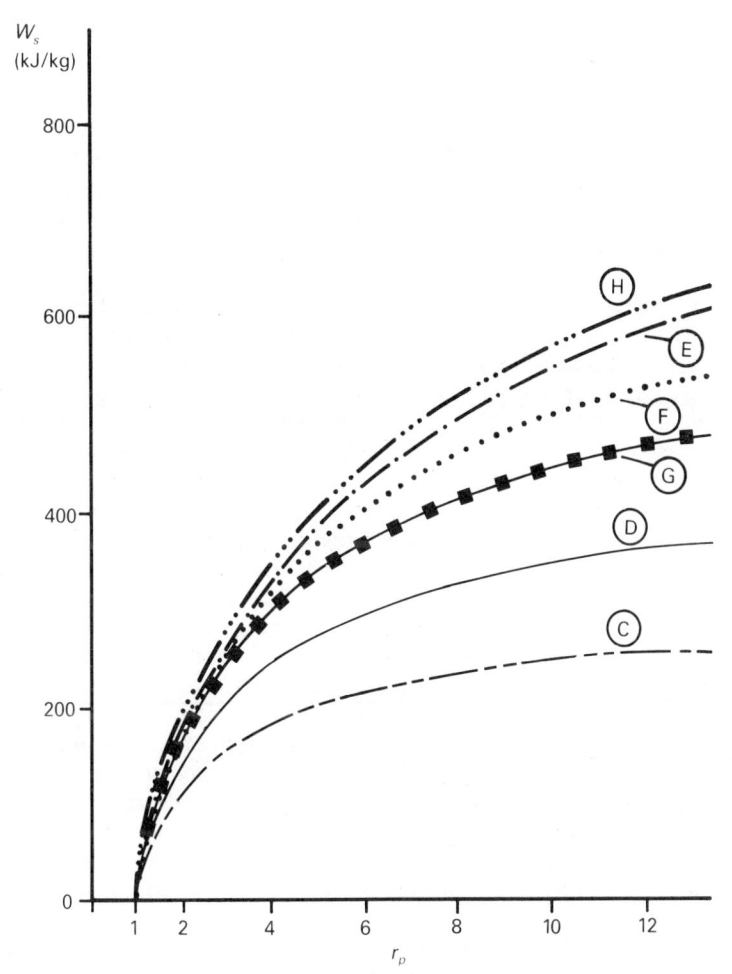

Fig. 7.13 The specific work output for a closed gas cycle with heat exchange, n stages of reheat and n stages of intercooling. The letters have the same meaning as in **Fig. 7.12**. (Air, $\theta = 4$, $T_1 = 302$ K.)

$$\eta \to 1 - 1/\theta \quad \text{and} \quad W_s \to 0 \quad (\delta \to 0) \tag{7.23}$$

The thermal efficiency is finally that for the Carnot cycle but there is zero work output. This result perhaps requires a little comment.

A total system is one at constant pressure, with heat exchange. There is an indefinite number of intermediate stages so the pressure ratio across each stage approaches unity as the number of stages increases. The work output of any gas cycle is zero when the pressure ratio is precisely unity (otherwise the First Law of Thermodynamics would be violated). If each component of the cycle is providing a zero specific work output then so will the total combination and so the total specific work output will vanish. The presence of the heat exchange is important from the point of view of the thermal efficiency. If the heat exchange is ideal the thermal efficiency at unit pressure ratio is always that of the Carnot engine for the corresponding temperature ratio (see Chapter 6). Our cycle, then, is at the pathological limit involving a succession of cycles with Carnot efficiency and no work output. It is interesting to notice, from eqn. (3.12), that the work output of the Carnot cycle is also zero for a unit pressure ratio (in eqn. (3.12), $\log(p_2/p_3) = \log(1) = 0$). It is perhaps not unfortunate that, after all this effort, the specific work output of even the ideally efficient cycle is zero because it would require an indefinitely large number of stages to function and this is hardly practical.

An essentially zero specific work output is also understandable even if the number of stages is very large though not indefinitely large. The very large number of compression stages would absorb all the work transfer provided by the very large number of expansion stages at low pressure ratios, with almost nothing left over. This limiting cycle with reheat, intercooling and perfect heat transfer and having the Carnot efficiency, is sometimes called the Ericsson cycle—in any practical application the number of stages would be large though not indefinitely large. This means that the Ericsson cycle is strictly a complete cycle rather than the total cycle in our terminology (see Question 7.7). The heat exchanger characteristics will be important in any application. It will be remembered from Chapter 6 that any inefficiency of heat exchange reduces the thermal efficiency of the cycle to zero at unit pressure ratio so in practice, when the pressure ratio is not far from unity, the cycle would have a very low thermal efficiency and a rather poor work output.

Our arguments have now come full circle. The Carnot cycle has no direct practical value but provides a standard against which a real cycle can be improved systematically. Such a systematic improvement allows a cycle to be devised as a limiting case which has the Carnot efficiency, but provides no work. In between there are cycles with a lower thermal efficiency than the Carnot ideal but which provide substantial work output and can be made and used: the main characteristics of the cycles are listed in Tables 7.1. and 7.2. It must be admitted, however, that this comparison between the ideal and the actual is not quite fair. The Carnot engine is the answer to the question 'Which is the most thermally efficient engine in principle?'. In the previous chapters we have asked the question 'Which is the practical engine, with heat receptance and rejection through conduction, that gives the maximum work output?'. We live in a non-ideal world but enjoy the bonus of being able to arrange matters so that we can make the most of it!

7.6 Comments on Practical Applications

A body requires energy to be accelerated from rest, that is to overcome the inertia of the stationary state. When it is moving through the surrounding medium it is acted on by a force of resistance (drag) and this must be precisely overcome by an energy source if a steady cruising speed is to be achieved and maintained, and at the end of the journey the drag force must be supplemented to bring the body ultimately to rest. It is the power unit that provides the energy to execute these requirements. Over the last twenty or thirty years air and marine units have more and more used gas turbine systems as the source of energy.

7.6 Comments on Practical Applications

Table 7.1. The Essential Characteristics of Constant Pressure Cycles Without Heat Exchange.

Cycle	Thermal Efficiency η	Work Ratio r_w	Specific Work Output W_s
Simple Joule Sect. 4.5.	$1 - \dfrac{1}{\rho_p} \equiv c$	$1 - \dfrac{\rho_p}{a}$	$\dfrac{C_p T_1}{\eta_c}(a - \rho_p)c$
Reheat (1 stage) Sect. 5.3.	$\dfrac{2a\left(1 - \dfrac{1}{\sqrt{\rho_p}}\right) - \rho_p + 1}{b - \rho_p + a\left(1 - \dfrac{1}{\sqrt{\rho_p}}\right)}$	$1 - \dfrac{\sqrt{\rho_p}(\sqrt{\rho_p}+1)}{2a}$	$\dfrac{C_p T_1}{\eta_c}\left\{2a\left(1 - \dfrac{1}{\sqrt{\rho_p}}\right) - \rho_p + 1\right\}$
Intercooling (1 stage) Sect. 5.4.	$\dfrac{a\left(1 - \dfrac{1}{\rho_p}\right) - 2\left(\sqrt{\rho_p} - 1\right)}{b - \sqrt{\rho_p}}$	$1 - \dfrac{2\sqrt{\rho_p}}{a(\sqrt{\rho_p}+1)}$	$\dfrac{C_p T_1}{\eta_c}\left\{ac - 2(\sqrt{\rho_p} - 1)\right\}$
Reheat and Intercooling (1 stage) Sect. 5.5.	$\dfrac{2\left\{a\left(1 - \dfrac{1}{\rho_p}\right) - \left(\sqrt{\rho_p} - 1\right)\right\}}{b - \sqrt{\rho_p} + a\left(1 - \dfrac{1}{\sqrt{\rho_p}}\right)}$	$1 - \dfrac{\sqrt{\rho_p}}{a}$	$\dfrac{C_p T_1}{\eta_c}\left\{2a\left(1 - \dfrac{1}{\sqrt{\rho_p}}\right) - 2(\sqrt{\rho_p} - 1)\right\}$
Isothermal Compression Sect. 5.7.	$\dfrac{\eta_t\theta(\rho_p - 1) - \rho_p\left(\dfrac{\gamma-1}{\gamma}\right)\log r_p}{\rho_p(\theta - 1)}$	$1 - \dfrac{\gamma-1}{\gamma} \cdot \dfrac{\rho_p}{\theta\eta_t(\rho_p - 1)}\log r_p$	$C_p T_1\left[\eta_t\theta c - \left(\dfrac{\gamma-1}{\gamma}\right)\log r_p\right]$

$r_p = p_2/p_1; \quad \rho_p = r_p^{\left(\frac{\gamma-1}{\gamma}\right)}; \quad a = \eta_t\eta_c\theta; \quad b = \eta_t(\theta - 1) + 1; \quad c = 1 - 1/\rho_p$

Table 7.2. The Essential Characteristics of Constant Pressure Cycles with Heat Exchange.

Cycle	Thermal Efficiency η	Work Ratio r_w	Specific Work Output W_s
Heat Exchange Sect. 6.4.	$\dfrac{ac - \rho_p + 1}{(b - \rho_p)(1 - e) - eac}$	$r_w = 1 - \dfrac{\rho_p}{a}$	$\dfrac{C_p T_1}{\eta_c}(ac - \rho_p + 1)$
Heat Exchange and Reheat (1 stage) Sect. 7.1.	$\dfrac{2a\left(1 - \dfrac{1}{\sqrt{\rho_p}}\right) - \rho_p + 1}{b - \rho_p + a\left(1 - \dfrac{1}{\sqrt{\rho_p}}\right) - ef_1}$ $\left[f_1 = a\left(\dfrac{1}{\sqrt{\rho_p}} - 1\right) + 1 - \rho_p\right]$	$1 - \dfrac{\sqrt{\rho_p}(\sqrt{\rho_p} + 1)}{2a}$	$\dfrac{C_p T_1}{\eta_c}\left\{2a\left(1 - \dfrac{1}{\sqrt{\rho_p}}\right) - \rho_p + 1\right\}$
Heat Exchange and Intercooling (1 stage) Sect. 7.3.	$\dfrac{ac - 2(\sqrt{\rho_p} - 1)}{b - \sqrt{\rho_p} - ef_2}$ $\left[f_2 = a\left(\dfrac{1}{\rho_p} - 1\right) - \sqrt{\rho_p} + 1\right]$	$1 - \dfrac{2\rho_p}{a(\sqrt{\rho_p} + 1)}$	$\dfrac{C_p T_1}{\eta_c}\{ac - 2(\sqrt{\rho_p} - 1)\}$
Heat Exchange and Reheat with Intercooling (n stages of each) Sect. 7.4.	$\dfrac{(n+1)\left[a\left(1 - \dfrac{1}{d_n}\right) - d_n + 1\right]}{b - d_n + an\left(1 - \dfrac{1}{d_n}\right) + ef_3}$ $\left[d_n = \rho_p^{\left(\frac{1}{n+1}\right)}\right]$ $f_3 = \eta_{lc}\left[\eta_t(d_n - 1) - \dfrac{\theta}{d_n} + 1\right]$	$1 - \dfrac{d_n}{a}$	$(n+1)\dfrac{C_p T_1}{\eta_c}\left\{a\left(1 - \dfrac{1}{d_n}\right) - d_n + 1\right\}$

$r_p = p_2/p_1$; $\rho_p = r_p^{\left(\frac{\gamma-1}{\gamma}\right)}$; $a = \eta_t \eta_c \theta$; $b = \eta_c(\theta - 1) + 1$; $c = 1 - 1/\rho_p$

Marine Units

Until thirty years ago ships (both military and civil) were almost universally powered by steam turbine systems fired by fossil fuels, first coal and later oil. The smallest vessels used diesel power, or for high speeds (for instance, naval motor gun boats) some form of petrol engine. The gas turbine was slow to enter the field primarily on grounds of reliability but also because of high fuel consumption. There was also the fear that marine applications would lead to excessive salt corrosion of the various components and therefore that maintenance would be a critical factor. These features can be of lower importance in some naval applications than in normal civil working and it was in naval areas that experience of the gas turbine applied to ship propulsion was first obtained.

The first exploration involved small naval craft where the requirement was high speed for short sorties from port in restricted waters (for instance, the Channel or the North Sea). The closed cycle was used, the working fluid (air in fact) being heated through a heat exchanger by an oil fuel whose waste gases were discharged to the atmosphere through a vent in the usual way. The expansion turbine was coupled to a propeller shaft through a reduction gear box to reduce the high revolutions of the turbine to the substantially lower revolution speeds of the propeller. Ship speeds of up to 51 knots were achieved for light torpedo boats under suitable conditions.

It was quickly realised that considerable saving of weight would result from using open gas cycles and this has become the norm. The air working fluid is drawn into the cycle through appropriate deck vents, the hot gases from the turbine being expelled to the atmosphere through a rear vent or, for larger ships, through a funnel. Reliability has now achieved fully acceptable levels and sea corrosion has proved controllable by washing through compressors and turbines with de-ionised water (not sea water).

High fuel consumption remains but it can be reduced by a suitable complex of power units. Early experiments involved a steam turbine unit for normal cruising supplemented for periods of high speed by a gas turbine unit which could be clutched into the propeller drive shaft. Such a combined gas and steam power unit (sometimes referred to as COGAS) proved a useful and flexible power unit. In some civil applications a diesel unit replaces the steam unit (then called COGAD–D for diesel). As gas turbine design developed the gas unit was used alone at high speeds, the steam or diesel unit being used for slow speed manoeuvering or slow cruising. When either the gas or the other unit is used separately of the other, the unit is a COGOS (combined gas or steam unit) or a COGOD (combined gas or diesel unit). More recently still the full propulsion unit involves only gas turbine power through the combination of one unit designed to have optimum working at the lower speeds while the other has optimum working conditions at higher speeds.

These gas units will achieve maximum thermal efficiency and work output by including all the components considered in the previous chapters. Heat exchange is essential but reheat and intercooling (perhaps two stages) will also be included. The lower pressure can even be below atmospheric if a degree of vacuum can be maintained. Some large warships are now powered in this way, for instance the aircraft 'through cruisers' of the Ark Royal type with a displacement of some 25 000 tons, which replaced the earlier Ark Royal class of steam turbine aircraft carriers with a displacement of 50 000

108 The Complete System

tons. The difference in displacement between the two types probably reflects the greater weight of the older steam units, even though the gas turbine and steam units provide comparable propulsive power. The naval designer's response to modern weapons systems, and particularly sea-to-sea and air-to-sea missiles, has led to a reduction in the use of protective armour for the ship, again reducing its overall displacement. The result is a capital ship of relatively high power to weight ratio, which means of high potential speed and manoeuvrability. Again, the power units can be started quickly from cold allowing rapid access to the sea from harbour, whereas steam units could take three, four or more hours to reach working pressure.

The complete success of gas turbine units for naval use has given a great fillip to civil applications and the gas system has considerable advantages in this connection. It has a high power density and is of low weight; it has great flexibility in use; it requires fewer personnel to run it and the maintenance is both easier and quicker; it is not necessary to build a specific power unit for a particular ship but rather a general unit for a particular range of ship and this reduces the cost of each unit (they are built as a 'run' and offered for sale internationally); it reduces the time of receiving the unit from the ordering stage; and power units can be selected almost from a catalogue with cheap 'off the shelf' replacements for maintenance. One possibility which is just beginning to receive attention is the automatic (computer) control of the unit and the possibilities here are great. All these features are attractive because they reduce costs both for purchase and running and lead to a more efficient ship overall.

The high power density of the gas turbine has also led to developments in hovercraft and hydrofoil transportation. The present restrictions on the general development of these systems lie in directions other than those involving the gas turbine propulsion units.

Aircraft Propulsion

It is sometimes popularly supposed that a major use of gas cycles is in aircraft propulsion but this is true only in a certain sense. It is, however, true that the increased reliability of such cycles has come about from aero-engine developments.

All aero-engine units working in an ambient atmosphere (assumes to be air) have one or more of three components: first, a fuel unit as a source of energy; second, a diffuser which appropriately channels the incoming air; and thirtly, a nozzle which directs the heated and compressed air away from the engine. This is also true for a piston engine, though here there is no diffuser or nozzle and the working unit is the piston engine coupled to the environment through a propeller. This coupling has its maximum effectiveness for air speeds a little below 600 miles per hour, which therefore provides an upper limit to propeller driven machines. For sub-sonic gas turbine flight the air is compressed by a compressor which must be driven and this is achieved by an expansion turbine. The output of the turbine needs, then, to be just sufficient to drive the compressor and no turbine work is, in fact, used directly to propel the aeroplane.

The incoming air is compressed and mixed with fuel; this is ignited, the resulting gases being forced rearwards through the turbine and then through the nozzle to be discharged to the atmosphere. It is the (unbalanced) force provided by the exhaust gases that is the motive power to drive the aeroplane. Increasing aircraft speed requires an increasing change of momentum (propulsive force) from the gas; this is associated with an increased gas speed although a reduced mass flow of air. For supersonic flight, the compression of the air entering the diffuser, through an associated shock wave, may provide sufficient compression without the use of a compressor at all. In that case (called a ram jet) there is neither a compressor nor a turbine unit in the engine, but the aeroplane must first reach an adequately high speed for the ram jet engine to function: one way of achieving this is to use rocket power initially.

Industrial Uses for Power
The gas cycle provides a highly attractive source of power for industrial uses, and the advantages listed for marine use (in Section 7.6.1) apply here. The cycle will usually be a closed cycle since, not needing to be mobile, weight is not an important consideration. A closed cycle has certain efficiency advantages over an open cycle in practice and it is possible to use gases other than air (for instance, helium) of higher γ value which will not be lost to the atmosphere because the cycle is closed. Apart from providing a low cost flexible power unit for low outputs, the gas cycle is also capable of providing larger power outputs for restricted time intervals. As one example, it is being used to boost the energy output of steam power stations for short periods of peak loading. The industrial applications would seem to be still in their infancy, not only for industrialised but also for under-industrialised countries.

7.7 Summary

1. The analyses of the last four chapters are joined together to provide a simple Joule cycle with components to enhance the thermal efficiency and the specific work output.
2. The cycle with heat exchange and one stage of reheat is considered separately from the cycle with heat exchange and one stage of intercooling. The characteristics of each are considered, including irreversibilities.
3. The complete system contains heat exchange, and at least one stage each of reheat and intercooling, but data for cycles with 1, 2, 5 or 10 reheat/intercooling stages are presented.
4. The total system contains an indefinitely large number of reheat and intercooling stages. In this case the thermal efficiency approaches the value for the Carnot cycle as the number of stages approaches infinity, the constant pressure cycle now showing, in this limit, isothermal heat acceptance and rejection even for a real working fluid.
5. Unfortunately, the specific work output of the cycle is zero in this limit.
6. The Joule gas cycle with some multi-pressure stages and regenerative heating is nevertheless found to present a highly attractive work source for practical application especially for mobile work production.
7. Some applications of the gas cycle are considered in the marine, aviation and industrial spheres.

7.8 Exercises

7.1 Referring to Section 7.4, the total gas turbine consists of an indefinitely large number of reheat and intercooling stages, with heat exchange, each component having unit efficiency.

The pressure ratio across each intermediate stage is indefinitely close to unity: write $r_p = 1 + \varepsilon$, where ε is a small quantity that vanishes in the limit of the number of stages becoming indefinitely great. Show that:

(a) the specific work output W_s can be written
$$W_s = ((\gamma - 1)/\gamma)C_p T_1(\theta - 1)\varepsilon$$
and so vanishes as $\varepsilon \to 0$

(b) if the heat exchange is not perfect ($e < 1$) then the thermal efficiency η vanishes in the limit $\varepsilon \to 0$.

Compare this behaviour with that of the Carnot cycle referring to Chapter 3, Section 3.2. Remembering that $\log(r_p) = \log(1 + \varepsilon) = \varepsilon$ when ε is vanishingly small, show that
$$W_s(\text{Carnot}) = C_v T_1[(\gamma - 1)/\gamma](\theta - 1)\varepsilon,$$
in this limit so that W_s (Carnot) $\to 0$ as $r_p \to 1$.

7.2 An Ericsson cycle is constructed with 4 reheat and 4 intercooling stages, and with heat exchange. If the isentropic efficiencies of the compressors and turbines are each 0.9 and the thermal efficiency of the heat exchange is 0.95, calculate the thermal efficiency and specific work output to be associated with the cycle. What is the effect if the efficiency of the heat exchange is (a) 0.9 and (b) 0.85? What implications do these results have for the practical application of the cycle?

7.3 A gas turbine system using air ($\gamma = 1.4$) as the working fluid contains reheat and heat exchange (but no intercooling). The overall pressure ratio is 8 and the temperature ratio is 4. What is the thermal efficiency and specific work output of the system if:

(a) there is one reheat stage and $\eta_T = \eta_c = 0.9$ and $e = 0.95$?;
(b) there are three reheat stages and $\eta_t = \eta_c = 0.85$ and $e = 0.9$?

7.4 A gas turbine system using air ($\gamma = 1.4$) as the working fluid contains intercooling and heat exchange (but no reheat). The overall pressure ratio is 8 and the temperature ratio is 4. What is the thermal efficiency and specific work output for the cycle if:

(a) there is one stage of intercooling and $\eta_t = \eta_c = 0.9$ and $e = 0.95$?;
(b) there are three stages of intercooling and $\eta_t = \eta_c = 0.85$ and $e = 0.9$?

7.5 Repeat the analyses of Questions 7.3 and 7.4 for closed cycles using helium ($\gamma = 1.66$) as the working fluid.

7.6 It is proposed to construct a closed gas turbine cycle with heat exchange but there is uncertainty whether there should be one stage of reheat with two stages of intercooling, or two stages of reheat with one of intercooling. Resolve the uncertainty by calculating the thermal efficiency and specific work output for each case using first air and then helium as the working fluid. You may take in each case $\eta_t = \eta_c = 0.9$ and $e = 0.92$.

8. NON-STEADY FLOW CYCLES FOR LOW POWER OUTPUT

In the previous chapters we have been concerned with cycles using a turbine/compressor system for work transfer, characterised by the working fluid being transported in steady flow whether the cycle is open or closed. Although such systems are well suited to a wide range of power outputs, they become unattractive in practice at low power levels (below, say, a few hundred kilowatts) because of the adverse behaviour of the turbine stages which suffer irreversible friction as they work. The associated friction force is to a large extent independent of the power output, therefore, the lower the power output the more significant the friction losses prove to be. For very low power outputs (such as those associated with motor vehicles) the overall thermal efficiency of the cycle involving turbines and compressors becomes so low as to be of no practical interest. A different approach to work and heat transfers is then required, not involving the use of a turbine as the mode of work transfer. This is the reciprocating cycle involving the combination of a number of non-flow processes and this approach forms the subject of the present short chapter. We should say at once that this does not involve purely constant pressure devices but it is of interest to see what alternative there is in practice to the rotary work transfer machinery associated with steady flow processes.

8.1 The Turbine and Compressor Have Gone

The removal of the turbine and compressor stages also removes the shaft mode of net work transport out of the engine and a different form of communication with the outside must be introduced. The general Carnot principles still apply, of course: the gas must be expanded and compressed and heat energy must be accepted and rejected. The pressure changes are achieved by the introduction of a moving piston in a cylindrical sleeve to provide a variable volume for the gas between the piston and the closed head of the cylinder. The heat transfer into the system is achieved by igniting fuel in the cylinder and the heat rejection involves the rejection of the spent fuel from the piston volume: this is the Internal Combustion Engine. The working fluid does not now move as a steady flow, as for the turbine devices, although the concept of a continually repeating cycle is maintained.

8.2 The Diesel Principle

Diesel invented the cycle which now bears his name during the search for a cycle that showed the Carnot efficiency (we should say at once that the Diesel cycle does not!). Heat transfer into the system is at constant pressure but the heat rejection is, instead, at constant volume. The p–V diagram for the cycle

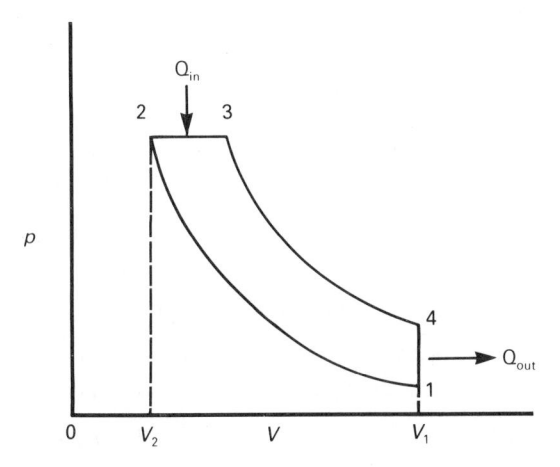

Fig. 8.1 The p–V diagram for the Diesel cycle.

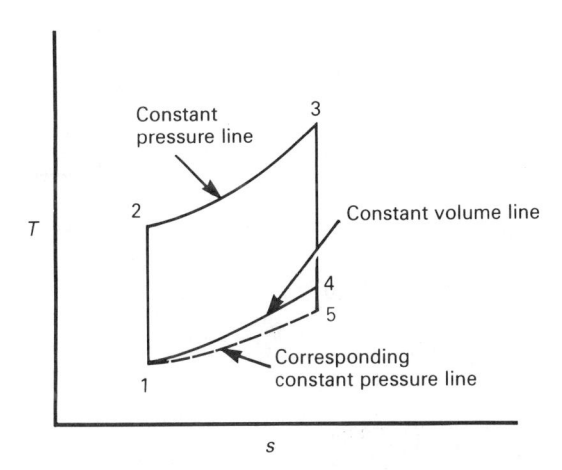

Fig. 8.2 The T–s diagram for the Diesel cycle. The dotted line refers to heat rejection at constant pressure (Joule cycle).

is shown in Fig. 8.1: the corresponding T–s diagram is shown in Fig. 8.2. The volume of the contained gas lies between the maximum value V_1 and the minimum value V_2. $(V_1 - V_2)$ is called the swept volume. The mass of fluid passing through the cylinder is $m(V_2 - V_1)$, where m is mass of the working fluid injected into the cylinder. As before, so now, we suppose the gas to behave as if it was ideal. We will use 1 kg of gas.

Thermal Efficiency and Work Output
Heat enters the cycle during the constant pressure stage 2–3 and is rejected during the constant volume stage 4–1: isentropic expansion is between 3–4 and isentropic compression between 1–2. Suppose we begin at the point 1 of the cycle. The initial compression between 1 and 2 involves the volume ratio $V_1/V_2 = V_r$ (>1): this is called the compression ratio. Heat is now added (in

114 Non-steady Flow Cycles for Low Power Output

practice the air contains the fuel which is ignited spontaneously at the high pressure p_2) and the gas expands to the volume V_3. The volume expansion associated with heat acceptance $V_3/V_2 = V_c$ (>1) is called the cut-off ratio. The isentropic expansion of the burnt gas is where the piston provides work. The expansion is stopped at the volume $V_4 = V_1$, which is also the initial volume, and the burnt gases are ejected. A new volume of free fuel and air is added and the cycle is repeated.

The heat transfer into the cycle is $Q_{23} = C_p(T_3 - T_2)$. The heat transfer out of the cycle is $Q_{41} = C_v(T_4 - T_1)$. The net heat transfer is

$$Q = Q_{23} - Q_{41} = C_p(T_3 - T_2) - C_v(T_4 - T_1)$$

The thermal efficiency, η_D, of the Diesel cycle is

$$\eta_D = \frac{C_p(T_3 - T_2) + C_v(T_1 - T_4)}{C_p(T_3 - T_2)} = 1 - \frac{(T_4 - T_1)}{(T_3 - T_2)} \frac{C_v}{C_p} \tag{8.1}$$

For the ideal gas we write $pV = RT$ so $V_3/V_2 = T_3/T_2 = V_c$. Under isentropic conditions $pV^\gamma =$ constant (A, say) so that $TV^{(\gamma-1)} = A/R$ where, as usual, $\gamma = C_p/C_v$. Then, $(V_1/V_2)^\gamma = V_r^{(\gamma-1)} = T_2/T_1$. This means we can write

$$T_3 = T_2 V_c \quad \text{and} \quad T_2 = T_1 V_r^{(\gamma-1)} \tag{8.2}$$

Also

$$T_4/T_3 = (V_3/V_4)^{(\gamma-1)} = (V_3 V_2/V_2 V_4)^{(\gamma-1)} = (V_3 V_2/V_2 V_1)^{(\gamma-1)} = (V_c/V_r)^{(\gamma-1)}$$

Then

$$T_4 = T_2 (V_c/V_r)^{(\gamma-1)} V_c \tag{8.3}$$

Inserting eqns. (8.2) and (8.3) into eqn. (8.1) gives

$$\eta_D = 1 - \frac{C_v}{C_p} \cdot \frac{(T_2 V_c^\gamma / V_r^{(\gamma-1)} - T_2 / V_r^{(\gamma-1)})}{(T_2 V_c - T_2)}$$

$$= 1 - \frac{1}{V_r^{(\gamma-1)}} \cdot \frac{(V_c^\gamma - 1)}{\gamma(V_c - 1)} \tag{8.4}$$

Apparently, the thermal efficiency depends both on the compression ratio, V_r, and on the cut-off ratio, V_c.

As the cut-off ratio approaches 1 we can write $V_c = 1 + \delta$, with $\delta \to 0$. Then also $V_c^\gamma = 1 + \gamma \delta$ so that $(V_c^\gamma - 1)/(V_c - 1)\gamma = 1$ when $\delta = 0$. It follows from eqn. (8.4) that $\eta \to 1 - 1/V_r^{(\gamma-1)}$ when $V_c = 1$. The significance of this result will become clear in Section 8.3. Because $V_c^\gamma - 1 > \gamma(V_c - 1)$, it follows that the thermal efficiency is increased as the cut-off ratio is reduced to unity.

The work output W is $(Q_{in} - Q_{out})$, that is

$$W = C_p(T_3 - T_2) - C_v(T_4 - T_1) \tag{8.5}$$

Using eqns. (8.2) and (8.3) we find successively

$$W = C_p T_2(V_c - 1) - C_v(T_2 V_c^{(\gamma-1)} - T_2/V_r^{(\gamma-1)})$$
$$= C_p T_2\{(V_c - 1) - (1/\gamma)(V_c^\gamma - 1/V_r^{(\gamma-1)})\}$$

or

$$W = C_p T_1 V_r^{(\gamma-1)}(V_c - 1)\left\{1 - \frac{1}{V_r^{(\gamma-1)}} \cdot \frac{(V_c^\gamma - 1)}{\gamma(V_c - 1)}\right\}$$

Remembering eqn. (8.4) we have finally

$$W = C_p T_1 V_r^{(\gamma-1)}(V_c - 1)\eta_D \tag{8.6}$$

The work output tends to zero as the cut-off ratio approaches its minimum value of unity. This is to be expected: reducing the cut-off ratio reduces the time available for energy to be transferred to the cycle and the limit $V_c = 1$ reduces this time to zero. No energy enters the cycle so no work emerges.

Comparison with the Joule Cycle

The Diesel cycle reduces to the Joule cycle when the heat rejection is made at constant pressure, the T–s diagram taking the form involving the dotted line in Fig. 8.2. The area under the closed curve 12351 is greater than that under the closed curve 12341: the former refers to the Joule cycle and the latter to the Diesel cycle. It follows at once that the thermal efficiency of the Diesel cycle is less than that for the Joule cycle under the same pressure ratio, and the work output is less.

The comparison with the Joule cycle can be taken further. Referring to Fig. 4.2, the isentropic compression stage for the Joule cycle is the line 1–2, linking the temperatures T_1 and T_2. The pressure ratio $p_2/p_1 = r_p$ is related to the temperature ratio T_2/T_1 through eqn. (4.5). The corresponding temperature ratio for the Diesel cycle is given in eqn. (8.2). It follows that r_p and V_r are related by $r_p = V_r^\gamma$ so that

$$V_r^{(\gamma-1)} = r_p^{(\frac{\gamma-1}{\gamma})} = \rho_p$$

The highest and lowest temperatures of the Diesel cycle are T_3 and T_1 respectively so we have

$$\theta = V_c V_r^{(\gamma-1)}$$

The expression (8.4) for the thermal efficiency can, then, be expressed alternatively as

$$\eta_D = 1 - \frac{1}{\rho_p} \cdot \frac{(V_c^\gamma - 1)}{(V_c - 1)} < 1 - \frac{1}{\rho_p} \tag{8.7}$$

This is the statement that the thermal efficiency of the Diesel cycle is inherently lower than that of the Joule cycle.

The work outputs of the two cycles can also be compared. From eqns. (4.17) and (4.15a) it follows for the Joule cycle that

$$W_s = C_p T_1 \rho_p (\theta/\rho_p - 1)$$

so the comparison is, in an obvious notation and using eqn. (8.6)

$$\frac{W(\text{Diesel})}{W_s(\text{Joule})} = \frac{\gamma(V_c^\gamma - 1)\eta_D}{(\theta/\rho_p - 1)\eta_J}$$

The value of the term on the right hand side increases as V_c increases but generally is less than unity.

116   Non-steady Flow Cycles for Low Power Output

Working Conditions

It was seen in Section 8.2 that the working characteristics of the Diesel cycle are inferior to those of the Joule cycle for a given compression ratio, and the question arises as to why the Diesel cycle is of any interest. The answer involves the different working conditions of the two cycles.

The thermal efficiency and work output of the Joule cycle both increase as the temperature ratio θ increases. Because the Joule cycle is a steady flow process, the temperatures in the Joule cycle are maintained continuously while the engine is working. Cooling of the whole engine becomes an important operating characteristic if high temperature ratios are to be used but the metallurgical limit is not to be surpassed. The Diesel cycle, on the other hand, is not a steady flow cycle. Energy is passed explosively (and so discontinuously) at the end of the compression stage and lastgs for a time determined by the cut-off ratio. The working fluid is now a fuel (an appropriate oil/air mixture) and is designed to fire spontaneously when the pressure in the cylinder exceeds a critical value. A very high temperature (perhaps 2500 K) is reached during ignition but this lasts only for a short time. Consequently, the metallurgical limits of the materials provide a less critical restriction on the conditions in the cycle and high pressure ratios can be safely used, especially if the metal in the region of ignition (the cylinder head in practice) is cooled appropriately. This allows the Diesel cycle to be worked with compression ratios far in excess of those possible for the Joule cycle and compensates for the inherently lower efficiency of the Diesel cycle. The cut-off ratio is generally not available to choice once the quantity of heat transfer into the cycle has been decided: both the time needed to transfer the heat and the dynamic ignition characteristics determine V_c. The fuel characteristics also set an upper limit to the compression ratio that can be achieved in the Diesel cycle as the fuel will ignite once the local pressure exceeds a characteristic value and this will be before the full fluid compression has been achieved if V_r is set too high. There is, then, a limit set by the working fluid to the degree to which the thermal efficiency and work output of the cycle can be increased.

8.3   Constant Volume Processes: The Otto Cycle

In the Otto cycle both heat reception and rejection are achieved at constant volume. Although this is entirely outside our brief, which is to constant pressure cycles, the constant volume condition is of interest for comparison as the limit of the Diesel cycle. (It should be pointed out that historically the Otto cycle pre-dated the Diesel cycle but history was rarely conducted in a logical fashion.)

Thermal Efficiency and Work Output

The p–V and T–s diagrams are shown in Figs. 8.3 and 8.4. The volume swept by the piston is again $V_1 - V_2$ ($= V_4 - V_3$). The region 1–2 contains the ignition of the fuel/air mixture, the gas mixture suffering isentropic compression. Heat enters the cycle during the constant volume stage 2–3 and is ignited at the most compressed point. The volume expands, again isentropically, between 3 and 4, and heat rejection occupies the constant volume region 4–1.

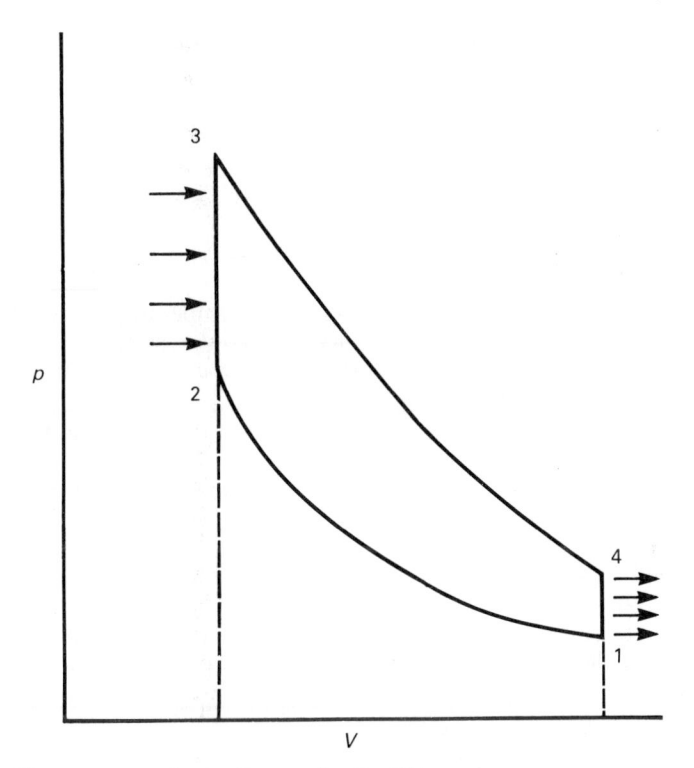

8.3 *Constant Volume Processes: The Otto Cycle* 117

Fig. 8.3 The pressure-volume diagram for the Otto cycle.

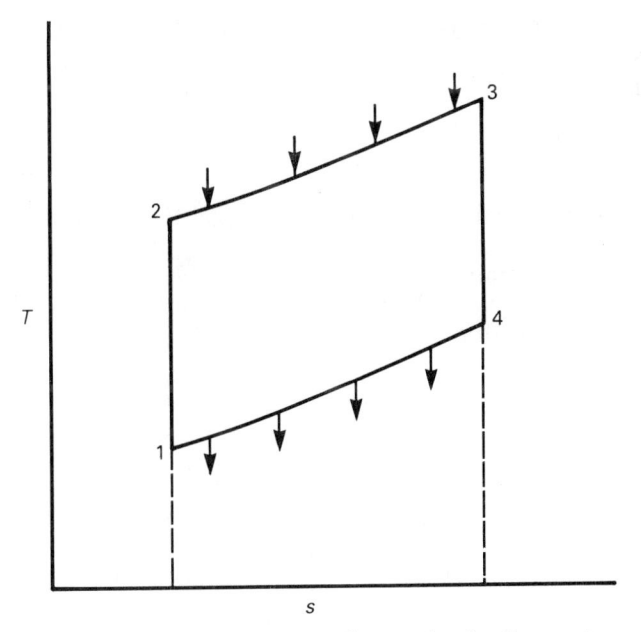

Fig. 8.4 The temperature-specific entropy diagram for the Otto cycle.

The Cycle Characteristics
The heat received and heat rejected by the cycle are
$$Q_{32} = C_V(T_3 - T_2) \qquad Q_{14} = C_V(T_4 - T_1)$$
The thermal efficiency η_O of the cycle is, therefore $\eta_O = 1 - (T_4 - T_1)/(T_3 - T_2)$. Also $T_2/T_1 = T_3/T_4 = V_r^{(\gamma-1)}$ so that $T_4 = T_3/V_r^{(\gamma-1)}$ and $T_1 = T_2/V_r^{(\gamma-1)}$. Therefore

$$\eta_O = 1 - 1/V_r^{(\gamma-1)} \tag{8.8}$$

The thermal efficiency does not depend on the temperature (in common with the simple Joule cycle) but increases with the compression ratio. A comparison between eqns. (8.4) and (8.8) shows that the thermal efficiency of the Otto cycle is greater than that of the Diesel cycle.

The regions of heat transfer are also regions of constant volume: it follows from the ideal gas law then that $T_3/T_2 = p_3/p_2 = r_p$, the pressure ratio for heat acceptance. Since $T_3/T_2 = T_4/T_1$ this is also the pressure ratio for heat rejection. Now, $V_r = T_2/T_1 = (T_2/T_3)(T_3/T_1)$. T_3 is the highest temperature and T_1 is the lowest temperature of the cycle. Denoting T_3/T_1 by θ, then $V_r^{(\gamma-1)} = \theta/r_p$. We can, it seems, formally rewrite eqn. (8.8) as $\eta_O = 1 - r_p/\theta$, even though there is not, in fact, an explicit dependence of η_O on the temperature. The highest temperature achieved in the cycle depends on the properties of the fuel rather than on the thermodynamics of the cycle.

The work output is
$$W = C_v(T_3 - T_2) - C_v(T_4 - T_1)$$
and this is readily arranged into the alternative form
$$W = C_v T_3 (1 - 1/r_p)(1 - 1/V_r^{(\gamma-1)})$$
$$= C_v T_3 (1 - 1/r_p)\eta_O \tag{8.9}$$

The work output depends on the highest temperature of the cycle so there is reason to run the cycle hot, but with adequate external cooling, to achieve the best working characteristics.

Working Conditions
That the Otto cycle has a greater thermal efficiency than the Diesel cycle for a given compression ratio allows its working conditions to be different. The air/fuel mixture is compressed in both cycles but the lower compression ratio for the Otto cycle precludes the usefulness of the mixture being self-igniting. Instead the fuel in the Otto cycle is ignited by a spark discharge, and this is the basis of the petrol internal combustion engine.

8.4 A Mixed Constant Pressure/Constant Volume Cycle

There is no necessity to restrict the heat acceptance stage to a constant pressure or a constant volume phase alone and many modern reciprocating engines employ a combination of these processes for heat input. Indeed, this combination is becoming the norm. Such dual or mixed cycles will have a p–V and a T–s diagram of the form shown in Fig. 8.5. Writing $V_c = V_3/V_2$,

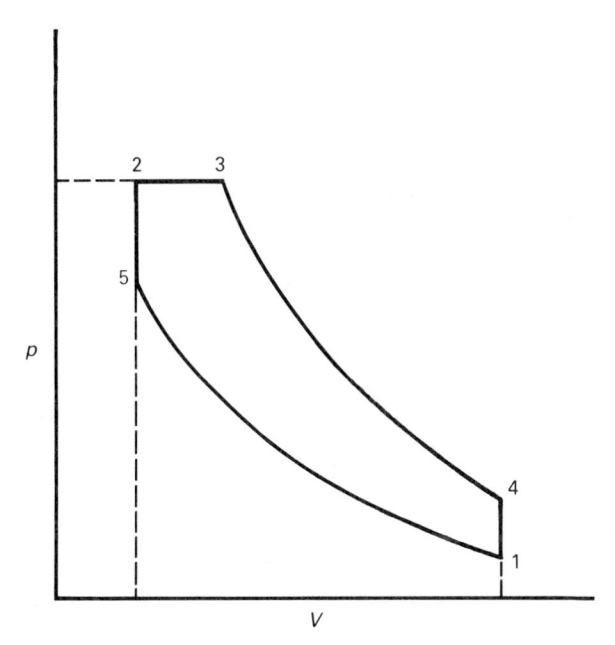

Fig. 8.5 The pressure-volume diagram for the mixed cycle.

$r_v = p_2/p_5$ and $V_r = V_1/V_5$, the thermal efficiency, η_M, of the mixed cycle is given by

$$\eta_M = 1 - \frac{1}{V_r^{(\gamma-1)}} \cdot \frac{r_v V_c^\gamma - 1}{\{(r_v - 1) + \gamma r_v (V_c - 1)\}}$$

For a given heat input and compression ratio, the efficiency of the mixed cycle is greater than that for the Diesel cycle but less than that for the Otto cycle.

8.5 Summary

1. The Diesel cycle has heat acceptance at constant pressure but heat rejection at constant volume. The fuel is ignited by pressure.
2. The general working characteristics of the engine are considered and are compared with those of the Joule cycle: the thermal efficiency of the Diesel cycle is less than that of the Joule cycle.
3. Heat acceptance and rejection at constant volume are the characteristics of the Otto cycle. The working characteristics and conditions for this cycle are briefly considered.
4. The concept is introduced of the mixed cycle with some heat acceptance at constant pressure and some at constant volume. The heat rejection is at constant volume. The working characteristics of this hybrid cycle are not explored.

PART II. CYCLES USING STEAM AS THE WORKING FLUID

Prologue

Gas cycles have many advantages as mobile power plants but their characteristics are not suited to large base power production in the megawatt range. Here steam is the appropriate working fluid at the present time. Although the Carnot cycle can be worked using steam (see Chapter 10) as with any other gas, steam is derived by heating water and the latent heat of vaporisation of water is not available for inclusion in the Carnot heat budget. The introduction of a change of phase into the power cycle must transcend the Carnot analysis but offers important new possibilities. The Carnot engine may well be the most efficient in an ideal world but we live in the real world where ideal fluids do not exist, inefficiencies abound and water is cheap and plentiful. Recognising this and acting on it was the achievement of Rankine. By accounting explicitly for the real properties of water and steam he constructed a work transfer cycle which is better than the Carnot cycle in practical application.

The simple Rankine cycle is not the last word in heat and work transfer and has been improved. The way this is done forms the subject of this Part. Lest, however, it be thought that the Carnot cycle has been eclipsed from our thoughts, the ideals of the thermodynamic cycle still offer a goal to be achieved in practice. The improvements of the simple Rankine cycle are, therefore, developed systematically with this goal in mind.

9. SOME PROPERTIES OF STEAM

Steam does not behave like an ideal gas nor does water behave like an ordinary liquid. For instance, almost all elements (germanium is also anomalous) contract on solidification from the liquid form so that the density increases and the solid sinks in the liquid. Ice is less dense than water and floats (witness icebergs or ice floating in a drink): this is extraordinary from a general point of view. The relation between p, V and T (that is, the equation of state) for water is not expressible by a simple formula covering a wide range of conditions. Instead, the equation of state is represented by a set of numerical values (obtained empirically by measurement) of the thermodynamic properties for a given pressure and temperature: these are expressed in the form of tables and the collection is called a set of steam tables (see Section 9.2). Some of the data (for instance the relation between specific enthalpy and specific entropy) are sometimes given alternatively in graph form for greater convenience in use. The correlation of the data has been overseen by an international commission so the data are recognised internationally: for this reason we sometimes refer to them as the International Steam Tables. The full steam tables are highly detailed and extensive and are usually necessary only for full design purposes. Abbreviated forms are more useful for many ordinary purposes and we shall presume in what follows that such secondary tables (for example, 'Thermodynamic and Transport Properties of Fluids' arranged by G. F. G. Rogers and Y. R. Mayhew) are available to the reader. If the full tables are available, however, this would be better still but the detail of the data will usually be unnecessary here.

9.1 The T–V, T–p and p–V Diagrams

Two independent thermodynamic properties of a fluid are sufficient to specify fully its state in equilibrium. Measurements are made of the temperature T, the pressure p and the volume V to specify the equation of state so any one of these can be expressed as a function of the other two. Different aspects of the behaviour of the fluid are highlighted by choosing different pairs of variables as the datum. The equation of state for a particular fluid is determined empirically and, although it could in principle be calculated from a knowledge of the properties of the molecules comprising the material and of the intermolecular forces, we accept it now as a relationship deduced from experimental data. The data are conveniently expressed as three graphs (often referred to as diagrams), a different variable being held constant in each case.

The T–V *Diagram*
Suppose a unit mass (1 kg) of ice is held in a container (it might be a cylinder with a piston) at constant pressure while heat transfer occurs from the outside. We might begin with atmospheric pressure, $p = 1.01325$ bar, for the

124 Some Properties of Steam

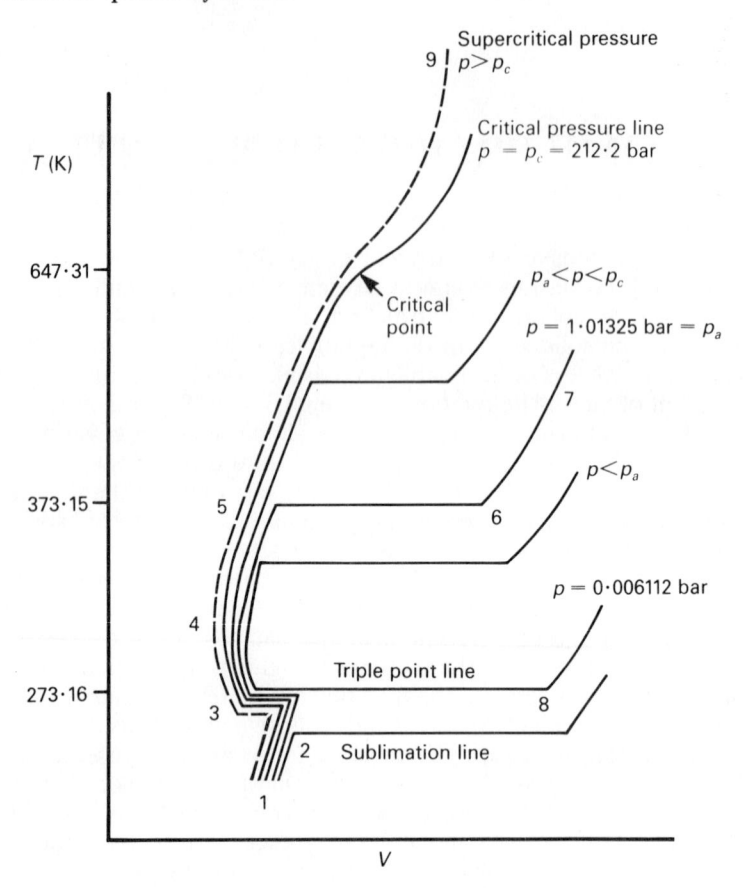

Fig. 9.1 The temperature-volume curves for steam (not to scale).

pressure in the cylinder. Values of T and V are plotted as a graph in Fig. 9.1 which we refer to now.

As heat is added the volume of the ice increases by a small fraction until the melting point is reached, $T = 273.16$ K (line 1–2 in Fig. 9.1). Adding more heat does not raise the temperature but more ice melts at constant temperature until all the ice has turned to water. The melting process at constant temperature is accompanied by a *reduction* of the volume and is represented by the constant temperature line 2–3. The quantity of heat required to melt the unit mass of ice is called the latent heat of fusion of water, measured in kJ/kg. The further addition of heat raises the temperature of the water (3–4 in Fig. 9.1) but at first the volume continues to decrease until the temperature is about 277.16 K (4 °C) when it reaches its minimum value corresponding to a maximum density. Raising the temperature above this point sees the volume increase until the point 5 is reached where the temperature is $T = 373.15$ K. The further input of heat now leaves the temperature of the water unaffected. The water begins to contain vapour (steam), the proportion increasing with further heat input. The vaporisation process (like melting) also occurs at constant temperature, the volume now increasing (for constant pressure): this is represented by the line 5–6. The heat required for the complete vaporisation of the water is called the latent heat of vaporisation. The distance along 5–6 relative to the total length 5–6 gives the proportion x of vapour that has formed at that point. Because x, by its definition, excludes the water it is often called the dryness fraction or the quality of the steam. The addition of further heat beyond the point 6 heats the steam and the temperature and volume increase (line 6–7). Constant pressure lines are called isobars and the T–V plot we are describing contains a set of isobars. The same behaviour is found for other pressures not far from atmospheric except that the change of volume on melting decreases as the constant pressure is made smaller, and the change of volume of vaporisation decreases as the pressure increases.

The boiling point decreases with decreasing pressure while the melting point increases by a small amount, the two points moving towards each other. At the pressure of 0.006 112 bar the melting and boiling points become identical. Under these conditions, and only under these conditions, solid, liquid and vapour water can exist together in full thermodynamic equilibrium. This is called the triple point. The corresponding line on the T–V diagram is the line 3–8 which is called the triple point line. For pressures below this the solid (ice) now forms vapour directly without passing through a liquid phase: this behaviour is called sublimation. The temperature at which the boiling and melting points are the same has been accepted internationally as a fixed point on the absolute temperature scale, being assigned the value 273.16 K. This region of the diagram is not of particular importance for steam cycles. Of greater interest from this point of view is when the ambient pressure is raised above atmospheric.

For pressures above atmospheric the isobar has the same general form as for atmospheric pressure except that there is a steady decrease of the change of volume on vaporisation as the pressure is increased. Eventually a pressure is reached where the volume change vanishes and the latent heat of vaporisation is zero. This is the critical point and marks the conditions where the distinction between liquid and vapour disappears. At the critical point $dT/dV = 0$ so there is here an inflection point in the isobar and not an extremum condition: for higher pressures $dT/dV > 0$. For water the critical point conditions are $p = p_c = 221.2$ bar, $T = T_c = 647.3$ K and $V = V_c = 0.003\ 17$ m^3/kg, these values forming the critical data for water. For supercritical pressures there is no vaporisation process and the isobar increases smoothly with increasing temperature. Such an isobar is the line 3–9 in Fig. 9.1.

The p–T Diagram

The same information can be expressed with p and T varying but V held constant. The plot is shown in Fig. 9.2. This form of presentation of the data shows the triple point clearly.

The p–V Diagram

The dependence of pressure on volume at constant temperature is information of practical utility and the curve for water is shown in Fig. 9.3. The lines of constant temperature are called isotherms. The separation between liquid and vapour below the critical point is well shown as is the merger of these two states above the critical point. The p–V diagram is useful for introducing one

126 *Some Properties of Steam*

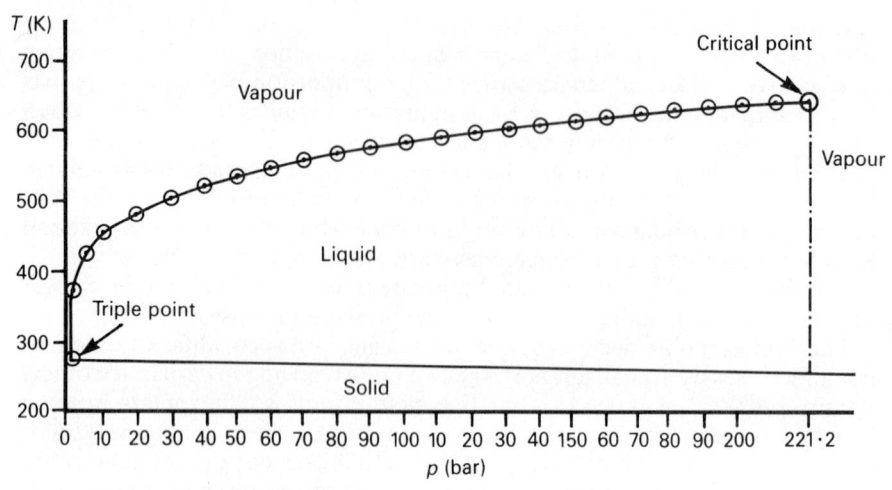

Fig. 9.2 The pressure-temperature relationship for water.

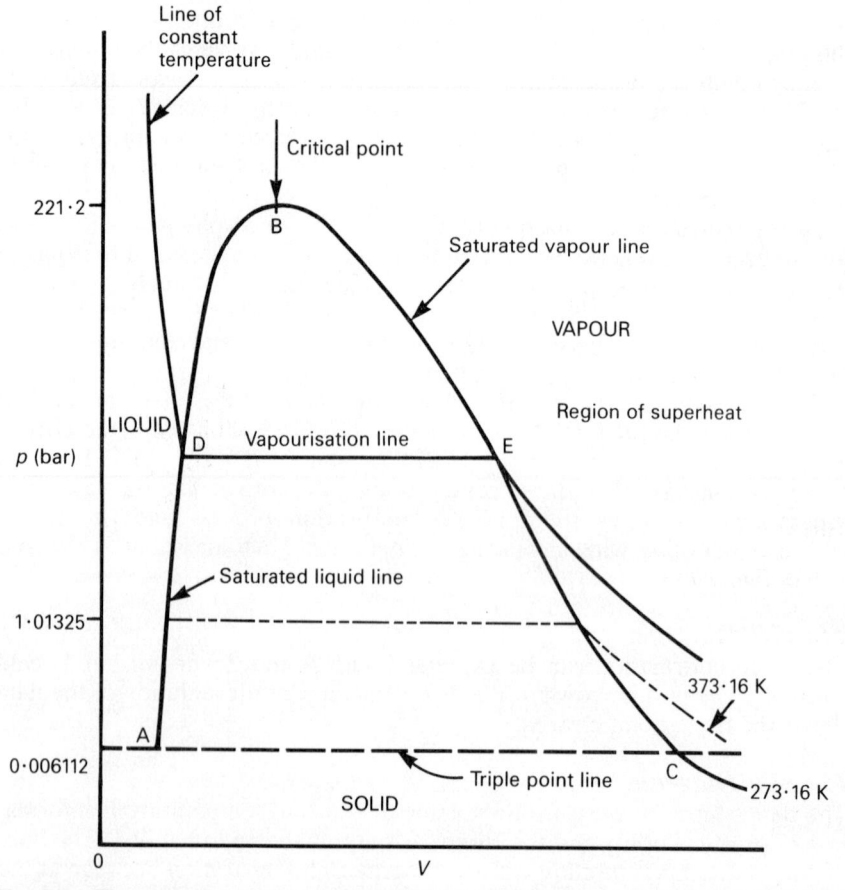

Fig. 9.3 The pressure-volume relationship for water, showing the critical point and the triple point line.

or two items of nomenclature. The line AB separating the liquid phase from the vapour is called the saturated liquid line. The line DE for vaporisation contains a liquid–vapour mixture and so is a two-phase region. This becomes entirely vapour along the line BC which is called the saturated vapour line, the steam being said to be dry saturated then. Steam lying to the right of the saturated vapour line is said to be superheated. The actual temperature of the superheated steam will be the same as that for some other isotherm (of lower temperature) as it meets the saturation line, and called the saturation temperature for that isotherm. The difference between the actual temperature of the steam and the saturation temperature for the associated isotherm is called the degree of superheat of the steam.

The state of a two-phase region is specified by a knowledge of both p and V, or both T and V. The state of a saturated vapour or a saturated liquid (one-phase regions) is specified by knowing either p or T.

Other Thermodynamic Variables

Once the equation of state is specified other thermodynamic quantities can be calculated using the various formulae of thermodynamics. We refer the reader to a treatise on thermodynamics for these formulae and the details of their derivation. It should be remembered, however, that these quantities are not given in absolute terms by the First and Second Laws of Thermodynamics but only relative to some chosen datum state. It is conventional to choose the properties at the triple point for the datum. This means that both u and s for the liquid are set equal to zero for that point.

Of particular importance now are the specific enthalpy $h = u + pV$ (where u is the specific internal energy) and the specific entropy s whose difference ds relative to a datum state is $ds = (du/T)du + (p/T)dV$. The specific heat capacities at constant pressure and constant volume have sufficiently small variations with both temperature and pressure that they are assumed to be constant throughout the changes of pressure and temperature of practical interest in power cycles.

9.2 Representing the Properties

The properties of steam, derived from empirical data which are also developed according to the relationships of thermodynamics, can equally well be presented in the form of tables or in the form of graphs: each form of representation has its particular convenience. The data are all given as specific quantities referring to unit mass of steam.

Steam Tables

These provide values of p, V, T, u, h and s between the sublimation point and the critical point. Data for the saturated liquid carry the subscript f while data for the saturated vapour carry the subscript g.

When both the liquid and vapour phases are present in some proportion in a two-phase region the thermodynamic quantity is an appropriate mixture of those for the two phases separately. For instance, suppose we consider the specific enthalpy of wet steam along the line DE in Fig. 9.3. 1 kg of the water–steam mixture can be regarded as composed of some quantity, x kg say, of saturated vapour and $(1-x)$ kg of saturated liquid, the total mass being 1 kg as required: x is the dryness fraction. This means we write

$$h = (1-x)h_f + xh_g$$

h is specified in terms of h_f and h_g when x is known. This expression can be written alternatively

$$h = h_f + x(h_g - h_f)$$

and the difference $h_g - h_f$ (>0 incidentally) needs to be calculated. The tables do this for us by listing also $h_g - h_f \equiv h_{fg}$ explicitly: this means we find h by looking up h_f, h_g and h_{fg} and, with x known, form

$$h = h_f + xh_{fg} \tag{9.1}$$

The dryness fraction has often to be found. In an isentropic expansion it is the specific entropy that is important. In that case s is known for the end of the expansion because it is the same as the value at the beginning and the dryness fraction follows from

$$x = (s - s_f)/s_{fg}$$

with s specified and s_f and $s_{fg}(= s_g - s_f)$ being read from the steam tables. In this way the dryness fraction is inferred for use in the calculation of the enthalpy according to eqn. (9.1).

The Mollier Diagram

This is the graph of specific enthalpy h against specific entropy s. For the SI system of units h is measured in kJ/kg and s in kJ/kg K. For ease of calculation various other data are also included in the diagram. Explicitly, the plot includes the set of lines of constant temperature (isotherms) from the saturation line usually up to about 800 °C; the set of lines of constant pressure (isobars) from the triple point pressure up to a little more than 1000 bar (100 MN/m^2); and lines of constant dryness fraction are given below the saturation line. There is here all that is necessary to calculate the working conditions of a steam cycle by following the changes of entropy and temperature that occur throughout the cycle.

The Mollier diagram is very valuable since it provides an alternative way of calculating the working conditions for a given cycle, separately from the direct use of steam tables. This alternative method of calculation is invaluable as a check on the direct use of steam tables. Errors will always appear in calculations—there is no shame in that—but they must be detected and eliminated (the shame is if they are not), and methods that provide a check on calculations are essential. It is wise to use both steam tables and the Mollier plot, independently, in any steam calculation.

The diagram supposes the system to be ideal. Isentropic efficiencies less than unity are accounted for by correcting the calculated enthalpy differences as they occur and moving to the appropriate new enthalpy location on the diagram, before proceeding with the next stage of the calculation.

The T–s Diagram

It is often useful to represent the cycle by using temperature and specific entropy as the independently variable quantities. The T–s diagram will include a set of isobars to show the variations of temperature throughout the

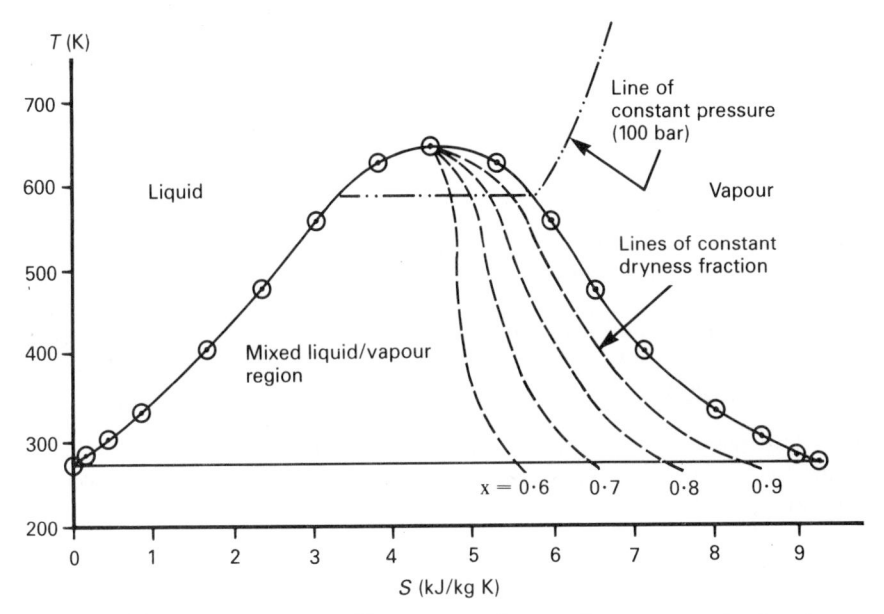

Fig. 9.4 The temperature-specific entropy diagram for water. x is the dryness fraction.

cycle directly. This is essentially a representation diagram and will be used extensively in the arguments to follow. The T–s diagram for steam is plotted in Fig. 9.4: the plot also includes some lines of constant dryness fraction x and the line of constant pressure for 100 bar.

Further Comment
Although concern here is with steam, the same methods are available for other vapours and the same plots can be made in appropriate forms. Not considered here, but of considerable interest in other connections, are data for refrigerants. Now the pressure–enthalpy diagram is of especial value.

9.3 Summary

1. Certain properties of steam are collected together and their representation is explained.
2. The three diagrams displaying the equation of state (p–V, p–T, T–V) are set down and used as a basis for defining the saturated liquid–vapour line, the mixed region of vaporisation, the triple point and the triple point line, and the critical point. What is meant by the dryness fraction, superheated steam and the degree of superheat is also explained.
3. The form and use of steam tables is explained as is the Mollier diagram and the T–s plot.

9.4 Exercises

9.1 Use the steam tables to plot the temperature–specific entropy diagram

130 Some Properties of Steam

for steam. Include the lines of constant dryness fraction $x = 0.95, 0.90, 0.85, 0.80, 0.75, 0.70, 0.65$ and 0.60.

9.2 Use the steam tables to plot the pressure–specific volume diagram for steam. Designate the regions of vapour, liquid and solid.

9.3 Use the steam tables to plot the temperature–specific volume diagram for steam. Mark the regions of vapour, liquid and solid. Mark the triple point line and the sublimation line.

9.4 Use mercury tables to plot the temperature–specific entropy diagram for mercury. Mark again the same lines of constant dryness fraction as for steam in Question 9.1.

9.5 Use the mercury tables to plot the temperature–specific volume diagram for mercury. In what way does this differ from the corresponding diagram for water?

10. THE CARNOT ENGINE WITH STEAM AS THE WORKING FLUID

10.1 Introduction

The Carnot cycle can be used with any working fluid although the advantageous properties arise only if the working fluid is ideal. In particular, the cycle can use steam and it is of interest to begin by finding the characteristics of the cycle using this real vapour. It must be noted that the steam must not be allowed to condense because the Carnot cycle does not explicitly include the possibility of the change of phase of the working fluid. The steam may well be wet due to partial condensation at some points of the cycle but it must not be allowed to condense entirely. The cycle then, having ideal expansion and compression stages though using a real working fluid, provides working characteristics that can be used as a datum against which the properties of other cycles can be judged.

10.2 The Working Characteristics

A steam plant which follows the Carnot principles closely is shown in Fig. 10.1. Water is heated in a heater H to form a steam at A which is compressed in the compressor C. The steam is then heated in the boiler B and expanded in the steam turbine T. Heat energy is rejected in the condenser D and the working fluid is returned to the heater H to be converted to dry saturated steam again. The boiler B is the high temperature reservoir (at temperature T_2) and the condenser D is the low temperature reservoir (at temperature T_1) of the cycle which acts as an *open* Carnot cycle. The T–s diagram is shown in Fig. 10.2.) The precise conditions of the high and low temperature junctions are determined by the pressures actually maintained there. The higher the pressure within the boiler the higher the temperature of the steam: the lower the pressure of the condenser, the lower the temperature there.

Although the concept of the Carnot cycle has been adhered to in Fig. 10.1, the practical application would probably not involve two separate sources of heat transfer into the cycle. The boiler H could be used to provide sufficient heat transfer to yield steam at an appropriate temperature after compression.

We start at the furnace end and use the notation of Fig. 10.2. Wet steam enters the boiler in the state 1, having come from the condenser. It is first compressed isentropically to the temperature T_h where its specific enthalpy is h_2 and its entropy $s_2 = s_1$. The work done on the steam is

$$W_{12} = h_2 - h_1 \qquad (10.1)$$

Heat Q_{in} enters the boiler to form saturated steam at state 3. The process is isothermal with temperature T_h and this is the hot end of the cycle. The specific enthalpy of the steam leaving the boiler is h_3 and its specific entropy is

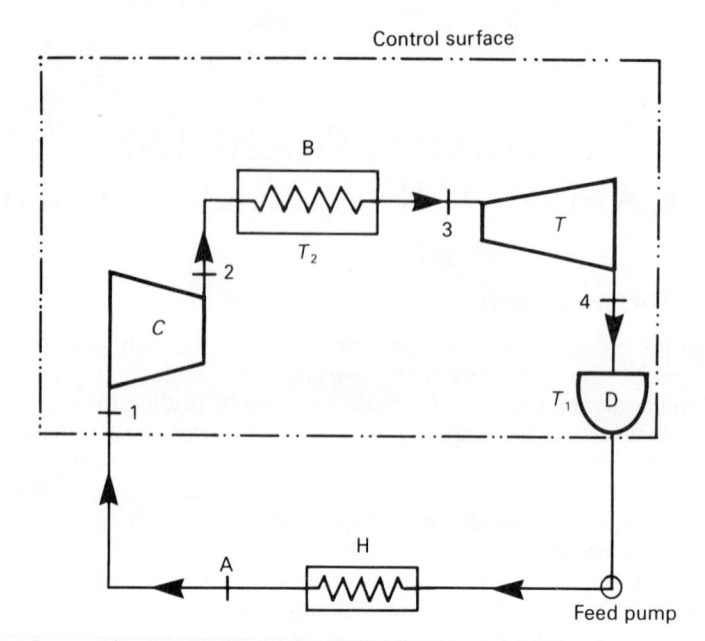

132 The Carnot Engine with Steam as the Working Fluid

Fig. 10.1 Layout for the Carnot cycle using steam as the working fluid.
The control surface involving steam is marked $-\cdot\cdot-\cdot\cdot-$. The maximum temperature is T_2 and the minimum T_1. The numbers are as in **Fig. 10.2**.

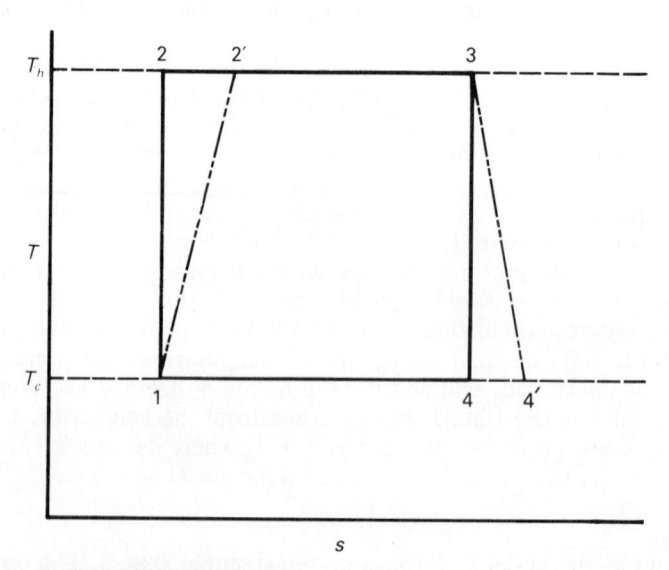

Fig. 10.2 Plot of temperature against specific entropy for the Carnot cycle using steam as the working fluid. The solid lines refer to the ideal cycle, while the broken lines refer to the cycle with inefficiencies.

s_3. The heat added to the system is

$$Q_{23} = h_3 - h_2 \qquad (10.2)$$

The steam is now expanded through a turbine (or reciprocating engine) to the state 4. This process is isentropic and $s_4 = s_3$. The work done by the steam, which is available for use, is

$$W_{34} = -(h_4 - h_3) \qquad (10.3)$$

With the expansion complete, the steam/water mixture now passes through the condenser at constant pressure and constant temperature T_c (the cold isothermal process) ending with the specific entropy s_1 and specific enthalpy h_1 with which the process started. The heat rejected (which cannot be used in the engine) is

$$Q_{41} = -(h_1 - h_4)$$

The thermal efficiency η of the process is, by definition

$$\eta = (W_{34} - W_{12})/Q_{23}$$

The work ratio W_r (see Section 2.6) is given by

$$W_r = (W_{34} + W_{12})/W_{34} \qquad (10.4)$$

The specific steam consumption (see Section 3.1.3) is

$$ssc = 3600/(\text{net work transfer from the system}) \text{ kg/kWh}$$
$$= 3600/W_{34} \text{ kg/kWh} \qquad (10.5)$$

A Specific Example

Suppose the boiler is held at 40 bar pressure while the condenser is held at 0.04 bar, providing a temperature about 10 °C above the ambient temperature of the environment.

The properties of steam are found from steam tables as explained in Chapter 9. We find

$$h_2 = h_f = 1087 \text{ kJ/kg and } s_2 = s_f = 2.797 \text{ kJ/kg K} \qquad \text{(steam tables)}$$

and

$$T_h = T_2 = T_3 = 523.6 \text{ K} \qquad \text{(steam tables)}$$

After the heat transfer to the steam the enthalpy and entropy are

$$h_3 = h_g = 2801 \text{ kJ/kg and } s_3 = s_g = 6.070 \text{ kJ/kg K} \qquad \text{(steam tables)}$$

The condenser pressure of 0.04 bar gives

$$T_c = T_4 = T_1 = 302.3 \text{ K} \qquad \text{(steam tables)}$$

The corresponding specific entropy is $s_4 = s_f = 0.422$ (steam tables)

Now (see Section 9.1, eqn. (9.1))

$$s_4 = s_3 = s_{f4} + x_4 s_{fg3}$$

so that (again from eqn. (9.1))

$$6.070 = 0.422 + x_3 8.051$$

giving $x_3 = 0.701$. Similarly,

$$s_2 = s_1 = s_{f1} + x_4 s_{fg1}$$

giving $x_4 = 0.295$. Then,

$$h_4 = 121 + 0.701 \times 2433 = 1826.5 \text{ kJ/kg}$$

and

$$h_1 = 121 + 0.295 \times 2433 = 838.7 \text{ kJ/kg}$$

We can now calculate the values of work done and the energy expended. From the values we have calculated

$$W = W_{34} - W_{21}$$
$$= (h_3 - h_4) - (h_2 - h_1)$$
$$= 727.2 \text{ kJ/kg}$$

and

$$Q = Q_{32} = (h_3 - h_2)$$
$$= 1714 \text{ kJ/kg}$$

The thermal efficiency is then $\eta = 727.2/1714 = 0.424$ or 42.4%. This is an interesting result because it shows that, under the conditions we have stipulated, even for the ideal case more energy is rejected (57.6%) than is available for useful work (42.4%). We shall find this feature to be common to all working engines (which could justifiably be called working heaters!).

The value of the thermal efficiency of the Carnot engine is obtained alternatively from eqn. (2.3), using the upper temperature (523.6 K) and the lower temperature (302.3 K): it is $\eta = 0.423$.

The work ratio for the engine is

$$W_r = W/W_{34} = 727.2/975.5$$
$$= 0.745$$

and the specific steam consumption is

$$ssc = 3600/727.2 = 4.95 \text{ kg/kWh}$$

Repeating this calculation for different boiler and condenser pressures allows Figs. 10.3 and 10.4 to be constructed showing (a) the effect of boiler pressure on the thermal efficiency and specific steam consumption for a fixed condenser pressure, and (b) the effect of condenser pressure on the thermal efficiency, the boiler pressure being held constant. For the effect of boiler pressure, the thermal efficiency increases steadily, fast at first but beyond about 130 bar is relatively insensitive to boiler pressure changes. On the other hand, the specific steam consumption has a minimum around 35 bar and then steadily increases to the critical pressure. Because the steam consumption dictates the size required for the boiler, the Carnot engine would best be constructed is practice to use a low boiler pressure where, unfortunately, the thermal efficiency is low.

10.2 The Working Characteristics

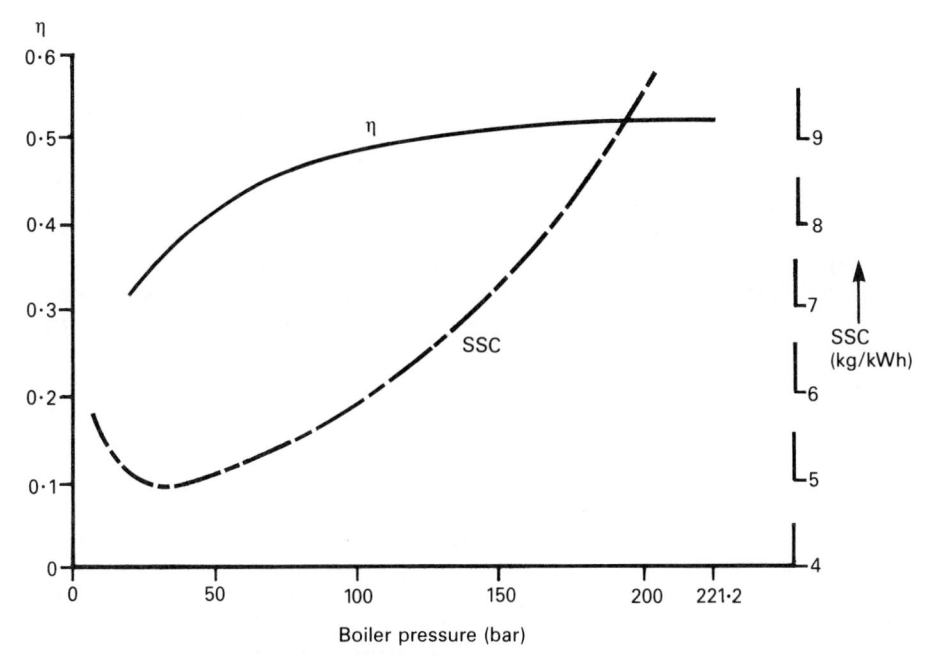

Fig. 10.3 The thermal efficiency and specific steam consumption (ssc) for a range of boiler pressures up to the critical pressure for the Carnot cycle using steam as the working fluid. The condenser pressure is held constant at 0.04 bar.

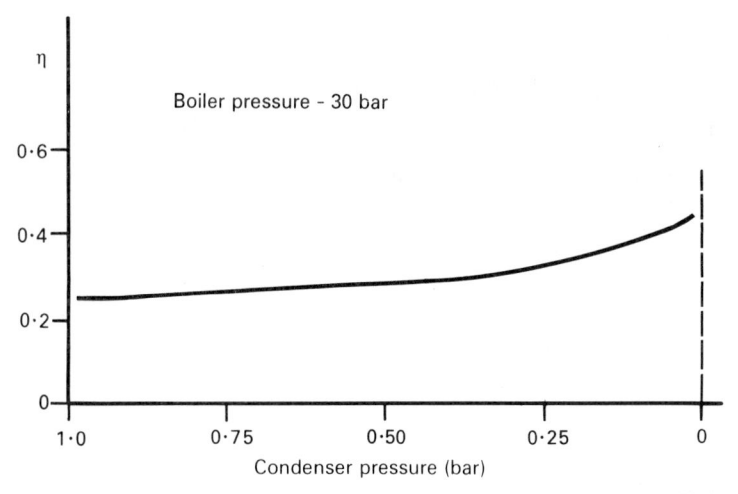

Fig. 10.4 The effect of changing the condenser pressure on the thermal efficiency of a Carnot cycle using steam as the working fluid. The boiler pressure is held constant at 30 bar.

10.3 Isentropic Efficiencies

The practical compressors and turbines are no more perfect for steam than for gas (rather less so if the steam is wet) and it is necessary to account for their irreversible behaviour. This is done by using the isentropic efficiencies introduced in Section 4.5 but we will use a slightly different notation to help us remember that it is steam that we are using: the isentropic efficiency for the expansion stage will be denoted by y and the isentropic efficiency for the compression stage by z. Explicitly, the isentropic efficiency for expansion is defined as the ratio of the change of specific enthalpy $(h_3 - h_{4'})$ for the stage including irreversibility as a fraction of the specific enthalpy change $(h_3 - h_4)$ for the ideal case:

$$y = \frac{[h_{3'} - h_4]}{[h_3 - h_4]} \qquad z = \frac{[h_2 - h_1]}{[h_{2'} - h_1]} \tag{10.6}$$

These definitions provide $y < 1$ and $z < 1$ for real systems with irreversibility. As before so now, the effect of irreversibility is to increase the entropy at each stage during the change of volume. With inefficiencies in the compression and expansion processes the Carnot cycle is represented in Fig. 10.2, the initial (ideal cycle) rectangle now being distorted by each process due to an increase of entropy. The 'ideal' cycle 12341 becomes 12'34'1. The work transfer on expansion is $W_{34'} = (h_3 - h_{4'})$ and on compression is $W_{12'} = (h_1 - h_{2'})$. The heat transferred to the cycle is $Q_{2'3} = (h_{2'} - h_3)$, which is rather less than for the ideal case. The heat rejected is $Q_{1'4} = (h_{1'} - h_4)$ which is rather more than before. The overall thermal efficiency must fall as the result of the effects of the non-isentropic components.

The Cycle Working Characteristics

The work $W_{12'}$ actually done in compression is related to that for the ideal case W_{12} by

$$W_{12'} = W_{12}/z$$

For expansion the corresponding work transfers are

$$W_{34'} = yW_{34}$$

The heat transferred into the working fluid is written

$$Q_{2'3} = h_3 - h_{2'}$$

But $-W_{12'} = -W_{12}/z = h_{2'} - h_1$, so that

$$h_{2'} = -W_{12}/z + h_1(1 - 1/z) + h_2/z$$

This gives

$$Q_{32'} = h_3 - h_2/z - h_1(1 - 1/z)$$

The thermal efficiency is then

$$\eta = [W_{34'} - W_{12'}]/Q_{12'}$$
$$= [yW_{34} - W_{12}/z]/[(h_3 - h_2/z) - h_1(1 - 1/z)]$$

in terms of the ideal work transfer values.

A Numerical Application

Let us consider again the example used for the ideal case, with boiler pressure 40 bar and condenser pressure 0.04 bar, but now with isentropic efficiencies less than unity. We might suppose that both the turbine and compressor have the same isentropic efficiencies of 0.8 so that $y = z = 0.8$. Using the values of the enthalpies for the ideal case given above the work transfer is found to be 469.31 kJ/kg and the heat transfer is found to be 1651.93 kJ/kg. This means that the thermal efficiency of the cycle is 28.41 %, which is substantially lower (by 32.84 % in fact) than for the ideal cycle. The work ratio is $W_r = 0.60$ and the specific steam consumption is $ssc = 7.67$ kg/kWh: the former is 19.45 % lower than for the ideal case while the latter is 54.95 % higher. Non-ideal components for compression or expansion affect the properties of the cycle significantly. Even if the cycle were constructed and used the non-ideal quality of the actual compression and expansion units would destroy many of its attractive thermal features.

10.4 Summary

1. The Carnot cycle is considered with steam as the working fluid. The working characteristics of the cycle are calculated for a particular pair of boiler and condenser pressures.
2. For a fixed condenser pressure, the thermal efficiency increases with increasing boiler pressure without passing through a maximum value. By contrast, the specific steam consumption passes through a minimum value at a lower boiler pressure after which it increases continuously to the critical point.
3. The effect of the condenser pressure at constant boiler pressure is less marked, the thermal efficiency increasing slowly as the condenser pressure is reduced.
4. The effects of non-isentropic pressure changes are investigated and the working characteristics of the cycle evaluated again. The adverse effects of irreversibilities are made clear.

10.5 Exercises

10.1 Calculate the heat and work transfers, cycle efficiency, work ratio and steam consumption of a steam driven Carnot cycle using a boiler pressure of 30 bar and a condenser pressure of 0.04 bar ($y = z = 1$).

10.2 Repeat the calculations of Question 10.1 for the different boiler pressures of (a) 50 bar and (b) 100 bar. The condenser pressure remains at 0.04 bar in each case.

10.3 Repeat the calculations of Question 10.2 for the different condenser pressures of (a) 0.02 bar and (b) 0.06 bar, the boiler pressure remaining at 30 bar.

10.4 Recalculate Question 10.2 with isentropic efficiencies of (a) 0.90, (b) 0.80 and (c) 0.70 for the compression and expansion processes, estimating the actual cycle efficiency and steam consumption in each case.

138 The Carnot Engine with Steam as the Working Fluid

10.5 Summarise the results of your calculations for Questions 10.1, 10.2, 10.3 and 10.4, together with the results of the main text, in graph form. Show particularly:

(i) the effects of (a) the boiler pressure, (b) the condenser pressure; and (c) the isentropic efficiency, on the thermodynamic efficiency of the cycle:

(ii) the effects of (a) the boiler pressure and (b) the condenser pressure on the work ratio;

(iii) the effects of (a) the boiler pressure, (b) the condenser pressure and (c) the isentropic efficiency on the specific steam consumption.

10.6 A Carnot cycle using steam as the working fluid has an initial pressure of 30 bar and a final pressure of 0.04 bar. The isentropic efficiency factors for the turbine and the compressor are y and z respectively. Show that the thermal efficiency η of the cycle can be expressed by

$$\eta = \frac{(904y - 215/z)}{(2010.5 - 215.5/z)}$$

Show further that the work ratio is given by

$$r_w = 1 - \frac{0.299}{yz}$$

and the specific steam consumption by

$$ssc = \frac{3600}{940y - 215.5/z}$$

Which component inefficiency has the greater effect on (a) thermal efficiency, (b) work ratio and (c) specific steam consumption?

Suppose it was decided to make such a plant with the requirement that $\eta > 0.30$, $r_w > 0.6$ and $ssc < 7$ kg/kWh. What constraints do these requirements place on the qualities of the turbine and compressor that can be used?

10.7 The effect of inefficiencies in the expansion and compression stages of a Carnot cycle reduce the mean thermal efficiency of the cycle. The effect can be reproduced by lowering the maximum temperature T_E of a hypothetical ideal cycle ($y = z = 1$) chosen specifically to represent the thermal efficiency of the inefficient cycle. If T_E is the effective maximum temperature of the hypothetical ideal cycle, use the results of Question 10.6 to derive the expression

$$T_E = \frac{[2010 - 215/z]}{[2010 - 904y]} T_1$$

for the effective upper temperature of the hypothetical ideal cycle, where T_1 is the (low) temperature for heat rejection.

10.8 Repeat the analyses of Questions 10.6 and 10.7 for the different initial pressures (a) 40 bar, (b) 60 bar and (c) 100 bar, the lower pressure remaining at 0.04 bar. Plot a graph to show how the effective temperature T_E depends on the value of the boiler pressure.

10.9 Repeat the analyses of Question 10.7 for the maximum pressure of 30 bar but with the low pressure having the values (a) 0.030 bar, (b) 0.035 bar, (c) 0.045 bar and (d) 0.05 bar. Plot a graph to portray the effect of the final pressure on the effective maximum temperature.

11. THE SIMPLE RANKINE CYCLE

An effective engine requires both a good heat capacity in the working fluid and an efficient conversion of that heat into work. The Carnot cycle provides the criterion for the second requirement but the first is not especially addressed. A vapour (with low density and low thermal capacity) does not have a very effective ability for absorbing and holding heat energy, unless the temperature is very high. Providing a high heat content for the given working fluid must, therefore, involve a large mass of vapour as working fluid and this is, for example, a characteristic of a gas turbine unit using air where the air mass is achieved by a high throughput. If the vapour were formed directly from the corresponding liquid, the latent heat of vaporisation could be added to the energy budget, and the heat content correspondingly increased. Steam was considered as the working fluid for the Carnot engine in Chapter 10, but it was important there not to allow the steam to condense during the cycle. If, however, a phase change is included (the water being heated to form dry steam), the latent heat of water is included in the heat budget, the high temperature fluid after expansion being condensed back to water to provide the low temperature end. There is another advantage in introducing the condensed phase of the working fluid. Liquids are highly incompressible and reach high pressures with very small decreases of volume. By compressing the *liquid* before it enters the boiler, the boiler pressure is made high without the use of a gas compressor unit. The work done during compression is very small (as we will find) in comparison with the expansion work output, so that the unavoidable inefficiencies associated with compression are effectively eliminated. The addition of the phase transition to the cycle was proposed by Rankine and the resulting cycle is named after him. Because a real working fluid (for instance, water/steam) is presumed from the outset, the Rankine cycle can form the basis for the practical conversion of energy into work for a vapour cycle.

11.1 The Basic Cycle

The basic Rankine cycle is closed and is shown schematically in Fig. 11.1. Heat is added to the working fluid over the range of temperature, in fact from the lowest, T_c of the condenser to the highest T_b of the boiler. The Carnot criterion for maximum thermal efficiency is that *all* the heat should be added at the highest temperature and the heat rejected should *all* be at the lowest temperature. Whilst the Rankine cycle meets the second requirement (steam is condensed at a constant low pressure), it does not meet the first. Consequently we can say at once that the thermal efficiency of the Rankine cycle must, as a matter of principle, be lower than for the corresponding Carnot cycle *between the same maximum and minimum temperatures, presuming all components behave ideally.* Since the real world is not ideal, and money constraints could make the components far from ideal, the thermal

11.2 The Features of the Cycle

efficiency of the Rankine cycle need not be lower in practice than that of the corresponding Carnot cycle if the inefficiencies are included in the tally for the overall thermal efficiency.

11.2 The Features of the Cycle

Referring to Fig. 11.1, water enters the boiler (point 5) and is first brought to the boiling temperature (point 2) appropriate to the pressure maintained in the boiler beyond which it is converted into steam. The steam leaves the boiler (point 3) in a dry saturated state and passes through a simple turbine system (T) where it is expanded, doing work (between points 3 and 4). The expanded steam leaves the turbine (point 4) and releases (rejects) heat in the condenser (C) at the low temperature maintained there by the vacuum. The now condensed steam, as water (point 1) is compressed and passes back to the boiler where the cycle begins again (point 5). The associated T–s diagram is shown in Fig. 11.2, using the number notation of Fig. 11.1.

The positive (turbine) work transfer is W_{34} while the negative (compressor) work transfer is W_{15}: the input heat transfer is Q_{35}. The thermal efficiency is, then,

$$\eta = W_{34}/Q_{35} = (h_3 - h_4)/(h_3 - h_5)$$

the work ratio and specific heat consumption are respectively

$$r_w = [(h_3 - h_4) - (h_5 - h_1)]/(h_3 - h_4)$$

and

$$ssc = [3600/(h_3 - h_4)] \text{ kg/kWh}$$

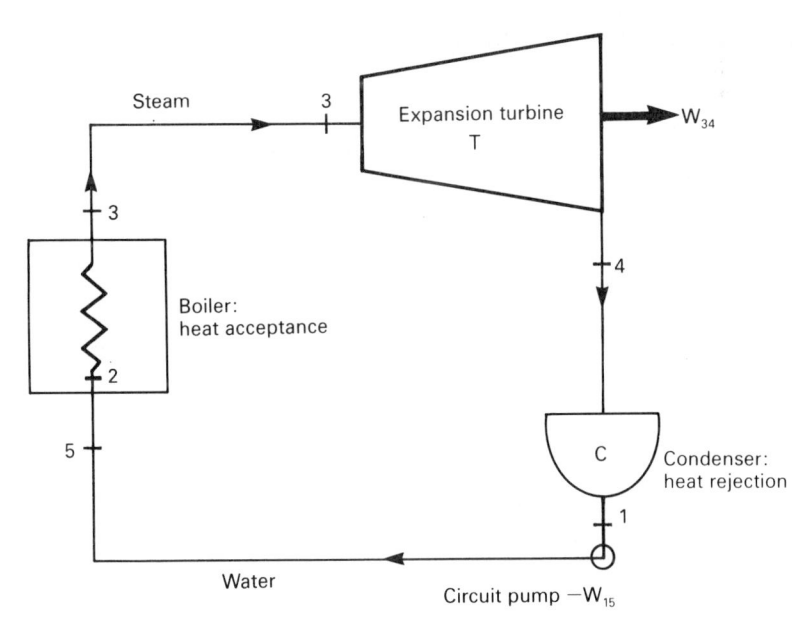

Fig. 11.1 Layout of the simple Rankine cycle. There are no losses in the circuit.

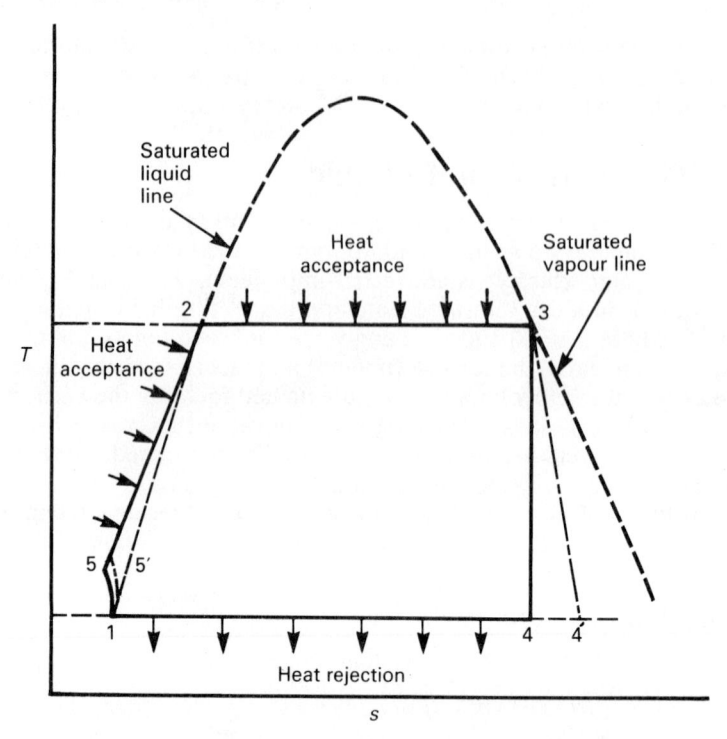

Fig. 11.2 The temperature-specific entropy diagram for the Rankine cycle.

11.3 The Compressor Work Transfer

The work done in compressing from condenser to boiler pressures is $W_{15} = h_1 - h_5$. The associated change of specific entropy follows from the thermodynamic relationship $T\mathrm{d}s = \mathrm{d}h - v\mathrm{d}p$, where v is the specific volume (the inverse of the density) of the liquid and $\mathrm{d}p$ is the accompanying change of pressure p. For the present case,

$$\mathrm{d}s_{15} = s_1 - s_5 \quad \mathrm{d}h = h_1 - h_5 \quad \text{and} \quad \mathrm{d}p = p_1 - p_5$$

But the compression is isentropic so there is no change of entropy, that is $\mathrm{d}s = 0$. The change of enthalpy $\mathrm{d}h$ in moving from state 1 to state 5 is related to the corresponding change of pressure by $\mathrm{d}h = v\mathrm{d}p$. Because of the extreme incompressibility of the liquid, its specific volume v can be treated as a constant throughout the compression to a very good first approximation and we have

$$\mathrm{d}h = h_1 - h_5 = v \int_5^1 \mathrm{d}p = v(h_1 - h_5)$$

The compressor work is consequently expressed by

$$W_{15} = h_1 - h_5 = v(p_1 - p_5)$$

This is negative because $p_5 > p_1$. The ratio of the positive (usable turbine) work output to the negative (compressor) work is then

$$W_{34}/W_{15} = [h_3 - h_4]/[h_1 - h_5]$$

Because $h_3 - h_4 > h_5 - h_1$ always, it follows that the compressor work is always less than the turbine work (a happy result otherwise this would be a machine to absorb work rather than produce it). The degree to which this is true depends on the pressure p_5 being held in the boiler: the higher the final pressure the more work that must be expended in achieving it, and this work increases with increasing pressure. For relatively low boiler pressures (up to a few tens of bar) it is true to about 1 part in 1000, but at higher pressures the accuracy may be somewhat less.

The neglect of the compressor work transfer in relation to the expansion work transfer will alway give a first approximation to the characteristics of the working of a Rankine cycle but account will need to be taken of the compressor work in accurate calculations at the higher pressures. This point will be confirmed in numerical calculations.

11.4 A Numerical Example

The workings of the simple Rankine cycle can best be judged by numerical results, and we will take two cases here; further analysis will follow from the examples in the Exercises (Section 11.8).

A Low Pressure System: ideal case with $y = z = 1$

A simple Rankine cycle using steam as the working fluid has a boiler pressure of 40 bar and a condenser pressure of 0.04 bar. The *T–s* diagram is that of Fig. 11.2. We shall use the notation displayed there.

The temperature in the boiler at 40 bar pressure is

$$T_2 = T_3 = 523 \text{ K} \qquad \text{(steam tables)}$$

while the condenser temperature is

$$T_4 = T_1 = 302.2 \text{ K} \qquad \text{(steam tables)}$$

For the specific enthalpies

$$h_1 = h_{f4} = 121 \text{ kJ/kg} \quad h_2 = h_{g2} = 1087 \text{ kJ/kg} \quad h_3 = 2801 \text{ kJ/kg}$$
$$\text{(steam tables)}$$

and for the specific entropy

$$s_3 = s_{g3} = 6.070 \text{ kJ/kg K} \qquad \text{(steam tables)}$$

This is the condition of the steam as it enters the turbine. It then suffers isentropic expansion, so that $s_3 = s_4$. At the end of the expansion there is a mixture of x % of steam and $(1-x)$ % of water so as usual we write

$$s_4 = s_{f4} + xs_{fg4}$$

But

$$s_f = 0.422 \text{ kJ/kg K} \quad \text{and} \quad s_{fg} = 8.051 \text{ kJ/kg K} \qquad \text{(steam tables)}$$

which gives $x = 0.701$. The enthalpy at point 4 then follows from

$$h_4 = h_{f4} + xh_{fg4},$$

144   The Simple Rankine Cycle

and the steam tables gives $h_{fg} = 2433$ kJ/kg. Therefore $h_4 = 1828$ kJ/kg.
The turbine work $W_{34} = h_3 - h_4$

$$= 2801 - 1828 = 973 \text{ kJ/kg}$$

The compressor work is, from Section 11.3,

$$W_{15} = -(h_1 - h_5)$$
$$= -v_f(p_1 - p_5)$$

since the water is now in liquid form. The pressure difference is from 40 bar to 0.44 bar so that $(p_1 - p_5) = 39.96$ bar (which is closely 40 bars to about 1 part in 1000). The physical properties of water provide $v_f = 10^{-3}$ m^3 so that, since 1 bar = 10^5N/m^2, $W_{15} = -10^{-3}$ (39.96×10^5)J/kg $= -10^{-3}$ (39.96×10^2)kJ/kg $= -3.996$ kJ/kg, or closely -4 kJ/kg (to 1 part in a 1000).

The net work $W = W_{34} + W_{15} = 973 - 4 = 969$ kJ/kg.
The heat input is $Q_{13} = h_3 - h_1 = 2679$ kJ/kg.

The thermal efficiency of the system is $\eta = W/Q_{12}$ or

$$\eta = \frac{969}{2679} = 0.362$$

The work ratio $r_w = W/W_{34} = 969/973 = 0.996$.
The specific steam consumption $ssc = 3600/969 = 3.71$ kg/kWh.

The comparison with the Carnot engine is interesting. From Section 10.2 the corresponding values for the Carnot engine working between the same upper and lower temperatures are

$$\eta = 0.424 \quad r_w = 0.745 \quad ssc = 4.95 \text{ kg/kWh}$$

Certainly the Carnot engine has a greater thermal efficiency than the Rankine engine working between the same temperatures. But the specific steam consumption for the Rankine engine is substantially lower and the work ratio substantially higher.

The Compressor Work
From the calculations in Section 11.4 we have found that $W_{34} = 973$ kJ/kg and $W_{15} = -4$ kJ/kg. This shows $W_{15} = 3.99 \times 10^{-3}$ W_{34}, and so to be negligible to 1 part in about 400. If this accuracy is sufficient for a particular purpose the compressor work can in practice be neglected. Certainly, there will be no loss of principle in our arguments but the length of the calculations will be somewhat reduced. Then

$$\eta = 973/2679 = 0.363 \quad r_w = 1 \quad ssc = 3600/973 = 3.70 \text{ kg/kWh}$$

A System at Higher Boiler Pressure
The boiler pressure will now be raised to 100 bar, the condenser pressure remaining the same at 0.04 bar. We neglect irreversibilities so that $y = z = 1$. From the steam tables, the boiler temperature is 311.0 °C = 584.3 K. The condenser temperature is 302.3 K as before. Again from the steam tables we find:

$$h_1 = h_f = 121 \text{ kJ/kg} \quad h_2 = h_f = 1408 \text{ kJ/kg} \quad h_3 = h_g = 2725 \text{ kJ/kg}$$

11.4 A Numerical Example

and $s_3 = s_g = 5.615$ kJ/kg K also $s_3 = s_4$. Consequently

$$s_3 = 5.615 = s_{g4} + xs_{fg4} = 0.422 + x8.051,$$

which gives $x = 0.645$. For the specific enthalpy

$$h_4 = 121 + 0.645 \times 2433 = 1690.29 \text{ kJ/kg}$$

The expansion work transfer is $W_{34} = h_3 - h_4 = 1034.71$ kJ/kg while the heat input is $h_3 - h_1 = 2604$ kJ/kg. This gives a thermal efficiency and specific steam consumption respectively

$$\eta = 1034.71/2604 = 0.397 \qquad ssc = 3600/1034.71 = 3.479 \text{ kg/kWh}$$

In comparison with the lower pressure cycle (Section 11.4) the thermal efficiency is higher and the specific steam consumption lower.

It is important to assess the contribution of the compressor work for this higher pressure system, which we were able to neglect for the lower pressure system. The compressor work transfer is

$$W_{15} = -v_f(p_1 - p_5) = -10^{-3}(0.04 - 100)10^2 = 9.996 \text{ kJ/kg}$$

Then

$$W_{15}/W_{34} = 9.996/1034 = 9.667 \times 10^{-3} \sim 10^{-2}$$

The compressor work is now 1 % of the turbine work and can still be regarded as small for many purposes and so neglected. It must be realised, however, that this may not be so for still higher boiler pressures especially if high accuracy is required of the calculations.

The thermal efficiency, specific steam consumption and dryness fraction for a Rankine cycle with condenser pressure 0.04 bar are shown in Fig. 11.3 over

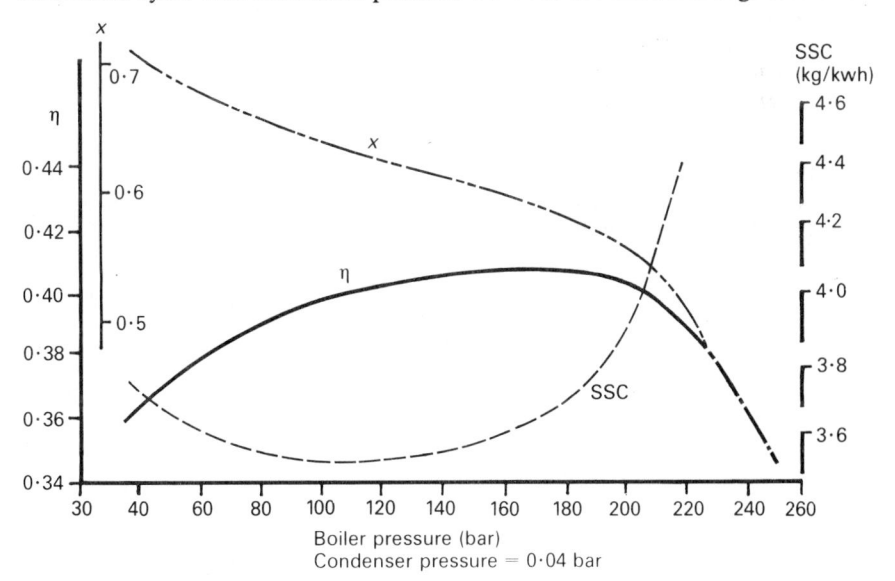

Fig. 11.3 Thermal efficiency, specific steam consumption and dryness fraction for the Rankine cycle, over a range of boiler pressures; the condenser pressure remains constant.

146 The Simple Rankine Cycle

the range of boiler pressures from 40 bar to 260 bar. It is seen that the thermal efficiency builds up to a maximum value in the region of 180 bar after which it falls quickly. On the other hand, the specific steam consumption falls at first, with a minimum at about 100 bar, after which it rises rapidly as the boiler pressure is increased. There is, apparently, a conflict if both maximum thermal efficiency and minimum specific steam consumption are required simultaneously because these conditions apply at different boiler pressures. The result must be a compromise but the cycle as it stands has a dryness fraction which is low even at the lower boiler pressures and steadily decreases as the boiler pressure is raised. The basic Rankine cycle needs further modifications, if only to improve the dryness fraction.

11.5 Imperfect Compression and Expansion

Real components show irreversible losses of energy that must be taken into account. The practical turbine will not achieve quite the expansion of the ideal and the compressor will not quite achieve the ideal compression. These effects are expressed in terms of appropriate coefficients of isentropic efficiency, introduced in Section 10.3 in connection with the Carnot cycle but also equally applicable here.

The Rankine cycle with less than ideal compression and expansion has a T–s diagram also included in Fig. 11.2 with the expansion and compression now shown dashed. Expansion has the specific enthalpy change $(h_3 - h_{4'})$ and compression has the enthalpy change $(h_{5'} - h_1)$. In calculating the characteristics of the system it is necessary to know $h_{4'}$ and $h_{5'}$.

The difference between $(h_3 - h_4)$ and $(h_3 - h_{4'})$ depends, by definition, on the efficiency of the turbine. The coefficient of isentropic expansion y is therefore defined as

$$y = \frac{(h_3 - h_{4'})}{(h_3 - h_4)} \tag{11.1a}$$

In practice a good (and in consequence usually expensive) turbine could have $y = 0.95$; less expensively, we might have $y = 0.90$ or 0.85, or even lower values if the steam is inordinately wet. The coefficient of isentropic compression, z, is defined analogously

$$z = \frac{h_1 - h_5}{h_1 - h_{5'}} \tag{11.1b}$$

We have seen that the compression work transfer is only a small fraction of the expansion work transfer so the quality of the turbine is of greater importance than of the compression in a Rankine cycle, which has this practical advantage over the Carnot cycle.

The effects of inefficiencies on the working of the cycle are expressed in terms of the corresponding ideal specific enthalpy values by way of the isentropic efficiencies y and z. Using the formulae of Section 11.2, the net work transfer $W = (h_3 - h_4)y - (1/z)(h_5 - h_1)$ and the thermal efficiency is expressed by the formula

$$\eta = y[h_3 - h_4]/[(h_2 - h_5/z) - h_1(1 - 1/z)]$$

11.5 Imperfect Compression and Expansion

The expressions for the work ratio and the specific steam consumption are

$$r_w = y - [(h_5 - h_1)/z]/(h_3 - h_4)$$
$$ssc = 3600/[y(h_3 - h_4)]$$

These formulae, of course, reduce to the corresponding forms of Section 4.2 when we set $y = z = 1$. It is readily seen that the effect of the inefficiencies on the cycle is to decrease the thermal efficiency, decrease the work ratio and increase the specific steam consumption. Interpreted another way, the cycle with inefficiencies will use more fuel than the ideal for a given work output, will give less work output for a given energy input and will require more steam for its working (and so a larger boiler) than the ideal. There is, in practice, a compromise to be drawn in any particular case on the degree of efficiency appropriate to the expansion unit, with the financial cost of buying a more efficient unit being balanced against the financial savings in running it.

Including Isentropic Efficiency

For the boiler pressure of 100 bar, $W_{34'} = yW_{34} = 1034y$. The heat transfer Q_{13} remains unaltered. The compression work transfer is neglected. The thermal efficiency and specific steam consumption become respectively

$$\eta = 0.397y \quad \text{and} \quad ssc = (3.479/y) \text{ kg/kWh}$$

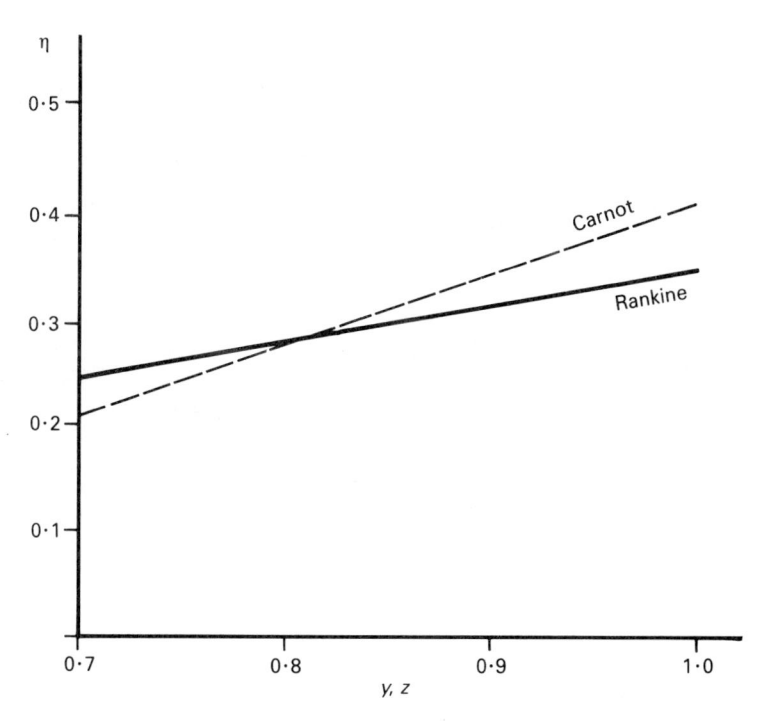

Fig. 11.4 Showing the dependence of the thermal efficiencies of the Rankine and Carnot cycles on the isentropic efficiencies of the expansion (y) and compression (z). Boiler pressure 30 bar, condenser pressure 0.04 bar.

148 The Simple Rankine Cycle

For $y = 0.9$ and $y = 0.8$ these values are changed to become

| $y = 0.9$ | $\eta = 0.357$ | ssc = 3.865 kg/kWh |
| $y = 0.8$ | $\eta = 0.318$ | ssc = 4.349 kg/kWh |

Apparently, the efficiency falls quite sharply as the quality of the expansion decreases while the specific steam consumption shows a corresponding rise.

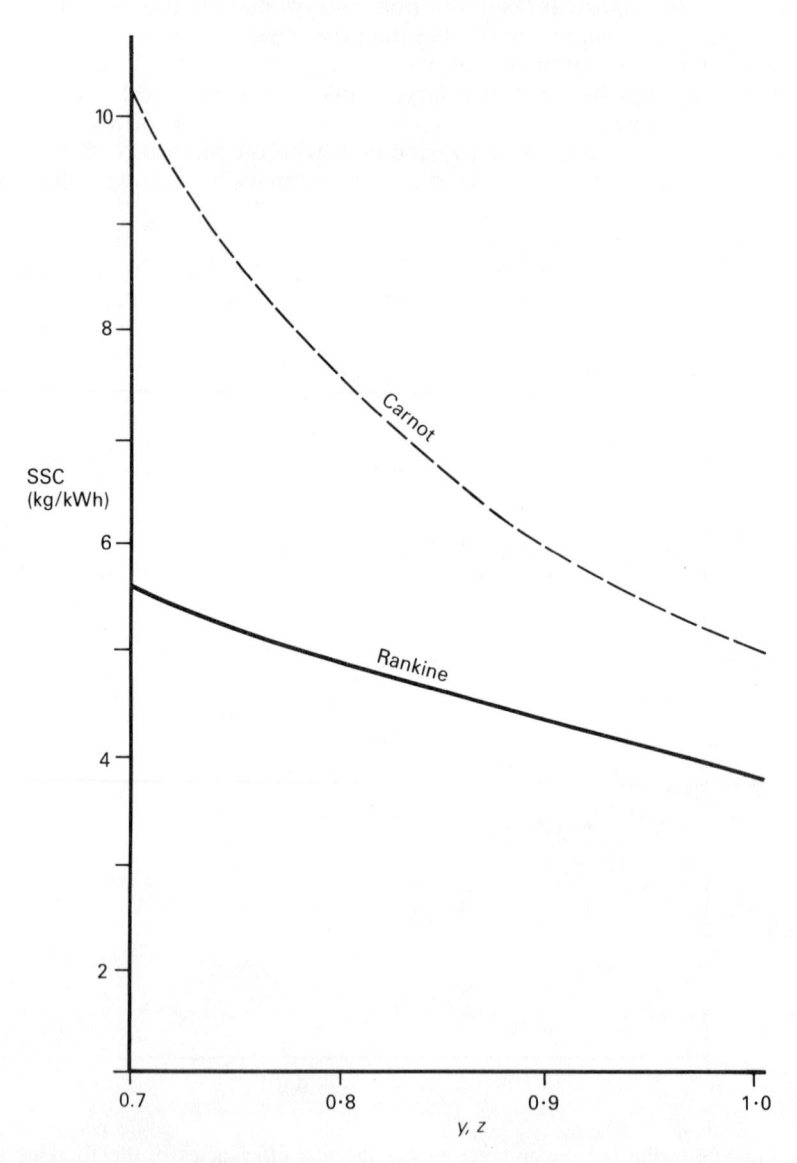

Fig. 11.5 Showing the dependence of the specific steam consumption (ssc) on the isentropic efficiencies of the expansion (y) and compression (z) stages for the Rankine and Carnot cycles.

A comparison of the effects of irreversibilities on the thermal efficiencies of the Carnot and Rankine cycles for the particular boiler pressure of 30 bar is shown in Fig. 11.4 (see Questions 10.5 and 11.8). The corresponding comparison for the specific steam consumptions is shown in Fig. 11.5. It is seen that the Rankine cycle is less susceptible to irreversibilities.

Comparison between the various results of Sections 11.4 shows that the effect of the increased boiler pressure is to improve the overall working conditions of the cycle. This is to be expected. Increasing the boiler pressure increases the receptor temperature and this will also reduce the steam consumption.

11.6 The Mean Receptor Temperature

As the working fluid is moved from an initial state with specific entropy s_1 and temperature T_1 to a final state with greater specific entropy s_2 and higher temperature T_2, the total change of specific enthalpy Δh_{12} is the area Δ 12341 in Fig. 11.6. The heat acceptance into the cycle is associated with the temperature increase $(T_3 - T_1)$. An equivalent isothermal change of the specific enthalpy, with temperature T_m (say), is represented by the area Δ 125641. The specific enthalpy acceptance will be the same in each case if Δ 12341 = Δ 125641. T_m is then a representative temperature for the enthalpy acceptance and is called the mean receptor temperature. Using T_m, heat reception by the Rankine cycle can be related to that in an equivalent Carnot cycle with upper temperature T_m. This hypothetical temperature is often useful in identifying the modifications necessary to improve the thermal characteristics of the Rankine cycle. A corresponding mean rejection temper-

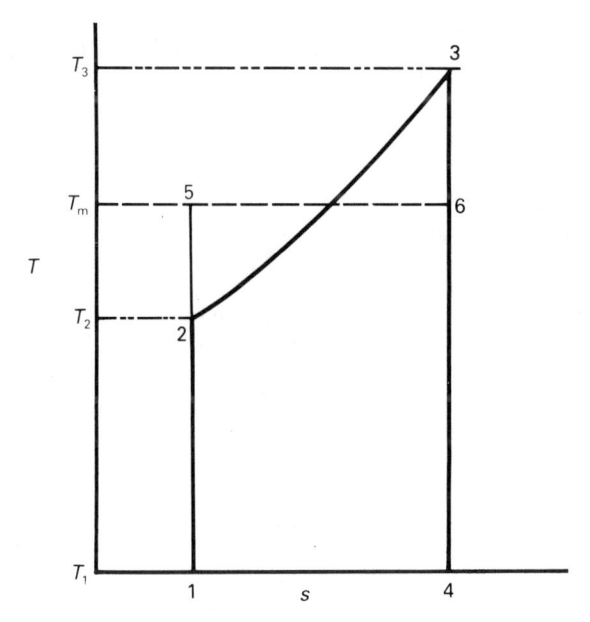

Fig. 11.6 The mean receptor temperature T_m in relation to the temperatures T_2 and T_3 marking vaporisation in the Rankine cycle.

ature can be defined in an analogous way but this is less useful because the isothermal nature of the heat rejection process in the Rankine cycle is assured by the condenser characteristics.

11.7 Summary

1. The Rankine cycle receives heat energy into the working fluid over a temperature range but rejects heat energy at a single temperature. This provides a characteristic thermodynamic behaviour for the system.
2. If water is used as the working fluid, the expansion process involves steam while the condensation process converts the steam back to liquid water.
3. The compression to the high temperature stage now takes place in the liquid phase. Liquid water has a very low compressibility coefficient. Compression is consequently a relatively efficient process.
4. Because compression takes place in the liquid form the compression work transfer is small in comparison with the expansion work transfer. In practical terms this reduces the effects of inefficiencies in the compression stage. Often the compression work transfer can be neglected in assessing the characteristics of the cycle.
5. The work ratio is now close to unity.
6. Isentropic efficiencies less than unity have relatively less effect on the working of the cycle than is the case for the Carnot cycle. This is important for the practical application of the cycle.
7. The practical Rankine cycle with less than ideal components can be more efficient than the corresponding Carnot cycle with the corresponding inefficiencies working between the same upper and lower temperatures.

11.8 Exercises

11.1 A steam turbine system works under the simple Rankine cycle. The boiler pressure is 30 bar and the condenser pressure is 0.04 bar. If the expansion and compression stages have ideal efficiency, find (a) the dryness fraction of the steam upon expansion; (b) the thermal efficiency of the cycle; (c) the work ratio, and (d) the specific steam consumption. Find the compression work transfer and establish it to be numerically negligible in comparison with the expansion work transfer.

11.2 The steam system is as in Question 11.1. The efficiency of the compression stage is perfect ($z = 1$). If the expansion stage has an isentropic efficiency y, show that the system then has the thermal efficiency $\eta = 0.350y$ and specific steam consumption $(3.84/y)$ kg/kWh. Calculate the values of η and ssc for $y = 0.97, 0.95, 0.90, 0.85, 0.80$ and 0.75. Plot a graph to show the behaviour of both η and ssc for increasing expansion efficiency.

11.3 A steam turbine system works under the simple Rankine cycle with a boiler pressure of 160 bar and a condenser pressure of 0.04 bar. If the expansion and compression stages each have ideal efficiency, calculate (a) the dryness fraction of the steam after expansion; (b) the thermal efficiency for the cycle; (c) the work ratio, and (d) the specific steam consumption for the system.

11.8 Exercises 151

Compare your results with those calculated in Question 11.1 and with those contained in Section 11.3 of the main text. Plot graphs to show the variations of the four quantities with change of boiler pressure, the condenser pressure remaining the same throughout.

11.4 Repeat the calculations of Question 11.1 for the different condenser pressures 0.015, 0.020, 0.025, 0.030 and 0.050 bar. Plot graphs showing how the thermal efficiency, work ratio and specific steam consumption vary with the condenser pressure, the boiler pressure remaining the same throughout.

11.5 Repeat the calculations of Question 11.4 for the boiler pressures 40 bar, 80 bar and 160 bar. Plot graphs to show how the variations of the dryness fraction, the thermal efficiency, the work ratio and the specific steam consumption with changing boiler pressure are themselves affected by changes of the condenser pressure.

11.6 The steam system treated in Question 11.4 has the expansion turbine replaced by one with isentropic efficiency y. By calculating the thermal efficiency, work ratio and specific steam consumption in each case, find the effect on these quantities of y having the different values 0.95, 0.90, 0.85 and 0.70, for each of the condenser pressures listed in Question 11.4. The boiler pressure is held throughout at 30 bar.

11.7 A steam turbine system working on the simple Rankine cycle is to be constructed within the Arctic Circle. A similar system (with boiler pressure 160 bar and condenser pressure 0.05 bar) was recently constructed in the tropics. If the same overall design is used for the new plant in what ways would it be desirable to modify the detail for the northern location? What might be a suitable pressure range for the new design?

11.8 Set down the essential thermodynamic differences between the Rankine and Carnot cycles.

A simple Rankine cycle, using steam as the working fluid, works between the boiler pressure of 30 bar and the condenser pressure of 0.04 bar. The turbine has an isentropic efficiency factor y and the compression work has isentropic efficiency factor z. Show that the thermal efficiency η of the cycle is given by the expression

$$\eta = \frac{(940y - 3/z)}{(2682 - 3/z)}$$

the work ratio r_w by the expression

$$r_w = 1 - 0.0032/yz$$

and the specific steam consumption by

$$ssc = 3600/(940y - 3/z) \text{ kg/kWh}$$

Compare these expressions with those for the corresponding Carnot cycle (Question 10.5) and construct the Figs. 11.4 and 11.5 of the main text comparing the working of the Carnot and Rankine cycles. In this way you will

152 The Simple Rankine Cycle

substantiate the statement that the Rankine cycle is less affected by irreversible behaviour than the Carnot cycle.

11.9 Accepting the expressions of the last question for the overall thermal efficiency, work ratio and specific steam consumption, derive the expression

$$\frac{\partial \eta}{\partial y} = \frac{\eta}{y - 0.0032/z}$$

for the variation of the thermal efficiency with the isentropic efficiency of expansion, the corresponding coefficient of compression remaining constant. Show further that the corresponding expression for the dependence of the specific steam consumption ssc on variations of the isentropic expansion coefficient y is

$$\frac{\partial}{\partial y}(ssc) = -\frac{3.825}{[y - 0.0032/z]^2} \text{ kg/kWh}$$

Show further that the effect of irreversibility in the compression alone, specified by the variation of the coefficient z, is described by

$$\frac{\partial \eta}{\partial z} = \frac{1-\eta}{z[1 - 0.0011z]}$$

and the specific steam consumption by

$$\frac{\partial}{\partial z}(ssc) = (ssc)[3/z^2] \text{ kg/kWh}$$

Use your formulae to deduce that η increases as either y or z increase, and that ssc decreases with increase of either y or z. Interpret these conclusions in terms of the operating characteristics of the system.

11.10 A simple Rankine cycle has a condenser pressure p_c and a boiler pressure $p_B = pp_c$, where p is the pressure ratio. The thermal efficiency is η, the net work output is W and the dryness fraction is x. It is required that the dryness fraction should not fall below the critical value $x_o = 0.88$.

By calculations involving the working characteristics of cycles with various values of p_B and p deduce that:
(a) η increases as p increases for every p_B
(b) W increases as p increases for every p_B
(c) the specific steam consumption (ssc) decreases as p increases for every p_B.

If η_o, p_o, W_o and $ssc(o)$ refers respectively to the values of η, p_b, W and ssc for $x_o = 0.88$ construct the following table:

Boiler pressure p_B bar	Pressure ratio p_p	Thermal efficiency η_o	Net Work output W_o kJ/kg	Specific steam consumption $ssc(o)$ kg/kWh
10	6.6	0.146	337	10.7
30	5	0.138	294	12.2
60	3	0.109	204	17.6
100	2	0.079	123	29.2

Taking account of the data used in constructing the table and also of the numerical data of the main text (including Fig. 11.3), comment on the implications of these data for the construction and use of a simple Rankine engine. (Under normal working conditions $p \sim 4000$.)

12. RAISING THE MEAN RECEPTOR TEMPERATURE: SUPERHEAT, REHEAT AND SUPERCRITICALITY

The simple Rankine cycle accepts heat over the range of temperatures spanned by the condenser and the boiler. The mean receptor temperature T_m (see Section 11.6) is always lower than the maximum temperature of the cycle, but becomes higher as the maximum temperature is increased. This is brought about by superheating the steam after it leaves the boiler.

The thermodynamic characteristics of the cycle improve as the maximum temperature is raised and the working characteristics are more attractive for improved dryness of steam throughout the turbine expansion system. Although it is clear from the T–s diagram that superheating must improve the dryness of the steam, the degree of dryness is insufficient for practical purposes (the dryness fraction is too low) except at the lowest boiler pressures (see Question 11.10). A higher degree of dryness is achieved by introducing a reheat stage into the cycle where the temperature of the steam is raised part way through the expansion phase. These matters are considered separately in the present chapter.

12.1 The Principle of Superheating

After leaving the boiler the dry saturated steam is passed through again, using a separate set of steam tubes, to further raise its temperature. The superheat pressure is the same as that of the boiler. The circuit is shown in Fig. 12.1 and the corresponding temperature/specific entropy diagram in Fig. 12.2.

The Enthalpy Changes

Referring to Fig. 12.2, the water is vaporised between the state points 5 and 2, becoming dry saturated at the boiler pressure p_b at the point 6. Maintaining the same boiler pressure, the steam is now superheated to the temperature T_3. The expansion work transfer (ideally at constant entropy) occurs between the state points 3 and 4. The steam is then condensed at the pressure p_c to be returned to the state point 1. Between the state points 1 and 5 the pressure is raised from p_c to p_b and the cycle starts again. It is the presence of the region from point 6 to point 3, when the steam is raised above its normal boiling temperature, that gives rise to the name superheating for the whole process.

Expressed in terms of the enthalpy, and using the notation of Fig. 12.2, the work transfer W is

$$W = (h_3 - h_4) - (h_5 - h_1)$$

12.1 The Principle of Superheating

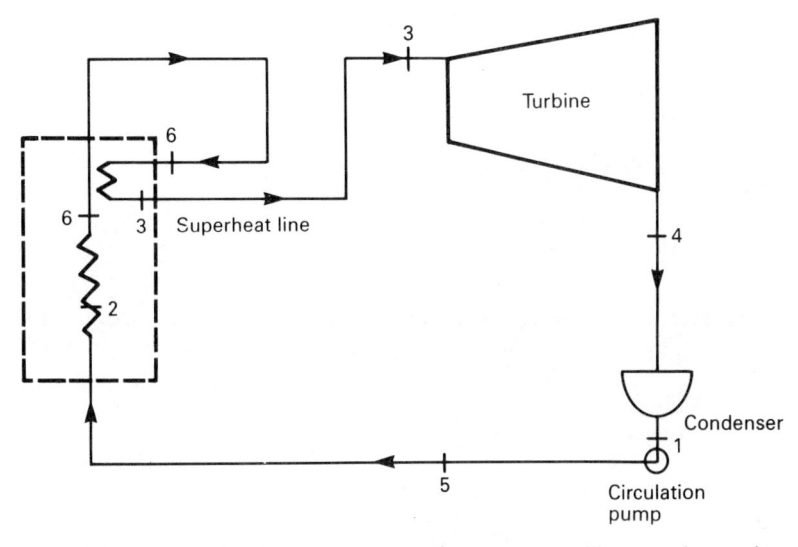

Fig. 12.1 The layout of the Rankine cycle with superheat. The superheat tubes are heated by furnace gases that have already vaporised the water.

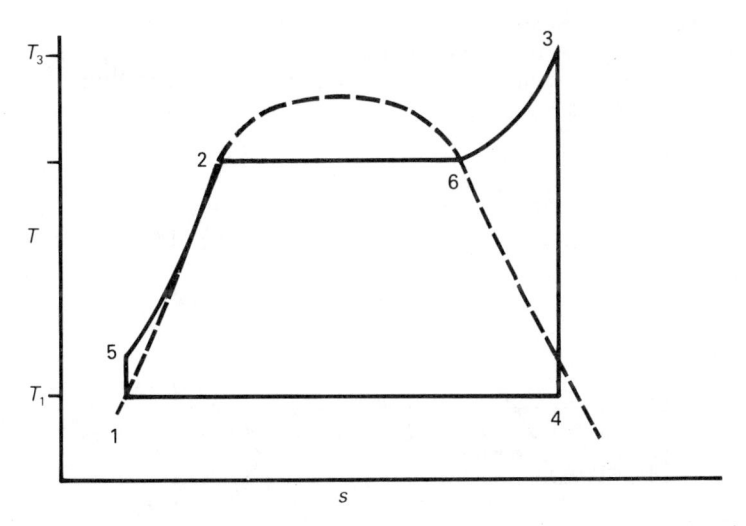

Fig. 12.2 The temperature-specific entropy diagram for a steam/Rankine cycle with superheat.

while the heat transfer Q is

$$Q = h_3 - h_5$$

The thermal efficiency η for the cycle is then

$$\eta = \frac{W}{Q} = \frac{(h_3 - h_4) - (h_5 - h_1)}{(h_3 - h_5)}$$

156 Raising the Mean Receptor Temperature

The work ratio r_w is expressed by

$$r_w = \frac{(h_3 - h_4)}{[(h_3 - h_4) - (h_5 - h_1)]}$$

and the specific steam consumption by

$$ssc = \frac{3600}{[(h_3 - h_4) - (h_5 - h_1)]} \text{ kg/kWh}$$

Also, $(h_5 - h_1) = v_p(p_b - p_c)$.

If it is sufficient to neglect the difference $(h_5 - h_1)$ in comparison with the difference $(h_3 - h_4)$, these formulae reduce to the form we will generally use

$$\eta = \frac{(h_3 - h_4)}{(h_3 - h_1)}$$

$$r_w = 1 \quad \text{and} \quad ssc = \frac{3600}{(h_3 - h_4)} \text{ kg/kWh}$$

The mean receptor temperature T_m is raised by superheating and consequently the thermal efficiency of the cycle must rise as is clear from Fig. 12.3. Heat transfer to the steam still spans a range of temperature so the thermal efficiency will still not be as high as that for the corresponding Carnot cycle working between the same maximum and minimum temperatures.

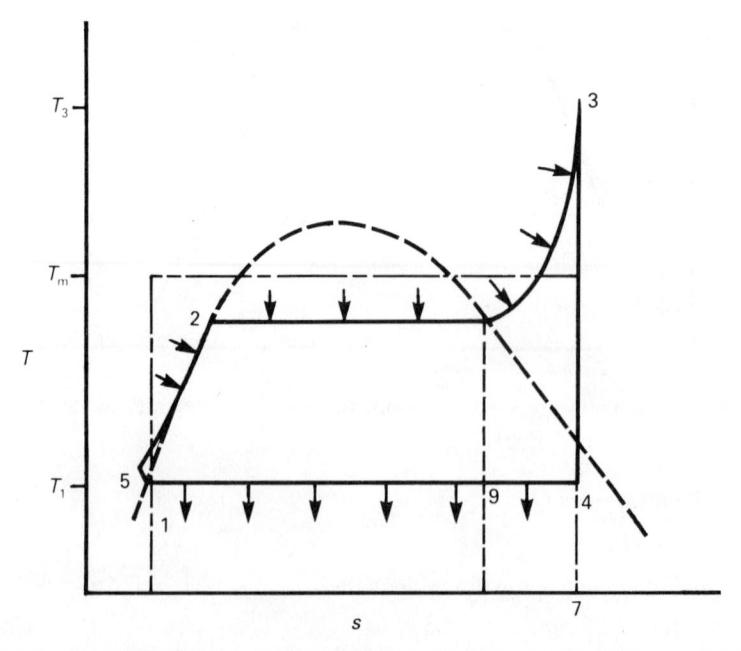

Fig. 12.3 The superheat cycle is composed of a simple Rankine cycle without superheat plus a supplementary cycle containing the superheat. The mean receptor temperature T_m is marked and $T_1 < T_m < T_3$.

12.1 The Principle of Superheating

These conclusions can be illustrated alternatively by regarding the superheated cycle as a simple Rankine cycle with no superheat (represented by the closed path 12691 in Fig. 12.3) supplemented by the superheat cycle 63496. For the non-superheat cycle, the heat received will be denoted by Q_n while that rejected will be Q_{on}: for the supplementary cycle the corresponding quantities are Q_s and Q_{os}. The thermal efficiency η of the complete cycle is

$$\eta = 1 - [(Q_{on} + Q_{os})/(Q_n + Q_s)] \qquad (12.1)$$

For the component cycles, where η_n denotes the thermal efficiency of the simple Rankine cycle and η_s that of the superheat cycle

$$\eta_n = 1 - [Q_{on}/Q_n] \quad \text{and} \quad \eta_s = 1 - [Q_{os}/Q_s] \qquad (12.2)$$

Simple manipulation of eqns. (12.1) and (12.2) allows η to be expressed in terms of the separate efficiencies η_n and η_s as

$$\eta = \eta_n\{1 + [(\eta_s/\eta_n) - 1]/[1 + Q_n/Q_s]\} \qquad (12.3)$$

It follows from eqn. (12.3) that $\eta > \eta_n$ when $\eta_s > \eta_n$, whatever the value of Q_n/Q_s may be. This, in fact, is always the case (see Question 12.6) so superheating will always increase the thermal efficiency of the cycle. Greater efficiency changes occur as Q_n/Q_s reduces and this means lower boiler pressure (see Question 12.6).

The net work output of the cycle is

$$W = (Q_n + Q_s) - (Q_{on} + Q_{os})$$

so the specific steam consumption is

$$ssc = 3600/[(Q_n + Q_s) - (Q_{on} + Q_{os})] \qquad (12.4)$$

Using the corresponding statement for ssc_n, referring to the non-superheated (Rankine) cycle, it follows easily that

$$ssc = ssc_n[1/(1 + p(Q_s/Q_n))] \qquad (12.5)$$

where $p = (1 - Q_{os}/Q_s)/(1 - Q_{on}/Q_n)$. It is found from the steam tables that $Q_s < Q_n$ always and $p \sim 1$ so that $p(Q_s/Q_n) > 0$. Consequently $ssc < ssc_n$ always, so superheating reduces the specific steam consumption of the cycle. The extent of the reduction decreases with increasing boiler pressure (see Question 12.6). The heat rejected by the full cycle is $(h_4 - h_1)$ while the heat rejected by the corresponding Rankine cycle is $(h_9 - h_1)$: it is readily seen that $(h_4 - h_1) > (h_9 - h_1)$. Consequently, the dryness fraction is increased by superheating.

The Effects of Irreversibilities

The thermal efficiency will be reduced by losses in the expansion and compression work transfers. We need only consider the expansion process explicitly here since we shall neglect the compression work transfer. For an isentropic efficiency y it readily follows that the formulae for the thermal efficiency and specific steam consumption are now

$$\eta = \frac{y(h_3 - h_4)}{(h_3 - h_1)}$$

and

$$ssc = \frac{3600}{y(h_3 - h_4)} \text{ kg/kWh}$$

It is seen from these formulae that, as before, inefficiencies decrease the thermal efficiency and increase the specific steam consumption but the general conclusions contained in eqns. (12.1) to (12.4) remain valid.

12.2 Numerical Example of Superheating

Ideal Cycle
Consider the cycle (also considered in Section 11.5) with boiler pressure 40 bar and condenser pressure 0.04 bar, the steam being superheated to 500 °C (773.3 K). This is only slightly below the metallurgical limit for the system. Now

$$h_1 = 121 \text{ kJ/kg} \quad h_3 = 3445 \text{ kJ/kg} \quad s_3 = 7.089 \text{ kJ/kg K} \quad \text{(steam tables)}$$

The expansion through the turbine proceeds at constant entropy, which means $s_3 = s_4$. But

$$s_{f4} = 0.422 \text{ kJ/kg K}, \quad s_{fg4} = 8.051 \text{ kJ/kg K} \quad \text{(steam tables)}$$

so that

$$s_3 = s_{f4} + x s_{fg4}$$

to determine the dryness fraction as $x = 0.828$. For the simple Rankine cycle (Section 11.4), $x = 0.701$.

At 0.04 bar

$$h_{f4} = 121 \text{ kJ/kg} \quad h_{fg4} = 2433 \text{ kJ/kg} \quad \text{(steam tables)}$$

so that

$$h_4 = 2135.8 \quad (\text{by } h_4 = h_{f4} + x h_{fg4})$$

These values give

$$W_{34} = (3445 - 2135.8) = 1309 \text{ kJ/kg}$$

and

$$Q_{31} = (3445 - 121) = 3324 \text{ kJ/kg}$$

leading to the thermal efficiency and specific steam consumption

$$\eta = 0.394 \quad ssc = 2.750 \text{ kg/kWh}$$

By comparison with the non-superheated Rankine cycle, both the thermal efficiency and the dryness fraction have been raised and the specific steam consumption reduced.

Suppose the boiler pressure is increased to 60 bar, the condenser pressure remaining unaltered at 0.04 bar. The superheat again will be to 500 °C. Then

$$h_1 = 121 \text{ kJ/kg} \quad h_3 = 3421 \text{ kJ/kg} \quad s_3 = 6.879 \text{ kJ/kg K} \quad \text{(steam tables)}$$

12.2 Numerical Example of Superheating

For isentropic expansion $s_s = s_4$, giving $x = 0.802$ (lower than for 40 bar boiler pressure) and $h_4 = 2072.29$ kJ/kg, using the same methods as before. The net work transfer is

$$W_{34} = (3421 - 2072.29) = 1348.7 \text{ kJ/kg}$$

and the heat transfer is

$$Q_{13} = (3421 - 121) = 3300 \text{ kJ/kg}$$

The thermal efficiency and specific steam consumption are

$$\eta = 0.409 \qquad ssc = 2.669 \text{ kg/kWh}$$

Increasing the boiler pressure increases the thermal efficiency and decreases the steam consumption. For comparison, the corresponding Carnot thermal efficiency for the same maximum and minimum temperatures is $\eta_{Carnot} = 0.654$. The wider gap between η and η_{Carnot} is an effect of the superheating.

The Work of Compression
The compressor work W_{15} is obtained, as before, from

$$W_{15} = v_s[p_1 - p_5]$$

For 40 bar and 60 bar we find respectively

$$W_{15}(40) = 3.996 \text{ kJ/kg} \qquad W_{15}(60) = 5.996 \text{ kJ/kg}$$

In each case these values are negligibly small when compared with the expansion work transfer and can be neglected to an accuracy of a few parts in a thousand.

General Results
The response of the thermal efficiency and the specific steam consumption to changes of the boiler pressure, for a condenser pressure of 0.04 bar and with superheating to 500 °C, are collected in Fig. 12.4 for the ideal case where $y = 1$ and the work transfer for compression is neglected. The thermal efficiency over the full range from 40 bar up to the critical pressure is seen to rise steadily with increasing pressure. In this respect the behaviour is different from that for the cycle without superheating (Fig. 11.3). The values of the efficiency are also higher by some 20 %. The specific steam consumption steadily decreases up to a boiler pressure of about 140 bar where it has a minimum value: for higher pressure the steam consumption increases. Although similar to the behaviour of the simple cycle (Fig. 11.3), the effect with superheating is less marked and the values of the steam consumption are also lower. The dependence of the thermal efficiency on the superheat temperature is shown in Fig. 12.5 for two boiler pressures, the condenser pressure remaining the same.

Inefficient Expansion
The effect of an isentropic expansion factor less than unity is important in practice. Again take the boiler pressure of 40 bar and the condenser pressure of 0.04 bar. Suppose $y = 0.9$. It is found that

160 Raising the Mean Receptor Temperature

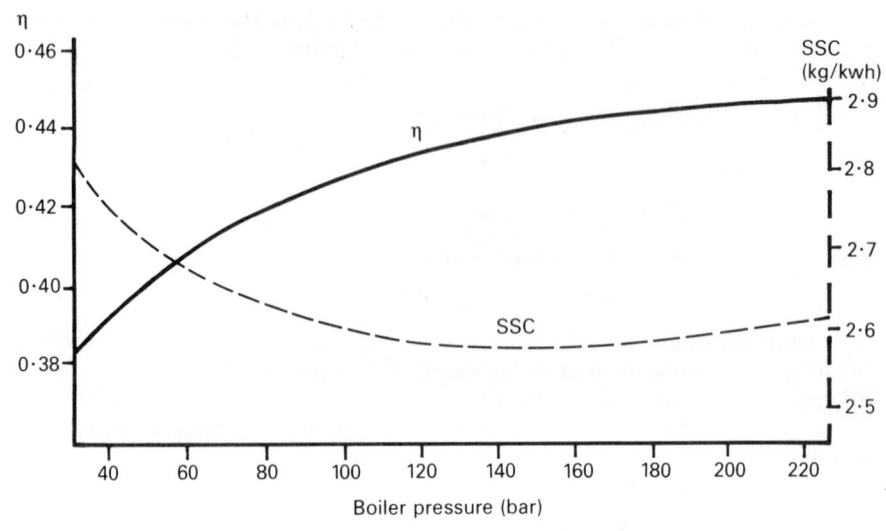

Fig. 12.4 Showing the effect of boiler pressure on the thermal efficiency and specific steam consumption for a Rankine cycle with superheat to 500 °C. The condenser pressure is held constant at 0.04 bar.

$$\eta_T(40) = 0.354 \qquad ssc(40) = 3.055 \text{ kg/kWh}$$
$$\eta_T(60) = 0.368 \qquad ssc(60) = 2.965 \text{ kg/kWh}$$

Irreversibilities decrease the thermal efficiency and increase the specific steam consumption quite significantly, showing the importance of employing a good quality turbine in the cycle. The dryness fraction is unaffected. It is also seen that increasing the boiler pressure by 50 % increases the thermal efficiency by 4 % but decreases the specific steam consumption by 3 %.

12.3 Reheating

The expansion of steam through the turbine involves a certain condensing of water as the enthalpy of the steam falls, the steam becoming quite wet towards the low pressure end, giving a relatively low dryness fraction (often below 0.75). This situation can be controlled by raising the steam temperature at a suitable point during the expansion, and so at a pressure much lower than the boiler pressure. This process is called reheat. Reheat is usually thermodynamically most effective up to the initial steam temperature: this gives the highest dryness fraction and usually leads to the best work transfer for the system. The steam circuit including reheat is shown in Fig. 12.6 and the corresponding T-s diagram in Fig. 12.7, for the simple system without superheat. The case where superheating is also present is shown in Figs. 12.8 and 12.9. Because superheat is an important component of modern steam plant we will suppose in what follows that superheat is present.

One stage of reheat involves two turbine units: a single high pressure unit covers the pressure range between the boiler and the reheat while a second

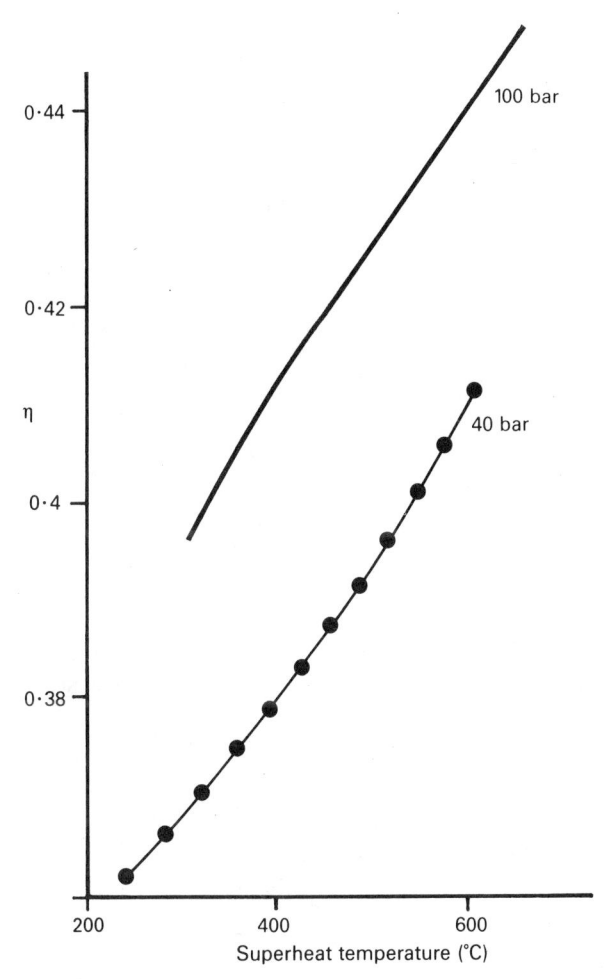

Fig. 12.5 Dependence of the thermal efficiency on superheat temperature for two pressures. The condenser pressure is 0.04 bar.

unit covers the range between the reheat and condenser pressures. The second unit usually comprises a separate intermediate and a low pressure component, to match the blading to the steam pressure and so achieve a higher isentropic efficiency than would be possible with a single unit.

General Criterion for Reheating
The essential feature is readily seen from the steam tables. The value of the specific entropy s_f increases while that of s_{fg} decreases with increasing pressure. Values of the dryness fraction x near to unity can be achieved when the specific entropy values after reheat are appropriately large. This will require a sufficient superheat temperature but it could also require a relatively low reheat pressure. This is clear by reference to Fig. 12.10.

162 *Raising the Mean Receptor Temperature*

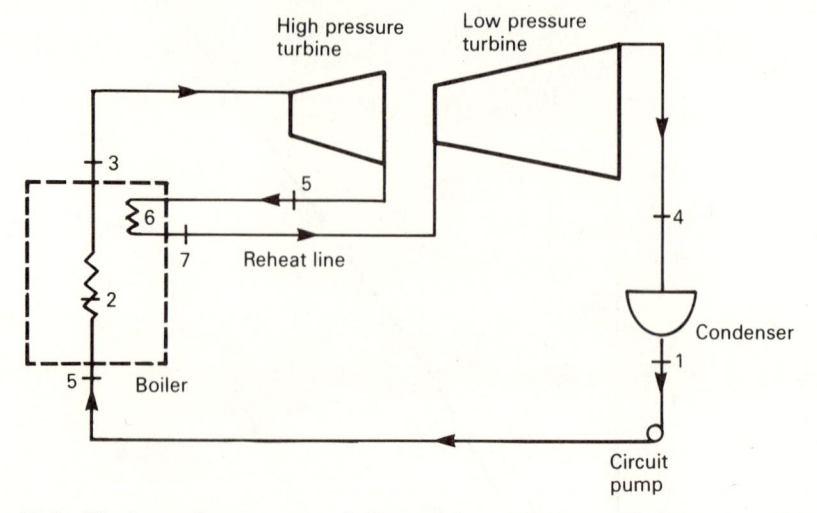

Fig. 12.6 The layout for one stage of reheat without superheat. The reheat is effected by flue gases after vaporisation of the water.

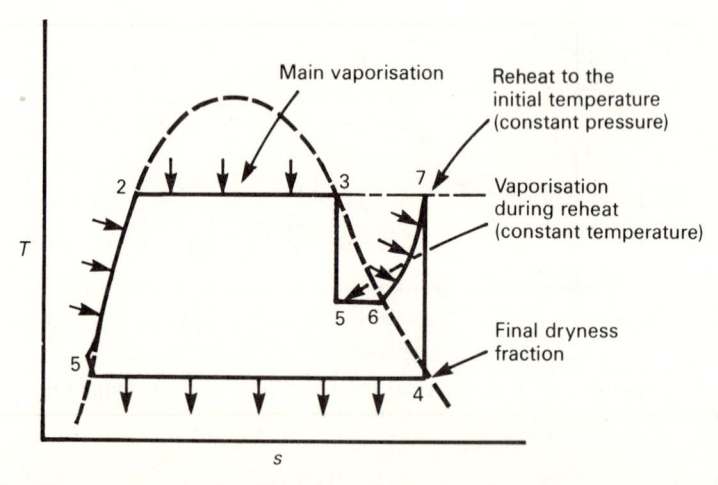

Fig. 12.7 The temperature-specific entropy curve for a Rankine cycle with one stage of reheat but no superheat.

Effect on η and ssc

Although reheat is applied primarily to control the dryness fraction the thermal efficiency cannot be overlooked entirely. Whether the thermal efficiency is increased by reheat or not depends on the particular cycle involved and it is possible for the thermal efficiency to fall. The criterion to tell what will happen in a particular case can be assessed as follows. The notation is as in Fig. 12.11.

12.3 Reheating

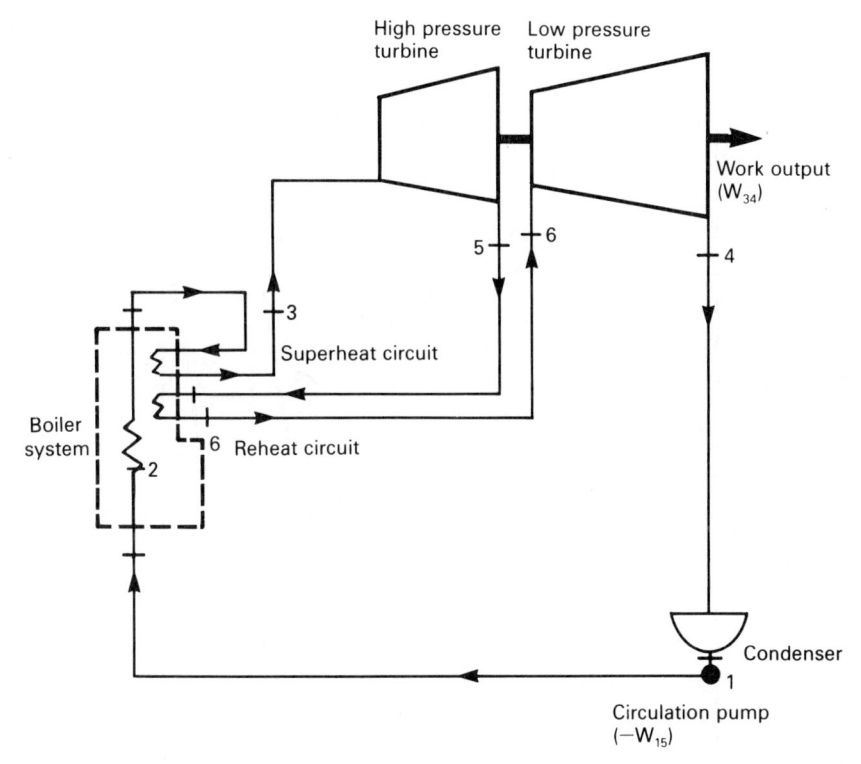

Fig. 12.8 The circuit layout including superheat and reheat. In practice, the second turbine would be divided into two – an intermediate pressure component and a low pressure component.

The total cycle is 1235641 and can be seen as being composed of the main cycle 123571 and the subsidiary cycle 56475. The corresponding Rankine cycle (without reheat but working between the same pressure ratio) would be 123571. The heat transfer into the main cycle is $Q_r = h_3 - h_1$ while that into the subsidiary cycle is $Q_h = h_6 - h_5$. The work transfer from the main cycle is $W_1 = h_3 - h_5$ and that from the subsidiary cycle is $W_h = h_6 - h_4$. We will further write $h_5 - h_7 = d$ and $W_r = W_1 + d$.

The thermal efficiencies for the total cycle η and for the subsidiary cycle η_h are compared with that of the associated Rankine cycle η_r by writing

$$\eta = \frac{W_1 + W_h}{Q_r + Q_h} \quad \eta_h = \frac{W_h}{Q_h} \quad \eta_r = \frac{W_r}{Q_r} \tag{12.6}$$

Then

$$\eta = \frac{W_r - d + W_h}{Q_r + Q_h} = \frac{W_r}{Q_r} \frac{[1 + (W_h/W_r) - (d/W_r)]}{[1 + (Q_h/Q_r)]}$$

Consequently the ratio of the thermal efficiencies for the total cycle including

164   *Raising the Mean Receptor Temperature*

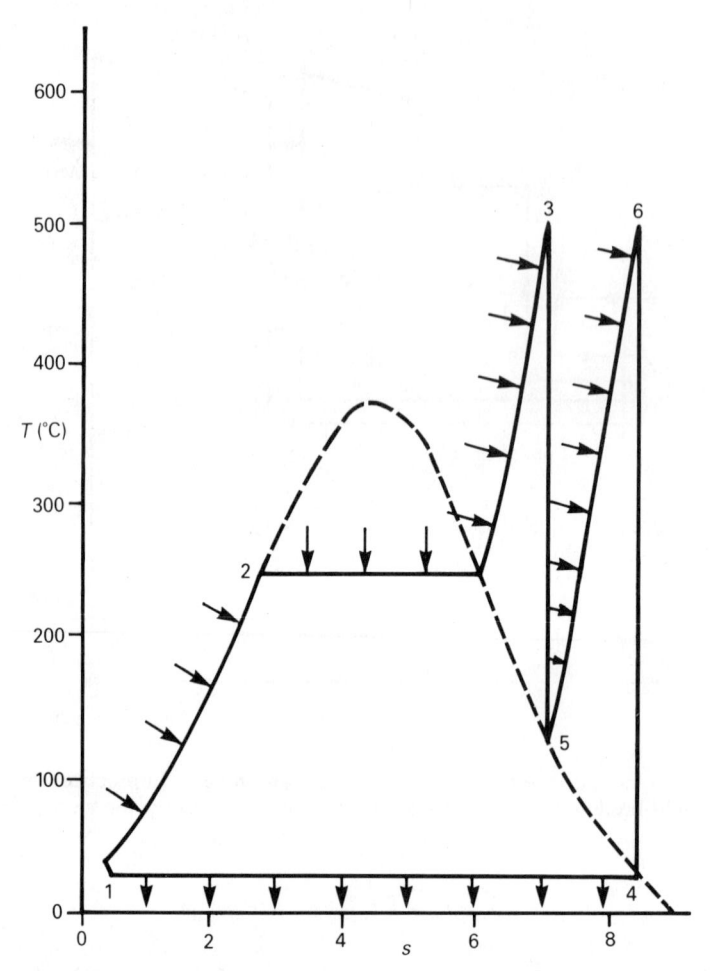

Fig. 12.9 The temperature-specific entropy diagram including superheat and reheat.

reheat and the associated Rankine cycle (which does not) is given by

$$\frac{\eta}{\eta_r} = \frac{1+(W_h/W_r)-(d/W_r)}{1+(Q_h/Q_r)}$$

$$= \frac{1+(Q_h/Q_r)(\eta_h/\eta_r)-(d/W_r)}{1+Q_h/Q_r} \qquad (12.7)$$

where we have used eqn. (12.6) to form the second form of the right hand side.

The thermal efficiency of the total cycle will be greater than that for the associated Rankine cycle without reheat, working between the same pressure range, if the numerator of (12.7) is greater than the denominator, so that

$$(Q_h/Q_r)(\eta_h/\eta_r) - d/W_r > Q_h/Q_r$$

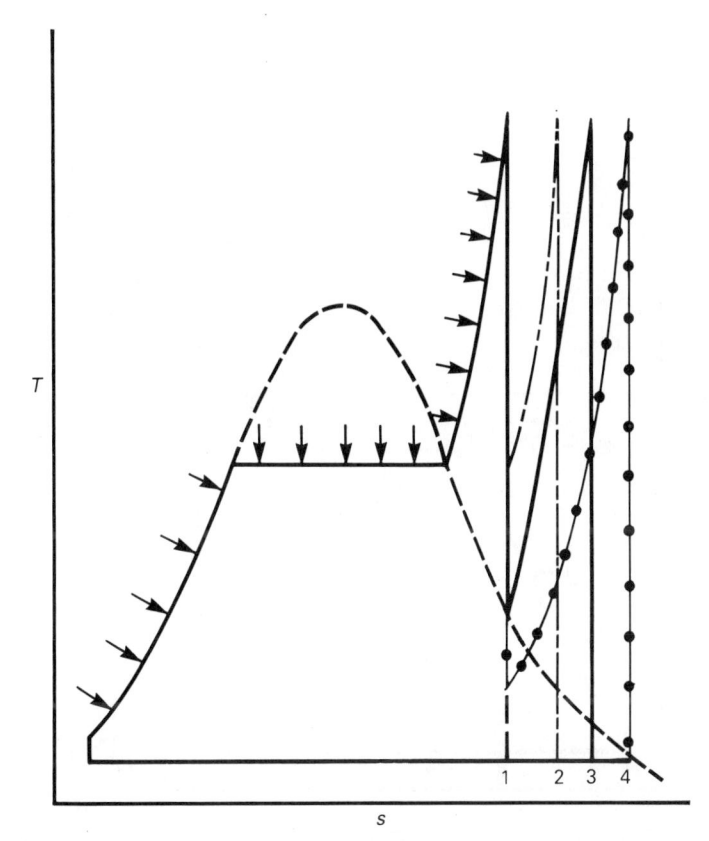

Fig. 12.10 Showing the effect of the reheat pressure on the dryness fraction of the steam leaving the turbine exit. For high dryness, the reheat pressure must be low. (See also **Fig. 9.4**). The points 1, 2, 3, 4 correspond to increasing dryness with decreasing reheat pressure.

that is if

$$\frac{Q_h}{Q_r}\left(\frac{\eta_h}{\eta_r} - 1\right) > \frac{d}{W_r} \qquad (12.8)$$

Because $Q_h < Q_r$, eqn. (12.8) requires that $\eta_h > \eta_r$, and by a sufficient margin. This need not be the case so the thermal efficiency of the total cycle need not be greater than that of the associated Rankine cycle (see Question 12.12).

The condition (12.8) is the more readily achieved for a smaller value of d for a given level of superheat. The smallest value of d (remembering that a reduction of wetness is the aim of the reheating) occurs when the reheat pressure is actually that for the saturation line. The heat acceptance for reheat Q_h is, unfortunately, also large then so the thermal efficiency can suffer. It is clearly necessary to reheat to the initial steam temperature if Q_h is to have a maximum value and so allow the condition (12.8) to be fully satisfied.

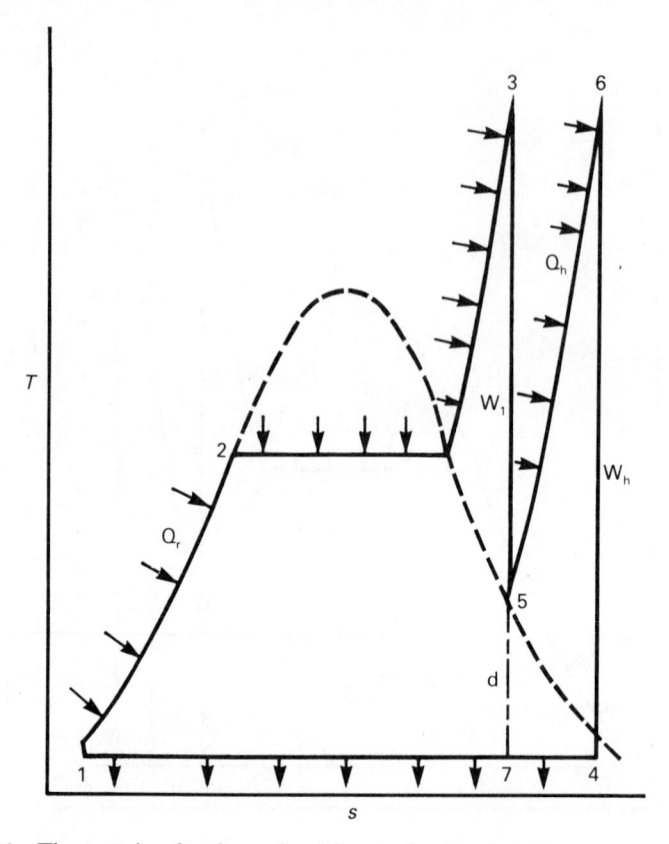

Fig. 12.11 The notation for the cycle with superheat and reheat.

Apparently, maximum dryness is achieved to some extent at the expense of the thermal efficiency.

Incidentally, it is readily seen that reheat will usually reduce the specific steam consumption of the cycle. For, in an obvious notation

$$ssc(\text{total}) = 3600/(W_1 + W_h) = 3600/(W_r - d + W_h)$$

$$= \frac{3600}{W_r} \frac{1}{(1 + [(W_h - d)/W_r])}$$

But $ssc(\text{Rankine}) = 3600/W_r$ so that

$$\frac{ssc(\text{total})}{ssc(\text{Rankine})} = \frac{1}{1 + [(W_h - d)/W_r]}$$

Providing $W_h > d$, it follows that $ssc(\text{total}) < ssc(\text{Rankine})$ and the steam consumption is reduced by the reheat. Indeed, d should be made as small as possible if reduction of steam consumption is especially important (it is almost always important in practice because reduced steam consumption allows for a smaller boiler). This condition also favours the largest value of the dryness

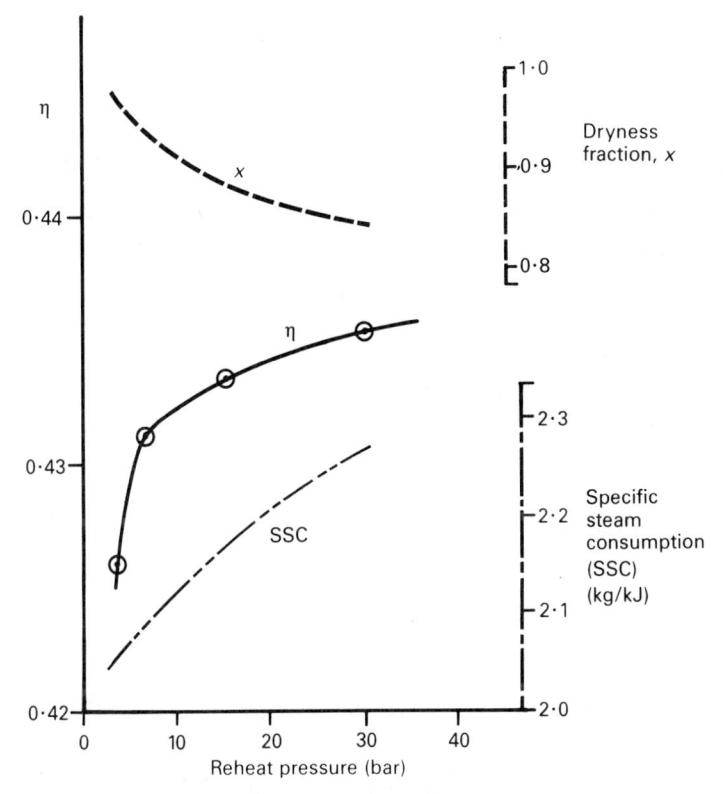

Fig. 12.12 The effect of the reheat pressure on the thermal efficiency, specific steam consumption and the dryness fraction for a cycle with reheat to 500 °C. Boiler pressure = 80 bar, condensed pressure = 0.04 bar.

fraction: this is shown in Fig. 12.12, which gives a quantitative expression for Fig. 12.10.

The precise reheat pressure which provides the optimum conditions with the greatest overall thermal efficiency for the cycle must be found from calculations. For a reversible cycle the optimum reheat pressure is often about 1/4 of the initial pressure. If irreversible effects are important, this 'best' reheat pressure is reduced, sometimes quite substantially. If reduction of the wetness of the steam is the central criterion (and it usually is) the reheat pressure should be low.

Numerical Examples

Consider again the Rankine cycle with boiler pressure 40 bar and condenser pressure 0.04 bar, with superheat to 500 °C. We will now add reheat to 500 °C. The notation is that of Fig. 12.9.

As before

$$h_1 = 121 \text{ kJ/kg} \quad h_3 = 3445 \text{ kJ/kg} \quad s_3 = 7.089 \text{ kJ/kg K} \quad \text{(steam tables)}$$

168   Raising the Mean Receptor Temperature

The expansion through the turbine will be stopped (point 5) when the dry saturation temperature is reached for some particular pressure, that is $s_3 = s_{g5}$ for some particular pressure. We find

$$s_{g5} = 7.089 \text{ kJ/kg when } p_5 = 3.0 \text{ bar}$$

with $h_{g5} = 2725$ kJ/kg   (steam tables)

The temperature is now raised, at 3.0 bar pressure, to 500 °C and

$$h_6 = 3486 \text{ kJ/kg and } s_6 = 8.324 \text{ kJ/kg}$$ (steam tables)

For the final expansion we require $s_6 = s_4$. This gives the relation involving the dryness fraction x

$$8.324 = 0.422 + x 8.051$$

so that

$$x = 0.981$$

This dryness fraction is much improved on that for the corresponding cycle without reheat (but superheated) considered in Section 12.2. But

$$h_4 = 121 + 0.981 \times 2433$$

so that

$$h_4 = 2509 \text{ kJ/kg}$$

Then, for the work and heat transfers

$$W_{34} = (h_3 - h_1) + (h_4 - h_6)$$
$$= (3445 - 2725) + (3486 - 2509) = 720 + 977 = 1697 \text{ kJ/kg}$$
$$Q_{31} = (h_3 - h_1) + (h_6 - h_5)$$
$$= (3445 - 121) + (3486 - 2725) = 3324 + 761 = 4085 \text{ kJ/kg}$$

The thermal efficiency and specific heat consumption are

$$\eta = 0.415 \qquad ssc = 2.121 \text{ kg/kWh}$$

Comparison with the results of section 12.2. shows that reheat has reduced the specific steam consumption by about 22 % and has also increased the thermal efficiency of the cycle by some 2 %.

It is interesting to notice that the work output of the reheated steam slightly exceeds that of the initial expansion while the heat acceptance by the reheat is substantially below that for the initial superheating.

To see the effect of increasing boiler pressure let us repeat the calculation for a boiler pressure of 60 bar, again with superheat to 500 °C and reheat to the same temperature. We find, by the same methods as before, the following data:

$$h_1 = 121 \text{ kJ/kg} \quad h_3 = 3421 \text{ kJ/kg} \quad s_3 = 6.879 \text{ kJ/kg K}$$

Expansion then follows until $s_3 = s_{g5}$: this provides the intermediate pressure $p_5 = 4.5$ bar. Then

$$h_{g5} = 2744 \text{ kJ/kg} \quad h_6 = 3484.5 \text{ kJ/kg and } s_6 = 8.139 \text{ kJ/kg K giving}$$
$$x = 0.959 \text{ and } h_4 = 2453 \text{ kJ/kg}$$

12.4 Supercritical Conditions

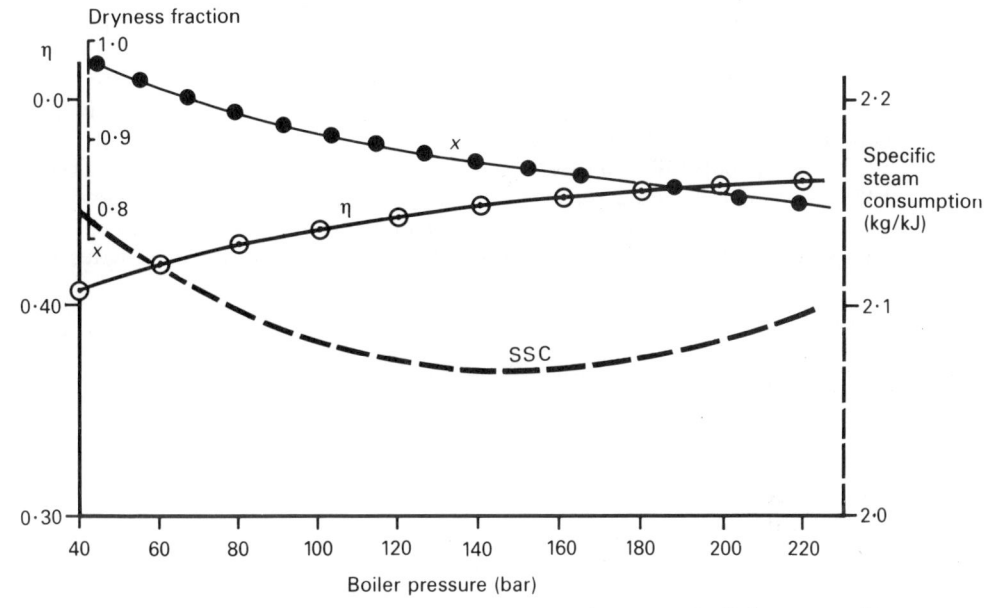

Fig. 12.13 Showing the effect of boiler pressure on the thermal efficiency, specific steam consumption and dryness fraction with superheat to 500 °C and one stage of reheat. The condensed pressure 0.04 bar.

The work transfer $W = W_{35} + W_{64}$
$$= (3421 - 2744) + (3484.5 - 2453)$$
$$= 1708.5 \text{ kJ/kg}$$

The heat transfer is $Q = Q_{13} + Q_{56}$
$$= (3421 - 121) + (3484.5 - 2744)$$
$$= 4040.5 \text{ kJ/kg}$$

The thermal efficiency and specific heat consumption are

$\eta = 0.423 \qquad ssc = 2.107 \text{ kg/kWh}$

Under these conditions involving a single reheat stage, an increase of 50 % in the boiler pressure reduces the specific steam consumption by about 2 % although it increases the thermal efficiency by about 1 %.

The effects of the boiler pressure on the thermal efficiency, specific steam consumption and dryness fraction from 30 bar to the critical pressure are shown in Fig. 12.13 for one reheat stage and superheating to 500 °C. The reheat pressure is that for which $s_3 = s_{g5}$ in the notation of Fig. 12.11.

12.4 Supercritical Conditions

As the boiler pressure is raised the critical point is eventually reached (at 221.2 bar pressure for water with the temperature 374.15 °C) and beyond this pressure there is no phase change line (see Fig. 9.1). The liquid phase passes

170 Raising the Mean Receptor Temperature

to the vapour phase directly and the specific enthalpy of the supercritical steam is as high as that achieved for a sub-critical pressure but with some degree of superheating. The move to high pressures could, therefore, be expected to lead, in principle at least, to a considerable reduction in the complexity of the boiler system.

There is, naturally, a less desirable aspect of supercriticality alone. Expansion of the steam through the turbine stage to enthalpies below the vapour saturation line of the T–s diagram is associated with an unacceptably high wetness of the condensed steam (see Fig. 12.14) and this complicates the

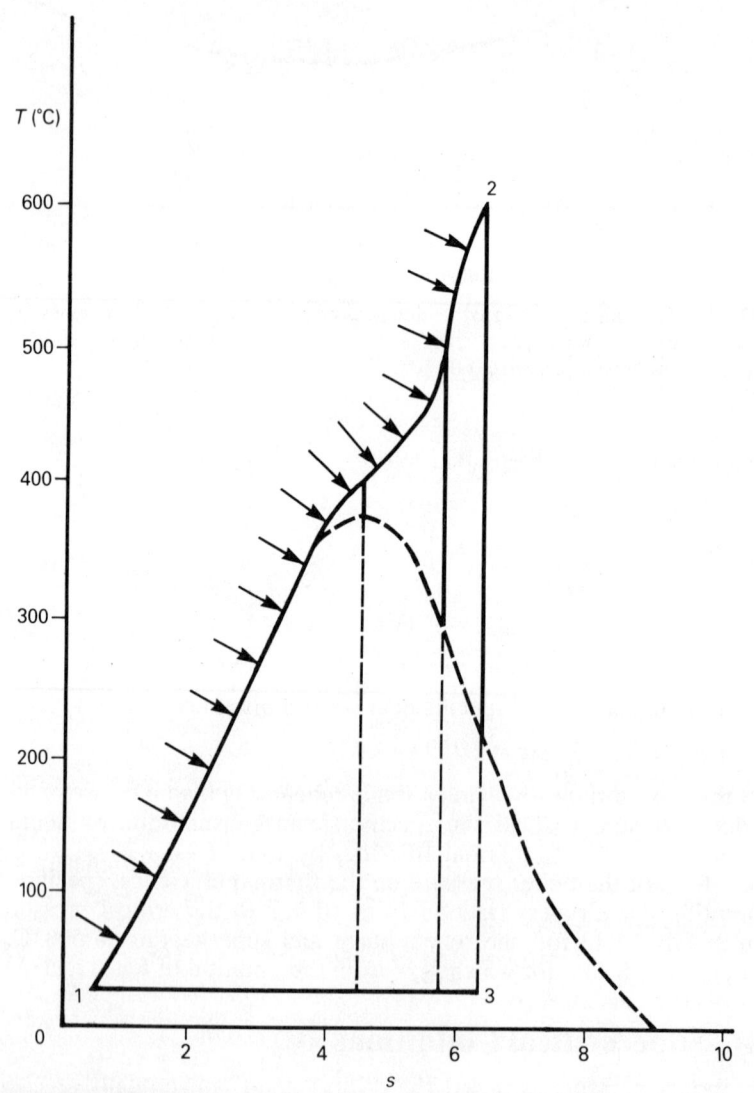

Fig. 12.14 The supercritical cycle with superheat. The dryness is small but is increased by the degree of superheat.

12.4 Supercritical Conditions

turbine design. But wetness can be controlled by reheating and a practical supercritical system must have substantial reheat in the cycle. It will also have superheat, which will alleviate the wetness problem to a small extent and will increase the thermal efficiency still further.

A Numerical Example

Suppose a boiler is working under supercritical conditions at 350 bar pressure, with superheating to produce steam at 600 °C. The steam passes through a turbine to be condensed at 0.04 bar (29 °C). The turbine expansion is entirely isentropic and there are no other losses in the system. The T–s diagram is that of Fig. 12.14. Then we have

$$h_1 = 121 \text{ kJ/kg} \quad h_2 = 3397 \text{ kJ/kg}$$
$$s_2 = 6.120 \text{ kJ/kg K} \qquad \text{(steam tables)}$$

The dryness fraction at the turbine exit after expansion x_3 is found from

$$6.120 = 0.422 + 8.051 x_3 \qquad \text{(steam tables)}$$

so that $x_3 = 0.708$. This is rather low for practical use but we will go on. In the usual way we find $h_3 = 1844$ kJ/kg. The thermal efficiency is

$$\eta = \frac{(h_2 - h_3)}{(h_2 - h_1)} = \frac{1553}{3276} = 0.474$$

and the specific steam consumption is

$$ssc = \frac{3600}{1553} = 2.318 \text{ kg/kWh}$$

Although the degree of wetness at the turbine exit is high, the thermal efficiency is highly attractive and the specific steam consumption is low in comparison with the non-critical cycles.

The wetness at the turbine exit can be controlled by introducing one stage of reheat into the system. Let us require a dryness fraction of at least 0.85 at the turbine exit, which we will achieve through reheat. The T–s diagram is now shown in Fig. 12.15, including one stage of reheat. There is now initial heating to critical conditions plus superheating between points 1 and 2, expansion between points 2 and 3, reheat between points 3 and 4 and the final expansion between points 4 and 5. As for no reheat, $h_1 = 121$ kJ/kg, $h_2 = 3397$ kJ/kg and $s_2 = 6.120$ kJ/kg K. The condition $s_2 = s_g$ defines point 3 and applies in the sub-critical region to fluid at a pressure of about 34 bar. Let the superheated steam expand isentropically to this pressure and it will lie close to the saturated vapour line. Then

$$6.120 = 2.710 + 3.426 x_3 \qquad \text{(steam tables)}$$

giving $x_3 = 0.995$, a very acceptable dryness fraction for that point in the cycle. For the specific enthalpy

$$h_3 = 1042 + 0.995 \times 1761 = 2795 \text{ kJ/kg} \qquad \text{(steam tables)}$$

The fluid is now reheated to the initial boiler temperature of 600 °C where

$$h_4 = 3677 \text{ kJ/kg} \quad \text{and} \quad s_4 = 7.441 \text{ kJ/kg K} \qquad \text{(steam tables)}$$

172 Raising the Mean Receptor Temperature

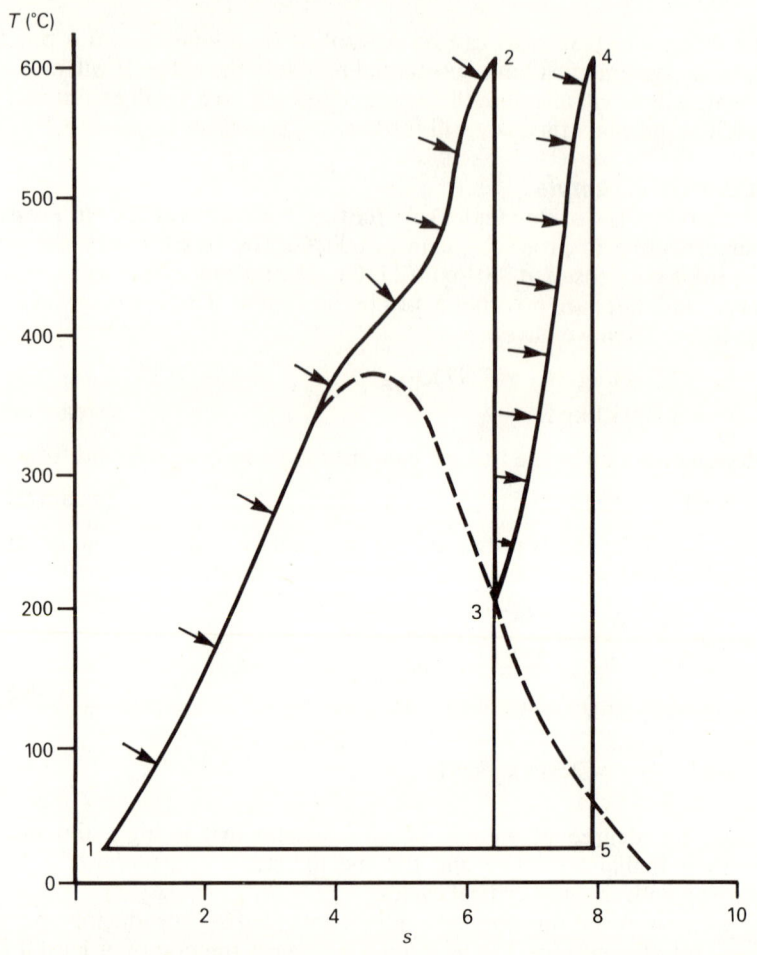

Fig. 12.15 The temperature-specific entropy diagram for superheat and reheat under supercritical conditions. In the text $x \not< 0.85$.

Instead of proceeding in our calculations from here in the usual way we go instead to the turbine exit point because we have prescribed the degree of dryness we wish to achieve at the turbine exit (point 4). The condenser pressure is 0.04 bar so we have from the steam tables, and remembering we require $x_5 = 0.85$

$$h_5 = 121 + 0.85 \times 2433 = 2189 \text{ kJ/kg}$$
$$s_4 = 0.422 + 0.85 \times 8.051 = 7.265 \text{ kJ/kg K}$$

The conditions for reheat are sub-critical and the specific entropy value of 7.265 kJ/kg K at 600 °C applies to a pressure of about 45 bar, so we will meet our target dryness with reheating at 34 bar. Explicitly, for the final isentropic expansion

$$s_4 = s_5 = 7.441 = 0.422 + 8.051 x_5 \qquad \text{(steam tables)}$$

giving for the dryness fraction $x_5 = 0.872$, or rather better than our lower limit. The enthalpy is

$$h_4 = 121 + 0.872 \times 2433 = 2242 \text{ kJ/kg} \qquad \text{(steam tables)}$$

The work transfer $W = (h_2 - h_3) + (h_4 - h_5) = 2037 \text{ kJ/kg}$
while the heat transfer $Q = (h_2 - h_1) + (h_4 - h_3) = 4158 \text{ kJ/kg}$. The thermal efficiency and specific steam consumption are therefore

$$\eta = 0.490 \quad \text{and} \quad ssc = 1.767 \text{ kg/kWh}$$

These are to be compared with the corresponding values above for the supercritical cycle without reheat of 0.474 and 2.317 kg/kWh.

Supercritical conditions are attractive from the points of view of thermal efficiency and specific steam consumption. With reheat the dryness fraction for the turbine stages is very good and, since the *ssc* is small, the boiler can be of small size. The disadvantage is the high pressure and temperature of the boiler, taking the current technology of the materials to the limit. For safety the boiler must be constructed from specially hardy steels and these are still very expensive.

12.5 Summary

1. The thermal efficiency of the Rankine cycle is increased if the mean temperature for the reception of heat transfer is raised. This requires the steam temperature at the inlet to the turbine to be raised well above the boiling point for the particular boiler pressure involved. The steam is superheated and the features of superheating are explored.
2. The wetness of the steam in the turbine is reduced by superheating over that for the simple Rankine cycle but the dryness fraction is still high judged by the best conditions in the turbine. The dryness fraction through the turbine can be raised by reheating the steam during the expansion, certainly once and perhaps two or three times. This is the process of reheat and some of its consequences are explored.
3. It is shown that superheating and reheating provides a combination of processes that both raises the thermal efficiency of the cycle and reduces the specific steam consumption.
4. Supercritical conditions are explored with superheat and reheat as an alternative way of raising the working level of the characteristics of the cycle.

12.6 Exercises

12.1 A steam turbine system has a boiler pressure of 40 bar and the emergent steam is superheated to 500 °C. The condenser pressure is 0.04 bar. There is no reheat stage. Calculate (a) the dryness fraction of the steam on leaving the turbine; (b) the thermal efficiency, and (c) the specific steam consumption for the system if the expansion is fully isentropic ($y = 1$) and the compression work transfer is neglected.

Repeat the calculations for the condenser pressures of 0.030, 0.035 and 0.050 bar. Plot graphs to show the dependence of dryness fraction, thermal efficiency and specific steam consumption on the condenser pressure.

174 Raising the Mean Receptor Temperature

12.2 The boiler pressure in Question 12.1 is raised to 60 bar. The superheat temperature remains the same and there is no reheat. Repeat the calculations of that question.

12.3 The boiler pressure of Question 12.1 is raised further to 100 bar, the superheat temperature remaining the same, and there is no reheat. Repeat the calculations of Question 12.1.

12.4 Use the results of Questions 12.1, 12.2 and 12.3, together with the results in the main text, to describe the way the dryness fraction, the thermal efficiency and the specific steam consumption of the cycle with superheat but no reheat are affected by boiler pressure and condenser pressure. Plot the results in graph form.

12.5 A Rankine cycle with superheat but no reheat operates between the boiler pressure of 30 bar and the condenser pressure of 0.04 bar, with superheating to 450 °C. The turbine and compressor stages have the isentropic efficiencies y and z respectively. Show that the thermal efficiency η of the cycle is given to good approximation by

$$\eta = \frac{(1210y - 3/z)}{(3222 - 3/z)}$$

the work ratio r_w by

$$r_w = 1 - \frac{0.0025}{yz}$$

and the specific steam consumption by

$$ssc = \frac{3600}{(1210y - 3/z)} \text{ kg/kWh}$$

How effective is superheating in improving the working characteristics of the system? To what extent can the deviation from unity of the isentropic efficiency of the compressor stage be ignored? Comment on the difference between this conclusion and that appropriate to the corresponding Carnot cycle.

12.6 Derive eqn. (12.3) of the main text from the eqns. (12.1) and (12.2). Show from (12.3) that the condition that superheat will increase the thermal efficiency can be expressed alternatively in the form $(Q_{os}/Q_s) < (Q_{on}/Q_n)$. Show that this result is always satisfied for steam by considering separately the cases of boiler pressure (a) 20 bar, (b) 50 bar, (c) 100 bar, (d) 150 bar and (e) 200 bar. In each case the condenser pressure is 0.04 bar. How does the ratio ssc/ssc_n depend upon the boiler pressure for fixed condenser pressure (eqn. (12.4) of the main text)? Plot graphs to show the effect of boiler pressure on the ratios η/η_n and ssc/ssc_n.

How applicable are these arguments to supercritical conditions? Is there a superheat temperature below which the arguments are not valid?

12.6 Exercises

12.7 A Rankine cycle with reheat to the appropriate dry saturation conditions but no superheat works between the pressure range of 30 bar at the boiler and 0.04 bar at the condenser. The final dryness fraction, where the expanded steam leaves the turbine, is to be 0.92. If the turbine and compressor stages each have ideal isothermal efficiencies, what is the thermal efficiency, work ratio and specific steam consumption for the cycle?

12.8 A Rankine cycle with superheat to 450 °C has one reheat stage to the superheat temperature. The boiler pressure is 30 bar and the condenser pressure is 0.04 bar. The turbine and compressor stages have ideal isothermal efficiencies. By choosing successive reheat pressures of (a) 15 bar, (b) 10 bar, (c) 7 bar and (d) 3 bar, construct a graph to determine the thermal efficiency, work ratio, specific steam consumption and dryness fraction at the turbine exit of the cycle.

12.9 What are the dryness fractions, the thermal efficiencies and the specific steam consumptions of the cycle considered in Question 12.8 if the isothermal efficiency of the expansion stage is (a) 0.90, (b) 0.85 or (c) 0.80?

12.10 What is the thermal efficiency, specific steam consumption and dryness fraction at the turbine exit of the cycle considered in Question 12.8 if the condenser pressure is changed to (a) 0.030 bar, (b) 0.035 bar or (c) 0.05 bar? Comment on the effect of the condenser pressure on the working of the cycle.

12.11 Repeat the calculations of Questions 12.8, 12.9, and 12.10 for a boiler pressure of (a) 40 bar, (b) 60 bar and (b) 100 bar. Comment on your results in relation to the working of the different plant conditions.

12.12 The condition that the thermal efficiency of a Rankine cycle with superheat should be increased by the addition of a reheat stage is eqn. (12.7) of the main text. Using the notation of the main text, show that

$$\eta/\eta_r = 1 + [(\eta_h/\eta_r) - 1]/[1 + (Q_r/Q_s)]$$

and that reheat will be associated with an increased thermal efficiency of the complete cycle if $\eta_h > \eta_r$, as deduced in the main text. Explore conditions under which this will *not* be the case.

12.13 It is proposed to construct a small mobile steam turbine power plant and there are two alternative designs. In each case, the isentropic efficiency of the turbine stage is $y = 0.95$ and the work transfer for compression can be neglected.

Design 1. Dry saturated steam leaves the boiler at 60 bar pressure and is expanded through the turbine to a condenser at 0.04 bar pressure. There is no superheat but the steam suffers one stage of reheat at a pressure which provides a dryness fraction at the turbine exit of 0.90.

Design 2. Dry saturated steam leaves the boiler at 40 bar pressure and is superheated to 500 °C. It then passes through the turbine stage without reheat and is condensed at 0.04 bar pressure.

176 *Raising the Mean Receptor Temperature*

Which of the two designs would offer (a) the highest thermal efficiency and (b) the lowest steam consumption? Discuss the practical and economic merits of each design remembering that the plant is to be both mobile and physically of small size.

12.14 If you were asked to offer a third design to satisfy the requirements of Question 12.11 what would this be? Give the reasoning behind your design, which should take account of the economics of the construction and of the running of the plant.

12.15 You are asked to adjudicate in the choice between two designs for a superheated steam turbine power plant.

Design 1. The boiler pressure is chosen to be 40 bar and the condenser pressure is 0.035 bar. The steam is to be superheated to 600 °C and expanded through a turbine of isentropic efficiency 0.92. The compression work transfer is *not* neglected and its isentropic efficiency is 0.95.

Design 2. The boiler pressure is now 30 bar and the condenser pressure is 0.04 bar. The steam is superheated to 500 °C and there is one stage of reheat to provide a final dryness fraction of 0.90 at the turbine exit. The isentropic efficiency for expansion is 0.92: the compression work transfer is included in the assessment and the compression efficiency is 0.95.

Deduce the expected working characteristics of the two systems and compare and contrast them. Which design would you choose if (a) high thermal efficiency or (b) low specific steam consumption was the particular requirement?

What would your advice be if you were asked to replace the designs to suggest a single improved version? Your considerations should also include the construction and running of the plant as well as its thermodynamic features.

13. REGENERATIVE HEATING

In the previous discussion we considered ways of increasing the maximum temperature of the cycle (and so of increasing the mean receptor temperature) and of controlling the degree of wetness of the steam during expansion. These needs have led to superheating and reheating, and improve the working of the cycle. However, none of these devices address the essential thermodynamic problem of reducing the range of temperature for heat reception into the system (indeed, superheating actually increases it). Only by reducing this temperature range, and so making the heat transfer more nearly isothermal, can the thermal conditions be moved significantly towards the Carnot ideal. Reducing the range of temperature is achieved if the feed water entering the boiler is above the temperature of condensation. This requires the feed water to be pre-heated between the condenser and the inlet to the boiler and using heat transfer within the cycle itself: external heat would not effect the reduction of heat input that we require. Such pre-heating of the feed water is termed regenerative heating (because it involves heat transfer within the cycle) or sometimes more simply feed heating (because it is the feed water that is heated). Feed heating is a vital part of any modern steam plant and the way such heating is achieved is the subject of the present chapter.

13.1 Preliminary Thermodynamic Considerations

Consider a simple thermal cycle with no superheating or reheating. If heat losses through piping and radiation are neglected (and these can be made small, at least in principle), the temperature drop through the cycle takes place between the entry to the turbine and the exit of the condenser. The feed water from the condenser to the boiler inlet is at the condenser temperature but its temperature can be raised through heat exchange with the turbine stage (see Fig. 13.1). If the feed water moving to the boiler were to be linked thermally to the turbine stage through a heat exchanger the feed water could be heated as it moves back to the boiler. The direction of water flow is that of increasing turbine temperature. The feed water can be arranged to traverse the full length of the turbine casing. To obtain the full thermodynamic efficiency, of course, the temperature difference across the heat exchanger between the expanding steam and the returning water must be indefinitely small at every point along the turbine. We consider first a method of achieving this requirement in principle; the application will be considered afterwards.

Superheat must be neglected at first because the feed water cannot achieve superheat conditions so the presence of superheated steam in the turbine will introduce a temperature difference between steam and water which cannot be infinitesimal even approximately. The solution to this problem will be explained later but the matter is best neglected at first until the principle of feed heating is clear.

178 Regenerative Heating

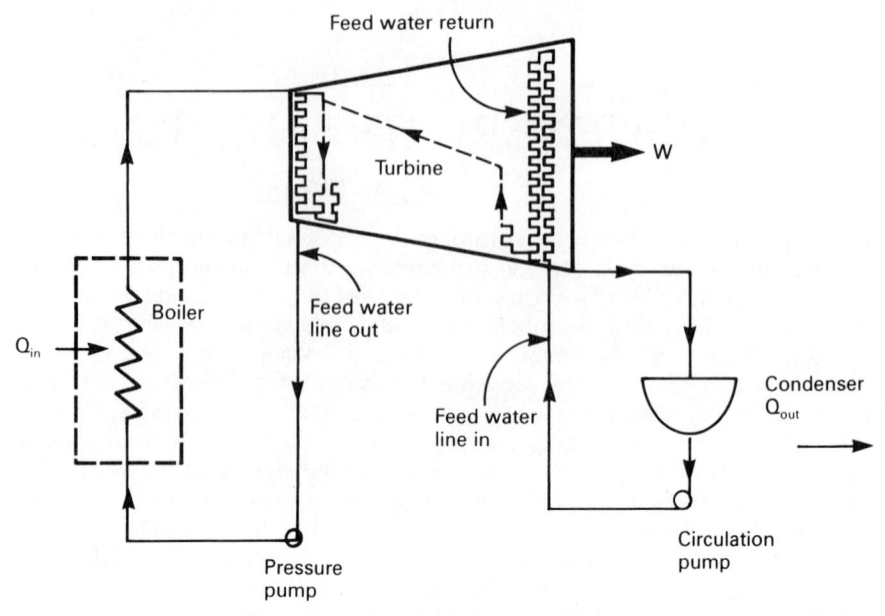

Fig. 13.1 Showing the principle of regenerative heating, the feed water to the boiler being brought into thermal contact with the turbine casing.

Isothermal and Adiabatic Transfers

The requirement of thermodynamic heat transfer between the steam and the feed water is that the process be a combination of isothermal and adiabatic changes. The expansion through the turbine is adiabatic and entropy transfer there will be isothermal. The turbine is recognised as the collection of separate blade stages that it actually is, each stage providing an infinitesimal contribution to the total expansion. Each stage is then characterised by an expansion of the steam at constant entropy and an entropy transfer from the steam to the water at constant temperature. The total entropy of the expansion/feed system remains unchanged and the actual expansion occurs at constant entropy at each stage. There is no mass transfer between the turbine steam and the feed water.

Suppose the steam between the stages is in perfect thermal contact with feed water through a heat exchanger so that an infinitesimal quantity of entropy $ds_s = dq/T$ (where dq is the quantity of heat transferred at the temperature T) is lost by the steam, passing isothermally to the feed water. The steam enters the next expansion stage downstream with entropy reduced by the infinitesimal amount ds_s, while the feed water passes to the next upstream expansion stage with the temperature increased by dT due to the absorption of the infinitesimal amount of energy dq ($= ds_w/T$), so that $dT = dq/\rho C_p$. There is no overall loss of entropy because the entropy ds_s lost by the steam is equal to that ds_w gained by the feed water, that is $ds_s + ds_w = 0$. The process for one stage is depicted in Fig. 13.2 in terms of the T–s diagram. Continuing the process over the full turbine stage leads to the T–s diagram of Fig. 13.3. The small regions at the entrance and exit to the

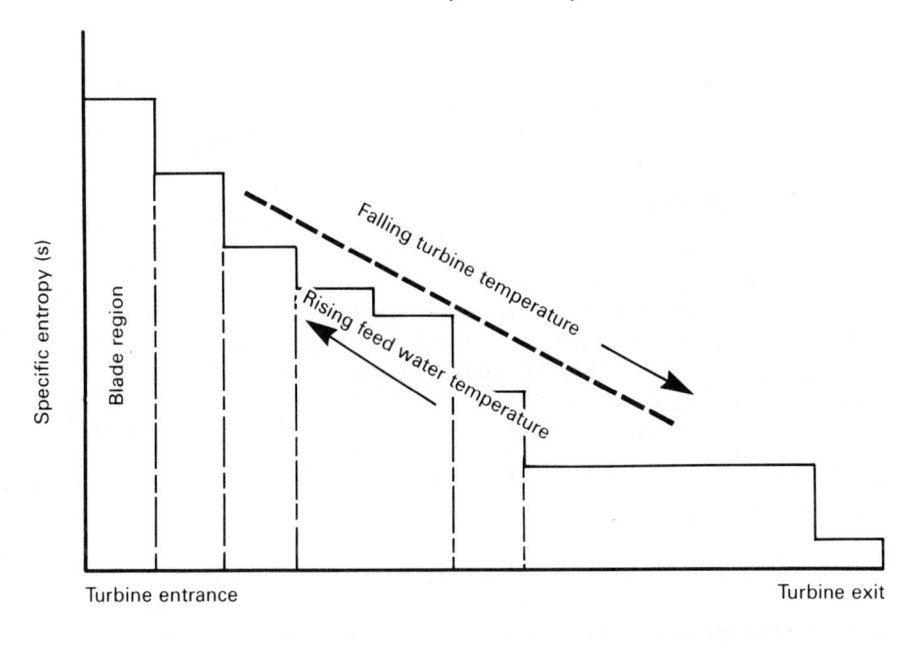

13.1 Preliminary Thermodynamic Considerations

Fig. 13.2 Showing the reciprocal fall of steam temperature in the turbine and rise of temperature in the feed water.

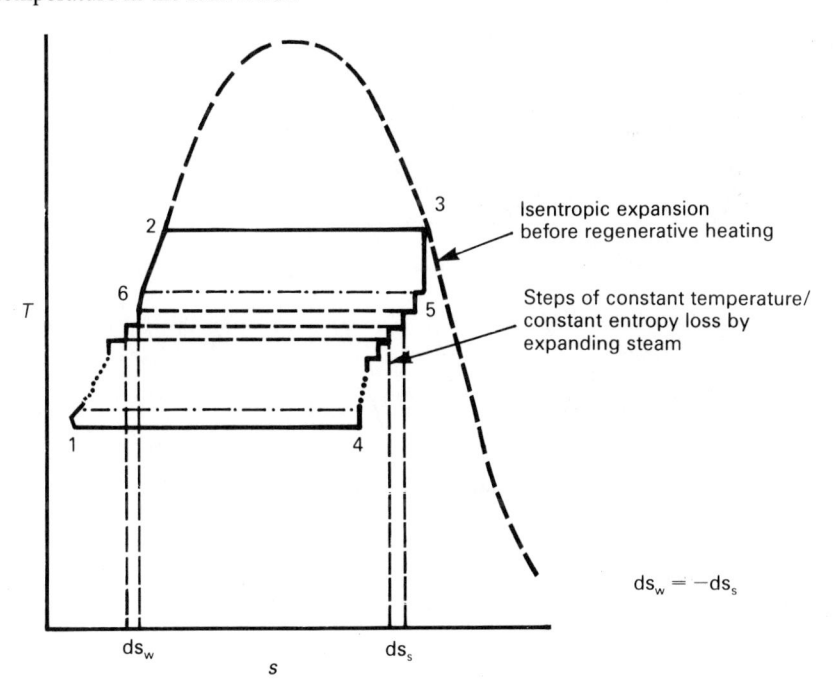

Fig. 13.3 Showing the elementary step form of the temperature-specific entropy diagram for a cycle with regenerative heating by conduction across the surface casing of the turbine.

180 Regenerative Heating

turbine where there is no regenerative heat transfer have isentropic expansion characteristics as before. The expansion work transfer has a step form which corresponds precisely with the step form of the heat transfer to the feed water. As the number of expansion stages becomes indefinitely large the step transfers degenerate to straight lines in this limit and the situation is shown in Fig. 13.4. Good thermodynamic practice is followed because all transfers of work have taken place adiabatically and all transfers of heat have taken place isothermally. This is clear from Fig. 13.4. It should be noticed that this is not the T–s diagram for a Rankine cycle.

The T–s diagram Fig. 13.4 has an indefinitely large number of transfer stages, presumed to traverse the entire length of the turbine from entrance to exit and the expansion and heating lines are smooth. In particular, the decrease of entropy ds_s of the steam at each stage is equal in magnitude to the increase of entropy ds_w of the feed water: this means that $ds_s + ds_w = 0$ or $ds_s = -ds_w$ at each point on the expansion/heating curves (respectively lines 3–4 and 1–2 in Fig. 13.4). It follows that the lines 1–2 and 3–4 are both parallel and of equal length. The heat reception occurs between states 2 and 3, the heat transfer being represented by the area 23872. The heat rejection occurs between the states 4 and 1, the heat transfer being given by the area 14651. The difference between these two areas (the area 23412) is, of course, a measure of the work output of the cycle.

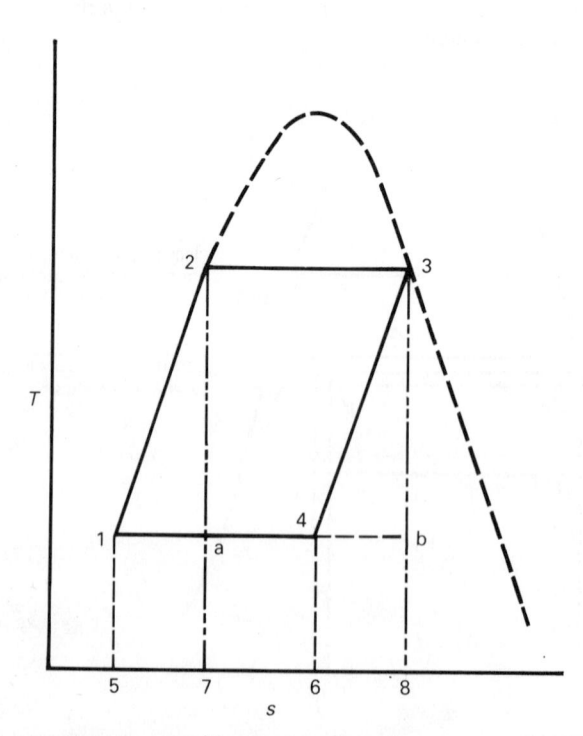

Fig. 13.4 The temperature-specific entropy diagram for the cycle with regenerative feed water heating. The energy transfer is by conduction through the turbine casing and the water/steam path is one continuous line.

Achieving the Carnot Efficiency

It will be realised that the T–s diagram for the Carnot cycle working between the same maximum and minimum temperatures is given by the curve 23ba2 of Fig. 13.4. The heat reception is along the line 23 (the same as for the regenerative cycle) and is represented by the area 23872. The heat rejection is along the line b–a with the representing area ab87a. The work output is represented by the area 23ba2. The geometry of Fig. 13.4 shows that

area 2a7512 = area 3b8643
area 1a751 = area 4b864,

from which it readily follows that the areas 23ba2 and 23412 are of equal magnitude. But these separately represent the work outputs of the regenerative and the Carnot cycles: the work output of the regenerative cycle is the same as that of the corresponding Carnot cycle. Because the heat input is the same for both cycles, their thermal efficiencies are the same. More than this, the two cycles have the same specific steam consumption. The regenerative heating system that we have described has achieved the thermal efficiency and specific steam consumptions of the ideal Carnot cycle.

It will be remembered that superheating has not been included in the discussion. Because it will spread the temperature at which heat is received by the cycle (and so break the rule of heat input at a single temperature), superheating will, in fact, reduce the thermal efficiency of the cycle from the Carnot ideal.

The Availability Function

Referring to the T–s diagram Fig. 13.3, the heat accepted by the cycle is the enthalpy difference $(h_3 - h_6)$ while the heat rejected is $T_c(s_4 - s_1)$, where T_c is the condensation temperature.

The thermal efficiency for the cycle is then

$$\eta = \frac{(h_3 - h_6) - T_c(s_4 - s_1)}{(h_3 - h_5)} \tag{13.1}$$

But $T_c(s_4 - s_1) = T_c(s_5 - s_6) = T_c(s_3 - s_6)$ because $s_5 = s_3$. Consequently eqn. (13.1) can be rearranged to the form

$$\eta = \frac{(h_3 - T_c s_3) - (h_6 - T_c s_6)}{(h_3 - h_6)} \tag{13.2}$$

The quantity $B = h - T_c s$ is called the availability function for the steady flow process (see Section 2.5). It refers to the lowest (condenser) temperature T_c and is a measure of the energy available for work transfer for temperatures of a heat reservoir higher than T_c. For a temperature $T(>T_c)$, $h = h(T)$ and $s = s(T)$. For several heat reservoirs at different temperatures, the availability functions for each reservoir, compared with T_c, are additive (algebraically). In terms of the availability functions referring to the highest and lowest temperatures for the heat acceptance (relative to the datum function B_1), eqn. (13.2) becomes

$$\eta = \frac{B_3 - B_6}{h_3 - h_6}$$

182 Regenerative Heating

This expression has the same form as for the Rankine cycle, Section 12.1 but is here expressed in terms of the availability function. B has a form reminiscent of the Gibbs free energy $G = u + pV - Ts$, and it is indeed closely related to it. But B relates the energy to the lowest temperature whereas G involves the temperature of the immediate environment of the reaction. The usefulness of B will be seen later (Section 14.2) when we consider some aspects of the furnace flue gases.

13.2 Practical Implementation of Feed Heating

The cycle we have described shows in principle the importance of regenerative heating for a cycle without superheat but is not immediately transferable to a practical design. The matching of the turbine expansion characteristics with the reversible transfer of heat to the feed water presents insuperable problems. Again, the reversible transfer of heat using the stages we have described would require an indefinitely large heat exchange system which is quite unrealistic. The transfer of steam enthalpy to the feed water is still valid but (the world being what it is) the transfer cannot fail to fall short of the ideal.

Finite Number of Exchange Points

The basic practical plant must still be the Rankine cycle. There will, however, be a finite number of heat transfer points linking the expanding steam to the feed water, but we will aim to make each such transfer as thermodynamically effective as possible. Rather than pass the feed water through the turbine casing (taking the feed water to the turbine) it is more practical to take a certain quantity of steam from the turbine to the feed water by bleeding steam off at appropriate points through the expansion stage. There is now a new element of the process in that a mass transfer is now maintained between the expanding steam and the feed water whereas in Section 13.1 no mass transfer was involved.

The steam bled off can provide no further work transfer in the turbine and is used entirely to raise the feed water temperature at that stage of the cycle. Thermal efficiency has obviously been lost to some extent and, since steam is being taken away from the expansion process before it is completed, the specific steam consumption must also suffer an increase. On the other hand, the mass flow through the turbine is reduced successively along its length and this will offer some advantage for the flow conditions at the low pressure end.

Comments on the Heat Exchange

Steam is bled from the turbine at selected points (the selection criteria will be considered later in Section 13.4) and heat exchanged between the bled steam and the feed water as shown in Fig. 13.5. The heat exchange must be as closely reversible as possible, and the way of achieving this will be considered in Section 13.5. There are two possible heat exchange systems.

In the first, called the open feed heater, the bled steam is introduced into the feed water directly. The combination of feed water and bled steam then moves upstream to the next heater stage and ultimately to the boiler. In this way all the bled steam is returned to the boiler without first having passed through the condenser. In the second method, called the closed feed heater,

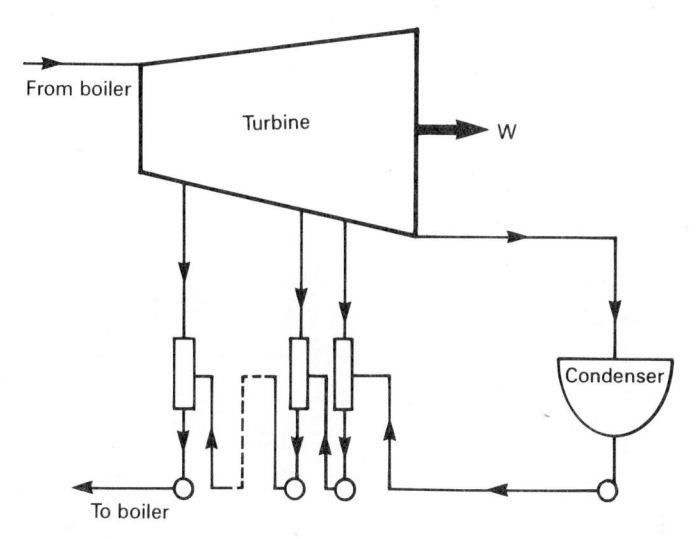

Fig. 13.5 Showing the flow diagram for the feed heat system of direct contact heaters. The steam entering the boiler has not been superheated.

the bled steam is contained in a pipe and is kept separate from the feed water during the heat exchange. Having lost enthalpy during this heat exchange, the bled steam is then passed downstream to the next lower heat exchange point and ultimately is returned to the condenser. All the bled steam now passes through the condenser and all the working fluid is reduced to the lowest temperature during the passage through the cycle.

The number of heat exchange points and where best they should be located along the turbine are problems of thermodynamic optimisation and we shall consider the criteria for them later (Sections 13.5 and 13.6). A practical design will usually include both types of heat transfer system, called collectively feed heaters. Quite generally, the closed feed heater is necessary for high pressure conditions (and especially when superheat is also included) while the open feed heater is useful at lower pressures. The open heater is useful as a method of removing air from the feed water and one open heater can in this way play the role of an aerator for the cycle. Air in the steam encourages cavitation in the turbine stage and so increases the wear on the turbine blades. Air also encourages corrosion in various parts of the system and is best excluded.

13.3 A Possible Cycle without Superheat

Before delving further into the theory of feed heating we will explore how the procedure might be applied in practice, and what special features will need further consideration. For this purpose we consider a Rankine cycle without superheat or reheat and will add one stage of feed heat to it. One stage represents the worst thermodynamic situation (almost no thermodynamic situation at all) and so the most severe test for the feed heat approach. The flow circuit is shown in Fig. 13.6 and the corresponding T–s diagram in Fig.

184 Regenerative Heating

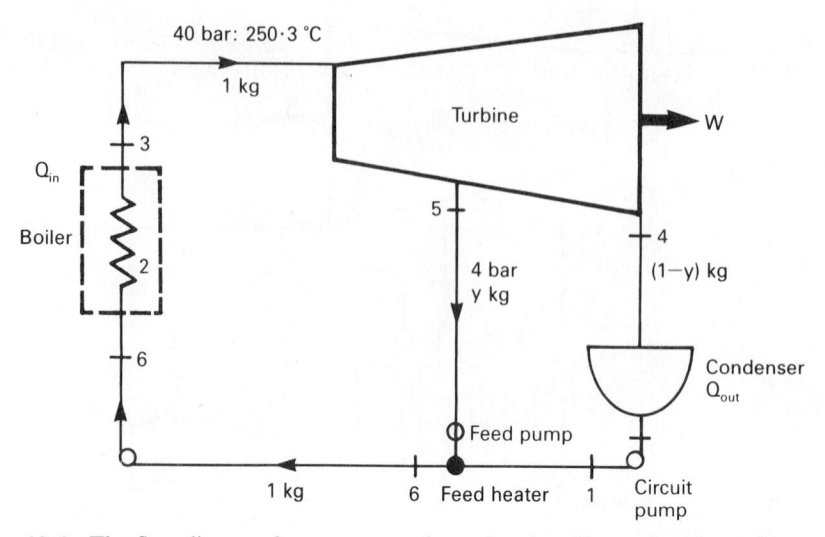

Fig. 13.6 The flow diagram for a non-superheated cycle with one feed heat unit.

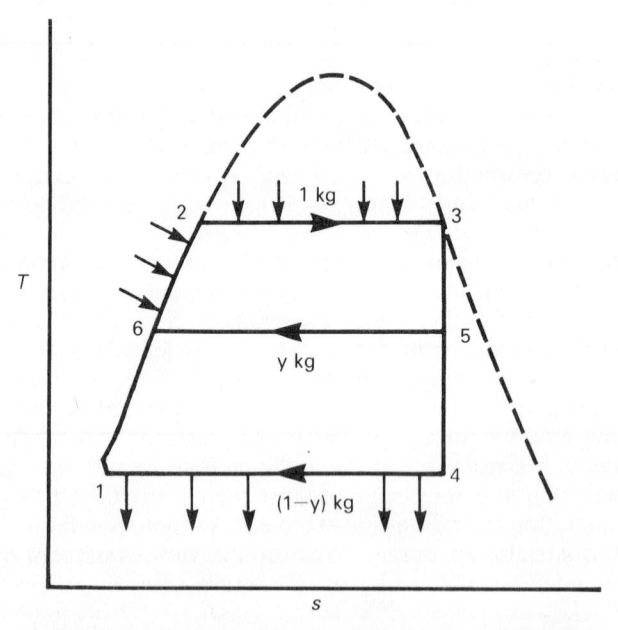

Fig. 13.7 The *T–s* diagram for a non-superheated cycle with one feed heat unit. The heat acceptance is between points 6 and 3 while the heat rejection is between points 1 and 4.

13.7. While Fig. 13.7 makes clear the heat and work transfers (and that concerns us now) it will be realised that it does not accurately portray the full conditions within the cycle. But those needed for our purposes are clearly shown and this will suffice.

13.3 A Possible Cycle without Superheat

Dry saturated steam leaves the boiler at 40 bar pressure and is condensed at 0.04 bar pressure. One open-type feed heater is introduced at a point along the turbine yet to be determined. Pump and other losses in the system are neglected in this example and it is supposed that the expansion and compression stages have an ideal efficiency ($y = z = 1$).

Initial and Final Steam Conditions

The boiler temperature T_B and condenser temperature T_C, together with the corresponding enthalpies and entropies are found from the steam tables to be

$T_B = 250.3\,°C = 523.6\,K \quad T_C = 29\,°C = 302.3\,K$
$h_1 = 121\,kJ/kg \quad h_3 = 2801\,kJ/kg \quad s_3 = 6.070\,kJ/kg\,K$

Selecting the Turbine Bleed Points

Returning to Fig. 13.6, steam is bled from the turbine at point 5 and introduced into the water feed line at point 6. It is necessary to choose the appropriate bleed point in terms of pressure and so temperature. One method is to choose different points along the turbine and select the one that shows the greatest thermal efficiency (this is done in Question 13.1). Alternatively, we could apply thermodynamic arguments to find the most effective bleed point (and this is done in Section 13.4). For the moment we use an heuristic argument.

The single bleed point taken too near the entry point to the turbine will have the greatest temperature difference between the bled steam and the feed water and is thermodynamically the least attractive condition. On the other hand, a bleed point chosen near the exit end of the turbine will involve steam containing the least enthalpy and so will have least effect on the feed temperature. As a compromise we take a bleed point with enthalpy mid-way between that of the saturated steam entering the turbine and the spent steam leaving it. We do not know what this enthalpy value will be at this stage but, since enthalpy increases with the temperature, we choose a point with temperature midway between the boiler and condenser temperatures.

This temperature is close to 145 °C and corresponds to a pressure of about 4 bar: the precise pressure is not, in fact, critical.

Enthalpies and Entropies

We select the bleed pressure as 4 bar and extract a steam/water mixture from the turbine after the initial steam has expanded to this point. If we suppose 1 kg of steam left the boiler to enter the turbine, we will extract Y kg of steam for feed heating. Then $(1 - Y)$ kg of fluid continues the expansion and is condensed at 0.04 bar pressure. We leave Y unknown for the moment. Then, at 4 bar pressure

$s_f = 1.778\,kJ/kg\,K \quad s_{fg} = 5.121\,kJ/kg\,K \quad h_6 = 605\,kJ/kg$ (steam tables)

The dryness fraction x_5 at point 5 follows from the condition for isentropic expansion $s_3 = s_5$, that is

$6.070 = 1.776 + 5.121 x_5$

giving $x_5 = 0.839$. This allows h_5 to be calculated from

$h_5 = 605 + 0.839 \times 2134 = 2394\,kJ/kg$

186 *Regenerative Heating*

After Y kg of steam has been bled off at point 5, $(1-Y)$ kg of steam continue the expansion to point 4 with $s_3 = s_4$. The corresponding dryness fraction x_4 is calculated from the expression

$$6.070 = 0.422 + 8.051 x_4 \qquad \text{(steam tables)}$$

giving $x_4 = 0.701$. The enthalpy at point 4 follows from

$$h_4 = 121 + 0.701 \times 2433 = 1828 \text{ kJ/kg}$$

As a check on our calculations, we can see that $[h_3 + h_4]/2 = 2315$ kJ/kg, which is close to the calculated value $h_5 = 2394$ kJ/kg, so our heuristic criterion for selecting point 5 has been satisfied to an adequate approximation.

Energy Balance at the Bleed/Feed Point

We must still determine the quantity Y of steam to be bled off from the turbine. This is done to satisfy energy balance at point 5 and yield 1 kg of water moving to the boiler at stage h_6. Y kg of bled steam with enthalpy h_5 is mixed with $(1-Y)$ kg of feed water with enthalpy h_1 to produce 1 kg of water with enthalpy h_6. Energy balance reuqires that the relation

$$h_6 = (1-Y) h_7 + Y h_5$$

shall be satisfied, giving $Y = 0.213$.

Work and Heat Transfers

Of the 1 kg of dry steam entering the turbine, 0.213 kg passes through the cycle 62356 and 0.787 kg passes through the condenser cycle 6346. All the steam is, therefore, expanded through the turbine between points 3 and 5 while $(1-Y)$ kg is expanded further through the points 5 and 4.

For the work transfer we have

$$\begin{aligned} \text{Work transfer} &= (h_3 - h_5) + (1-Y)(h_5 - h_4) \\ &= (2801 - 2394) + (1 - 0.213)(2394 - 1828) \\ &= 852 \text{ kJ/kg} \end{aligned}$$

For the heat transfer we have

$$\text{Heat transfer} = (h_3 - h_6) = 2801 - 606 = 2195 \text{ kJ/kg}$$

The thermal efficiency of the cycle is then

$$\eta = \frac{852}{2195} = 0.388$$

The specific steam consumption is

$$ssc = \frac{3600}{852} = 4.22 \text{ kg/kWh}$$

For comparison, the corresponding values for the simple Rankine cycle without feed heat are $\eta = 0.362$ and $ssc = 3.71$ kg/kWh. The single stage of feed heat has increased the thermal efficiency by a little more than 7 % and increased the specific steam consumption by nearly 14 %. It seems that even one stage of feed heating has an important effect on the efficiency of the cycle.

The increased thermal efficiency arises even though the work transfer is reduced because the heat transfer to the cycle is reduced even more: the increased specific steam consumption arises because steam is now bled from the turbine when it has only partially expanded.

Two Feed Heaters
The calculations can be repeated for more than one feed heater; suppose there are two. We will again suppose the bleed points to be placed to give equal enthalpy intervals along the turbine. The intermediate pressure points for the bleed points are then at 9 bars and 0.9 bar. The dryness fraction at the first bleed expansion is 0.878, at the second is 0.784 and at the final expansion is 0.702. Of the 1 kg of steam entering the turbine, 0.159 kg is bled off at the first point and 0.116 kg (rather less) is bled off at the second. The work transfer during the cycle is 820.73 kJ (rather less than for one stage) but the heat transfer into the system is 2058 kJ (again less than for one feed heat stage). The thermal efficiency and specific steam consumption for the two-feed heat cycle are now $\eta = 0.399$ and $ssc = 4.386$ kg/kWh. The thermal efficiency has again been raised, but so also has the specific steam consumption. More feed heaters can be added with the same result but the effect on the thermal efficiency will be proportionately less than for one or two heaters. The thermal efficiency and specific steam consumption change very little after eight or nine feed heat stages have been included, although there is always some further improvement as the number of stages is increased. The improvement has, however, to be weighed against the increased complexity and cost of the plant and there is usually no advantage in practice in including more than eight feed heat stages. This number is associated with a substantial complexity of flow pipes adding to the volume requirements of the plant.

13.4 The Thermodynamics of the Bleed Point Locations

It was seen in the last section that feed heating does in fact increase the thermal efficiency of the cycle if applied to the correct thermodynamic conditions of the bleed point. There it was guessed that the bleed points should be equi-spaced on the enthalpy scale, so that the temperature difference between successive bleed points is the same across the turbine. This guess will now be developed and substantiated theoretically.

Steam and Feed Water Flows
A boiler supplies dry saturated steam directly to the turbine, and the steam is taken out through a number of bleed points to heat the feed water on its return from the condenser to the boiler. The circuit is shown in Fig. 13.8, where three successive representative feed heater points are shown and labelled $j-1$, j and $j+1$ in the direction of the feed water flow. It will be supposed that 1 kg of steam passes through the condenser as the result of S_B kg of steam entering the turbine.

The mass of steam extracted at the jth turbine bleed point is written S_j with

188 Regenerative Heating

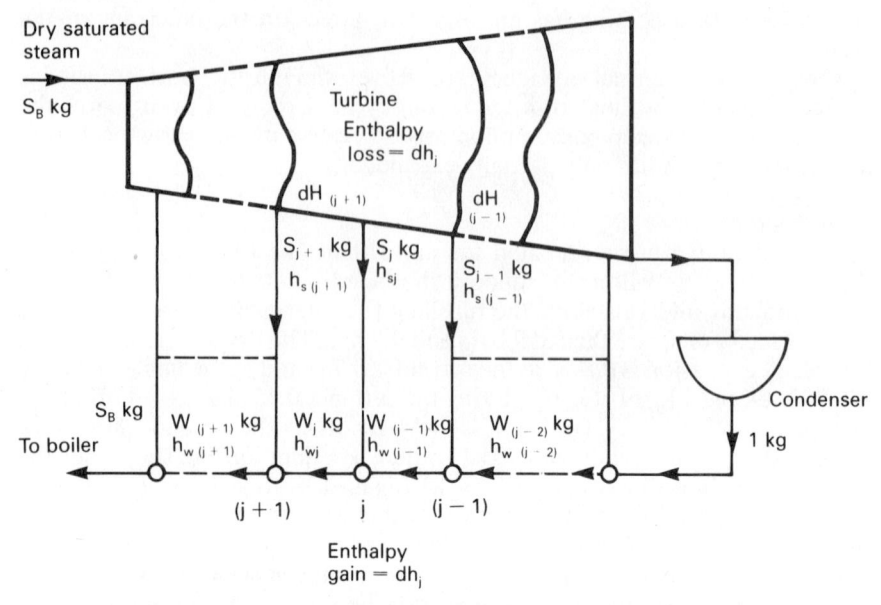

Fig. 13.8 Schematic representation of mass and enthalpy distributions in the turbine and feed heat line for three heaters taken as representative of the infinite set.

enthalpy h_{sj}. This is mixed with the feed water from the $(j-1)$th feed heater downstream (that is, nearer the condenser): let the mass of feed water from the $(j-1)$th heater be $W_{w(j-1)}$ and the specific enthalpy $h_{w(j-1)}$. The rise in the specific enthalpy on passing through the jth feed heater is $dh_j = h_{wj} - h_{w(j-1)}$. The corresponding loss of specific entropy by the steam in the turbine is $dH_j = h_{sj} - h_{wj}$. The energy balance between the steam and water sides requires the relation

$$S_j[h_{sj} - h_{wj}] = W_{(j-1)}[h_{wj} - h_{w(j-1)}] \tag{13.3}$$

to be valid. Also, because the mass of steam/water is constant

$$W_j = W_{(j-1)} + S_j \tag{13.4}$$

For convenience write

$$dh_j = h_{wj} - h_{w(j-1)} \quad \text{and} \quad dH_j = (h_{sj} - h_{wj}) \tag{13.5}$$

so that dh_j represents the gain of enthalpy by the feed water due to the action of the jth heater and dH_j represents the loss of enthalpy by the steam at the jth heater. Then it follows, by combining eqns. (13.3), (13.4) and (13.5), that

$$\frac{W_j}{W_{(j-1)}} = 1 + \frac{dh_j}{dH_j}$$

For the three heaters we can write correspondingly

$$\frac{W_{(j+1)}}{W_{(j-1)}} = \left(1 + \frac{dh_{(j+1)}}{dH_{(j+1)}}\right)\left(1 + \frac{dh_j}{dH_j}\right) \tag{13.6}$$

13.4 The Thermodynamics of the Bleed Point Locations

For the full set of n feed heaters we must sum all the contributions of the form (13.6) to obtain

$$S_B = \sum \left(1 + \frac{dh_j}{dH_j}\right) \tag{13.7}$$

This relates the mass flow into the turbine, for unit mass flow through the condenser, to the enthalpy changes at the various stages through the cycle.

Optimum Feed Heat Transfer
Suppose we now vary the position of the jth bleed point, leaving all the others constant with a view to maximising the steam flow S_B through the turbine. Using the expression (13.7) means that this will be done within the requirements of energy balance in the feed heat system. Because we can suppose that the effect of varying the jth heater will leave conditions upstream unaffected, we can take the conditions at the $(j-1)$th heater point as the datum for enthalpy. If K is the enthalpy rise between the $(j-1)$th and $(j+1)$th heaters, let k_j be the enthalpy difference between the $(j-1)$th and jth heater points. The enthalpy difference between the jth and $(j+1)$th heaters is then $dh_{(j+1)} = K - k_j$. We can write eqn. (13.7) in the form

$$S_B = A\left(1 + \frac{K - k_j}{dH_{(j+1)}}\right)\left(1 + \frac{k_j}{dH_j}\right) \tag{13.8}$$

where A accounts for the combined effect of all the remaining heaters which remain fixed. The enthalpy change between the steam and water sides will be approximately the same at each bleed point and we will suppose this to be the case now, so that we can set $dH_{(j+1)} = dH_j = dH$ (say).

To maximise the steam flow in the turbine we require

$$\frac{dS_B}{dk_j} = 0 \quad \text{and} \quad \frac{d^2 S_B}{dk_j^2} < 0 \tag{13.9}$$

Then, using eqn. (3.8)

$$\frac{dS_B}{dk_j} = A\left\{\left(-\frac{1}{dH}\right)\left(1 + \frac{k_j}{dH}\right) + \left(1 + \frac{K - k_j}{dH}\right)\left(\frac{1}{dH}\right)\right\}$$

$$= A(1/dH)^2[K - 2k_j] = 0 \tag{13.10}$$

This expression is zero if $k_j = K/2$, that is if the enthalpy difference between the heater points $(j+1)$ and j is the same as the enthalpy difference between the heater points j and $(j-1)$. The same argument can be applied to each bleed point in the system and leads to the approximate rule that the bleed points should be distributed such that the enthalpy difference between any consecutive pair is the same for all pairs. Because the enthalpy difference is related to the corresponding temperature difference, the bleed points can alternatively be distributed such that the temperature differences in the turbine between consecutive pairs is the same for all such pairs. This result substantiates the criterion used in Section 13.4 and is shown in practice in Question 13.1.

To show that the steam flow is a maximum under these conditions we must follow the requirements of eqn. (13.9) and find the second derivative of S_B

190 Regenerative Heating

with respect to k_j to confirm that this is a negative quantity. Explicitly, differentiating eqn. (13.10) with respect to k_j gives

$$\frac{d^2 S_B}{dk_j^2} = -2(1/dH)^2$$

This is negative because $(1/dH)^2$ is always positive. S_B indeed has a maximum value with respect to variations of k_j.

13.5 Feed Heating Using a Cascade

The analysis of the last Section has substantiated the heuristic procedure used in Section 13.3 for the selection of bleed points along the turbine expansion line. The use of open feed heaters is expensive on feed pumps since each line will need to have one. Although we have agreed to neglect the pump energy in the calculations this is not necessarily an acceptable approximation when many feed heaters are employed. A method which avoids this difficulty is the cascade method, using closed feed heaters for the heat transfer between the bled steam and the feed water line. One advantage is a substantial reduction in the number of feed pumps necessary to service the heater lines: only one is now involved.

A feed heat system involving two closed feed heaters in a cascade

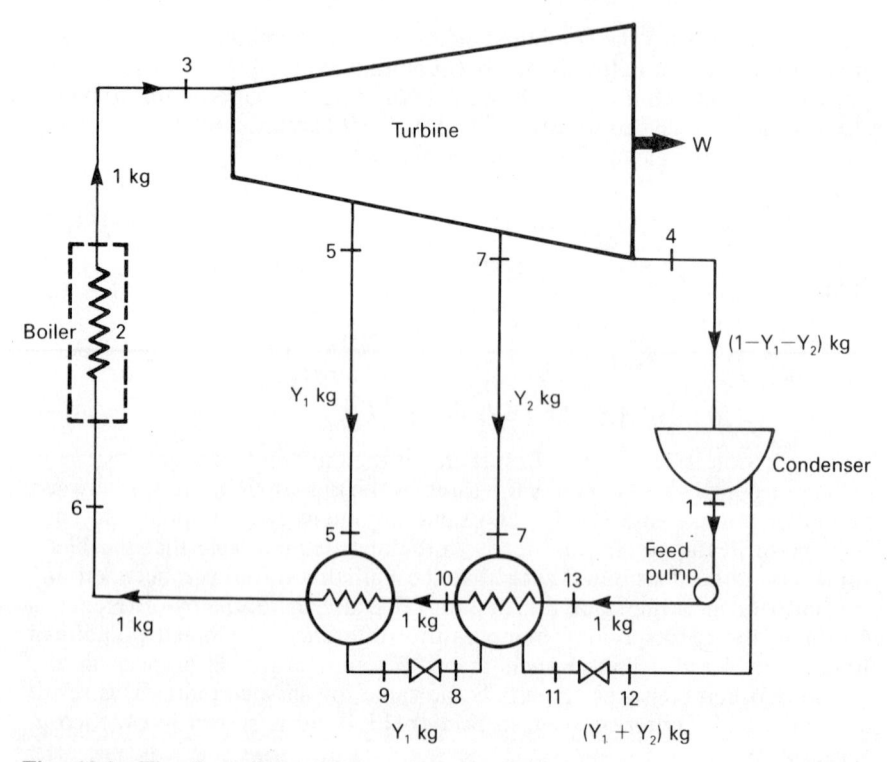

Fig. 13.9 The flow diagram for a non-superheated cycle with two feed heaters cascade, including the associated throttling points.

13.5 Feed Heating Using a Cascade

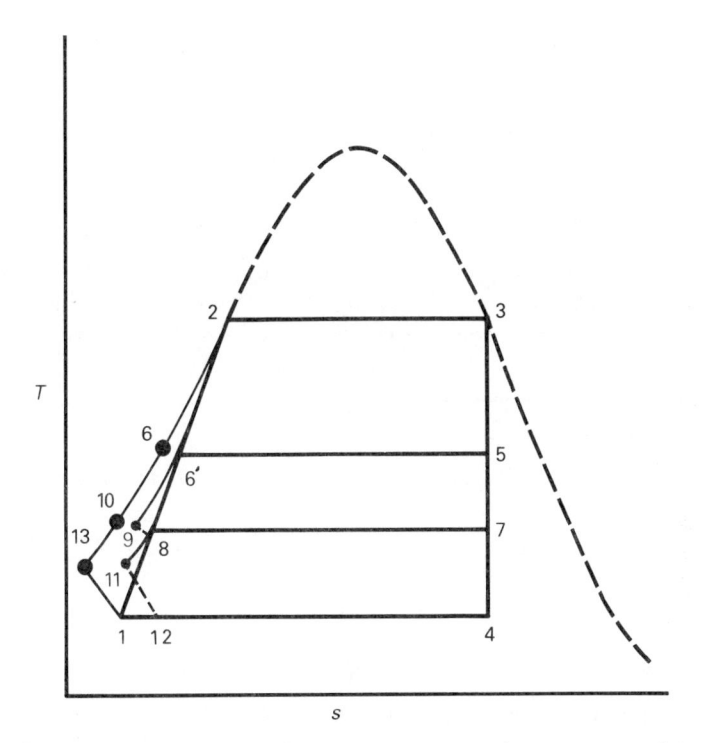

Fig. 13.10 The temperature-specific entropy diagram for a non-superheated cycle with two feed heaters in cascade.

arrangement is shown in Fig. 13.9. The associated T–s diagram is shown in Fig. 13.10. It will be seen that the circulation of working fluid can be achieved now with the use of one feed pump, at least in principle. All the bled steam is condensed and passed through the condenser rather than joining the feed water line directly at the point of heat exchange. The use of closed heaters presents practical problems which we will avoid now. The heat exchange linkage will be strong only if the heat exchangers have a large volume and throttle points (two are shown in Fig. 13.9) will be needed to achieve the appropriate steam/water conditions in the bleed lines. These issues will not be considered here.

Rather we will make two assumptions:

(a) that the throttling of the condensed bled steam brings the fluid close to the saturation line in Fig. 13.9, which means that $h_8 = h_9 = h_{10} = h_f(p_7)$, which is the value the enthalpy of the liquid corresponding to the pressure p_7 of the bleed line 7;
(b) that the enthalpy in the compressed liquid feed heat line is closely the same as that for the saturated liquid at the same temperature, so that is $h_9 = h_8$.

We shall also neglect the work of pumps and suppose the expansions are fully isentropic with no other losses in the system. This means setting the feed pump work $W_{1,13} = 0$ and supposing $h_1 = h_{11} = h_{12} = h_{13}$.

192   Regenerative Heating

With these assumptions we can form the following data table, using the steam tables and the results of Section 13.3.

$h_1 = 121$ kJ/kg $\qquad h_3 = 2801$ kJ/kg
$s_3 = 6.070$ kJ/kg K $\qquad h_5 = 2527$ kJ/kg
$h_7 = 2181$ kJ/kg $\qquad h_4 = 1828$ kJ/kg
$h_8 = 405$ kJ/kg $\qquad h_6 = 743$ kJ/kg

The heat exchange at the two heaters gives relations involving Y_1 and Y_2, the quantities of steam to be extracted from the turbine flow. Remember that all the fluid is now condensed and returns to the boiler along the feed water line. Then we have

$$-h_6 + Y_1 h_5 - Y_1 h_8 + h_8 = 0$$
$$-h_8 + Y_2 h_7 + h_8 Y_1 - (Y_1 + Y_2) h_1 + h_1 = 0$$

These relations lead to the values $Y_1 = 0.159$ and $Y_2 = 0.116$, the same values as for the open feed heaters of Section 13.3. The work transfer is found to be 821 kJ/kg and the heat transfer to be 2058 kJ/kg, as for the open heater system of Section 13.3. The thermal efficiency is then $\eta = 0.399$ and the specific steam consumption is $ssc = 4.386$, as before.

13.6   Feed Heating in a Superheated System

The introduction of superheat into the cycle raises the mean receptor temperature but does this at the expense of an increase of the range of temperature for heat reception. The principles of feed heating can be applied to a superheated system to reduce the range of heat reception but its introduction raises a new problem.

The Thermodynamics for Expanding the Bled Steam

Superheating introduces the problem of mixing superheated steam from the turbine bleed points (at least at the high pressure end) with condensed water along the feed line while maintaining thermodynamically reversible conditions. Consider one bleed point as an example, with a *T–s* diagram given in Fig. 13.11.

A quantity of steam which has been expanded to some point along the turbine is extracted to mix with feed water that has arrived from the next feed heater point downstream. The temperature of the bled steam is required to be T_s (say) represented by the line FF' in Fig. 13.11. The dry saturation point for this temperature is point F, so it is necessary to adjust the bled steam to the thermodynamic point F on the *T–s* diagram if the heat transfer with the feed water is to be reversible. The expansion line for the steam is AA' and the appropriate temperature would be achieved at the point F' but unfortunately the entropy difference between the points F and F' is not infinitesimal: the heat transfer between bled steam and feed water cannot occur reversibly at that point. It is necessary, then, to bring the steam to the true thermodynamic conditions represented by the steam saturation point F.

This requires a number of steps. First, the expansion line is stopped at the point F' which is on the constant pressure line p_s and has the temperature T_s which has been chosen. The steam temperature is now reduced by an

13.6 Feed Heating in a Superheated System 193

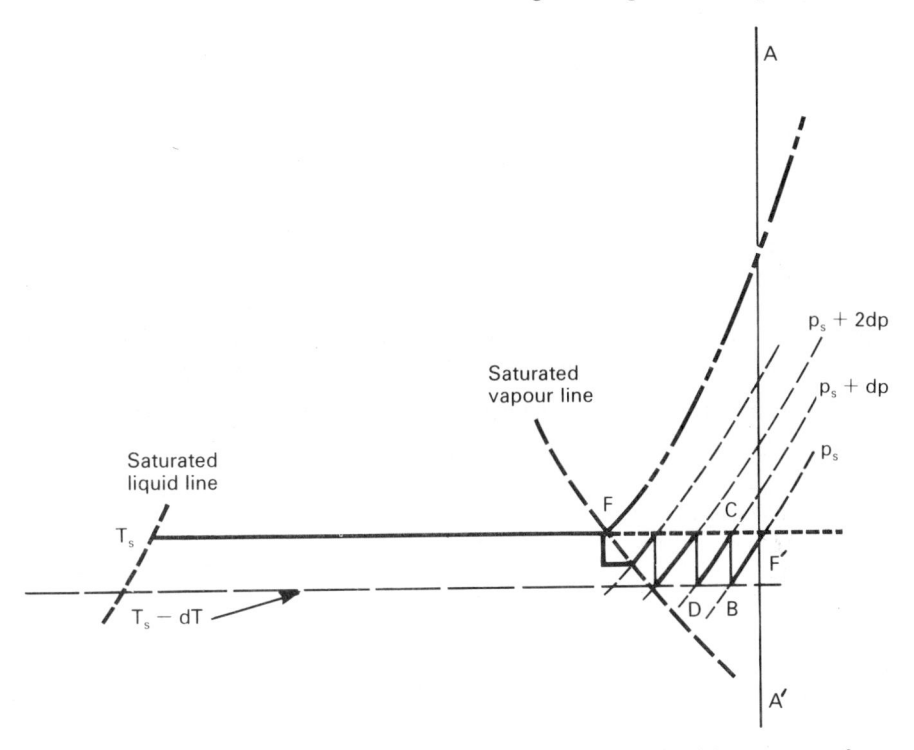

Fig. 13.11 Showing the thermodynamic details of the ideal feed heater stage for a cycle using superheated steam.

infinitesimal amount dT by cooling at constant pressure along the line F'B to the temperature $T_s - dT$ (point B). The steam is still superheated and the amount of superheat is reduced by first heating through the temperature range dT at constant entropy (line BC) to achieve the higher pressure $p_s + dp$ at point C, and then cooling along the constant pressure line CD, reaching again the temperature $T_s - dT$ at point D. This process of elementary cooling and isentropically raising the pressure is continued until the point F is ultimately reached, as shown in Fig. 13.11. The saw-toothed path shown there will be reduced to a smooth line as the variations of temperature, entropy and pressure are made vanishingly small, that is as the number of steps is increased indefinitely. The adjustment of the conditions of the bled steam to those of the feed water become thermodynamically reversible in this limit.

To achieve this in practice (even approximately) requires the use of a variable pressure direct contact heater that has been constructed to introduce both the isentropic and isobaric changes required: but the design of the feed heater will not be considered here. While such closed variable pressure feed heaters will certainly be necessary for bleed points in the high pressure turbine region, simple open heaters are often sufficient when the expansion line from the turbine crosses the dry saturation line. Consequently, a practical system with several feed heaters will be a mixture of types, the closed heater controlling the bleed conditions at the high pressure end and open heaters controlling the conditions near the exit to the turbine. It will, however, not be

194 Regenerative Heating

necessary in our calculations to draw a distinction between the two heater types.

The Effect of Including Superheat

To see the effect of including superheating in the feed heat system consider again the cycle with a boiler pressure of 40 bar with superheating to 500 °C, the final expansion being to 0.04 bar in the condenser. One open feed heater is included in the cycle. The associated *T–s* diagram is shown in Fig. 13.12. This extends the case considered in Section 13.3 and is related also to the system considered in Section 12.4.

The bleed point pressure h_6 will be chosen so that $(h_2 - h_6) = (h_6 - h_1)$. From the steam tables we find

$$h_1 = 121 \text{ kJ/kg} \quad h_2 = 1087 \text{ kJ/kg} \quad h_3 = 3214 \text{ kJ/kg} \quad s_3 = 6.769 \text{ kJ/kg K}$$

Then, $[h_2 + h_1]/2 = 604$ kJ/kg and this corresponds very closely to the pressure $p_5 = 4.0$ bar: the bleed pressure is then taken as 4 bar for which $h_6 = 605$ kJ/kg. Then from the Mollier diagram we find $h_5 = 2819$ kJ/kg. The remaining steam in the turbine is expanded to the exit so that

$$s_3 = s_4 = 6.769 = 0.422 + 8.051 x_4$$

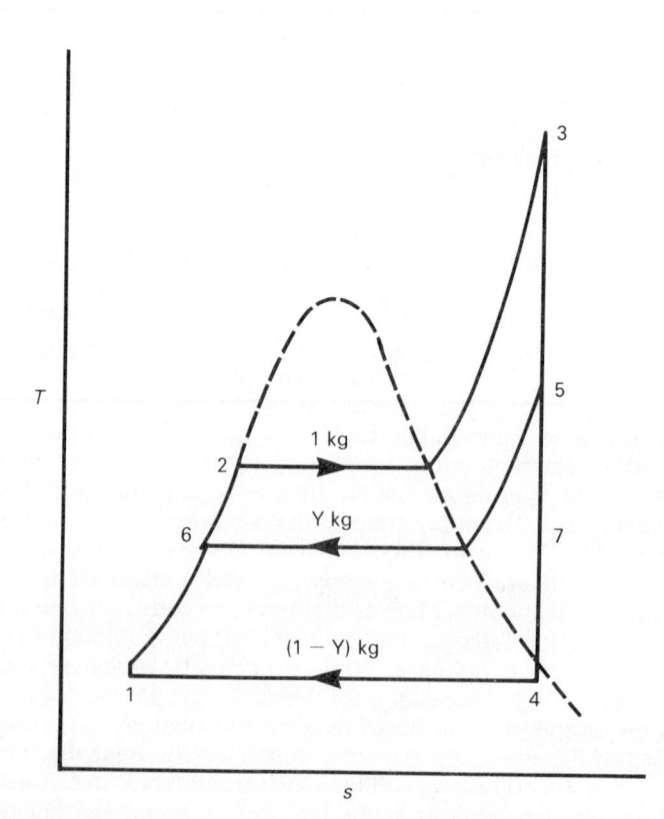

Fig. 13.12 The temperature-specific entropy diagram for a superheated cycle with one feed heater.

The final dryness fraction is then $x_4 = 0.788$ and

$$h_4 = 121 + 0.788 \times 2433 = 2039 \text{ kJ/kg}$$

The mass of water Y through the feed heater is found from the energy balance expression

$$Y = \frac{(h_6 - h_1)}{(h_5 - h_1)} = \frac{(605 - 121)}{(2819 - 121)} = \frac{484}{2698} = 0.1794$$

The work transfer $= (h_3 - h_5) + (1 - Y)(h_5 - h_4)$

$$= (3214 - 2819) + 0.8206(2819 - 2039) = 1068 \text{ kJ/kg}$$

the heat transfer $= (h_3 - h_6) = (3214 - 605) = 2609 \text{ kJ/kg}$

The thermal efficiency and specific steam consumption are

$$\eta = \frac{1068}{2609} = 0.409 \quad \text{and} \quad ssc = \frac{3600}{1068} = 3.371 \text{ kg/kWh}$$

Comparison with the results for superheating alone (Section 12.2.4, $\eta = 0.354$, $ssc = 3.055$ kg/kWh) shows that the single feed system has increased the thermal efficiency by rather more than 15 % but at the expense of increasing the specific steam consumption by some 10 %.

The comparison of feed heating with and without superheating (specifically for the pressure range 40 bar to 0.04 bar) can be made by remembering the results of Section 12.6 where we found that, with superheat to 500 °C, $\eta = 0.399$ and $ssc = 4.386$ kg/kWh; with superheat to 500 °C, $\eta = 0.409$ and $ssc = 3.371$ kg/kWh. Superheat has raised the thermal efficiency by about 30 % and has reduced the specific steam consumption by more than 20 %. The dryness fraction is also increased by superheating, although it is still low for practical purposes.

The effect of increasing boiler pressure is to increase the thermal efficiency but also to increase the specific steam consumption. Suppose the boiler pressure is increased to 100 bar, with one feed heater and superheat to 500 °C. The bleed point pressure will now be at 10 bar to provide the best thermal conditions. The thermodynamic flow and state is shown in Fig. 13.13(a), where the work transfer $W = 1194$ kJ/kg and the heat transfer $Q = 2610$ kJ/kg. This gives the thermal efficiency $\eta = 0.457$ and specific steam consumption $ssc = 3.015$ kg/kWh.

The addition of a further feed heat stage increases the thermal efficiency further. For a two feed heat cycle, again with superheat to 500 °C, the bleed pressure are now chosen to be 26 bar and 2.8 bar. The various thermodynamic and flow data are shown in Fig. 13.13(b). The work output $W = 1130$ kJ/kg and the heat input $Q = 2401$ kJ/kg, giving a thermal efficiency $\eta = 0.471$ and a specific steam consumption $ssc = 3.186$ kg/kWh.

13.7 Summary

1. Regenerative feed heating in the cycle, in which steam bled from the expansion turbine is used to heat feed water going to the boiler from the condenser, allows the range of temperature associated with heat reception in the simple Rankine cycle to be decreased.

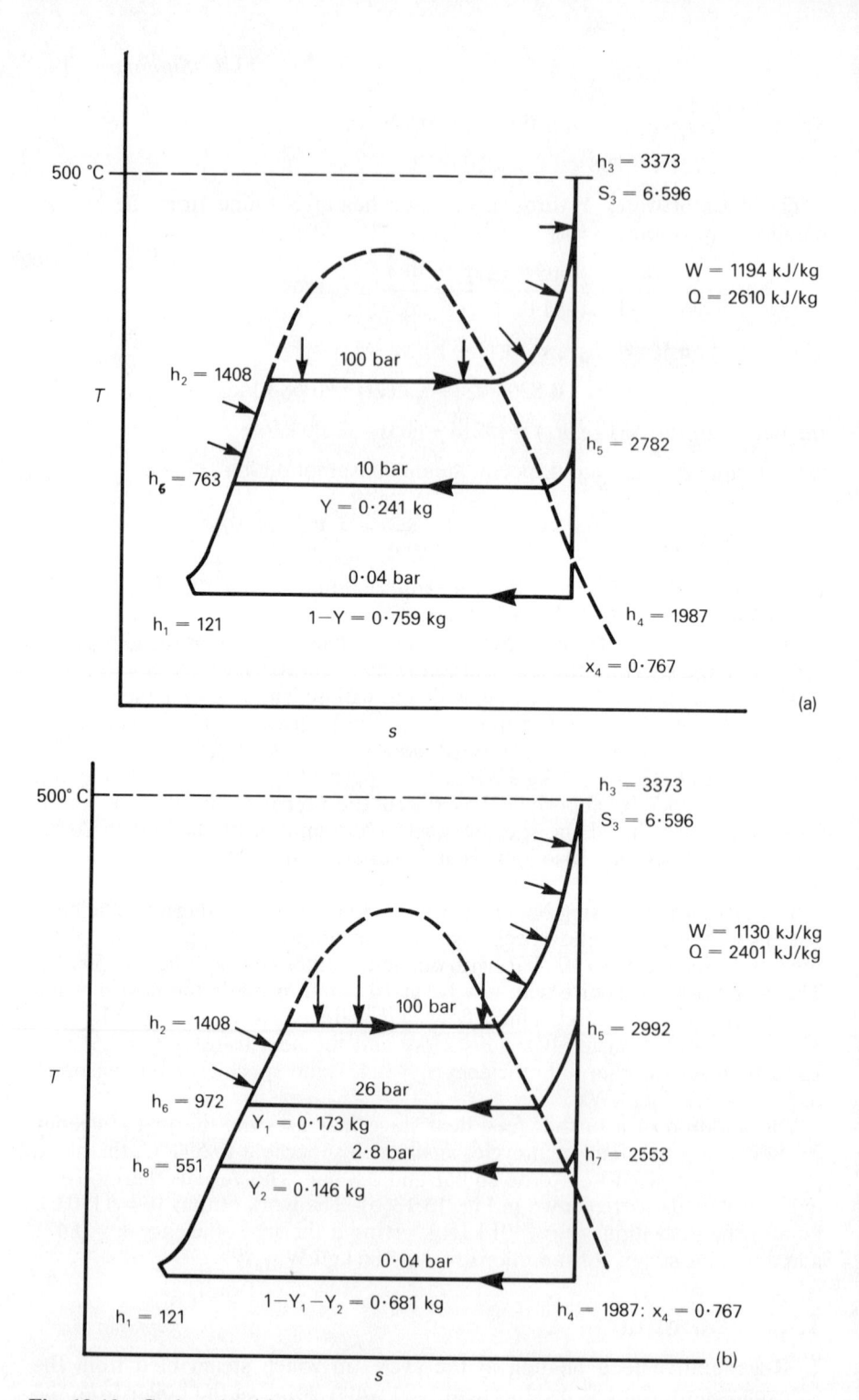

Fig. 13.13 Cycles with (a) one and (b) two feed heat stages. h is in units of kJ/kg; s is in units of kJ/kg K.

2. For a Rankine cycle without superheat the procedure is straight forward and the heat reception can be made essentially isothermal in principle and in practice to a good approximation.
3. For a cycle with superheat the thermodynamic conditions for the bled steam must be adjusted to allow the transfer of heat from the steam to the feed water to take place reversibly. The process necessary to achieve this is considered.
4. The introduction of a feed heat system into the cycle increases the thermal efficiency of the cycle but also increases the specific steam consumption.
5. Because steam is extracted at points along the turbine, the mass flow of steam through the turbine is reduced along the expansion line.
6. Even an approximate representation of thermodynamic equilibrium will require the use of more than one feed heater in the cycle: in modern steam plant as many as eight may be used.
7. Open and closed feed heater arrangements are recognised as necessary to cover the full range of superheat conditions in practice and a system with several feed heaters will usually contain a mixture of these two heater types, closed near the entrance to the turbine and open near the exit.
8. The importance of an availability function $B = h - T_C s$ has been recognised, where T_C is the condenser temperature. This relates to the useful work that can be extracted from a cycle for given conditions of the cold reservoir.

13.8 Exercises

13.1 A Rankine cycle with boiler pressure 40 bar and condenser pressure 0.04 bar has one stage of feed heat at the turbine bleed temperature T. There is no superheat or reheat in the system. The turbine expansion and boiler compression have ideal isentropic efficiencies ($y = z = 1$) and there are no other losses in the system.

Calculate the thermal efficiency and specific steam consumption for the different values of the bleed temperature $T = 80\,°C$, $100\,°C$, $160\,°C$ and $200\,°C$. Plot graphs to show the dependence of the thermal efficiency and the specific steam consumption on the intermediate (bleed) pressure and substantiate the statement of Section 13.2 of the main text that the best thermal efficiency is found when the intermediate pressure is about 4 bar, with a temperature of about 145 °C.

13.2 A Rankine cycle works between the pressures of 40 bar and 0.04 bar. There is no superheat or reheat, but there are open feed heaters in the circuit. Calculate the dryness fraction, the thermal efficiency and the specific steam consumption for the cases of three, four and six feed heater stages placed for optimum cycle efficiency. In each case the turbine can be supposed to have ideal efficiency ($y = 1$) and the compression work transfer can be neglected.

Plot graphs to show the dependence of the thermal efficiency and specific steam consumption of the system on the number of feed heaters used. What conclusions about the possible practical working of the system can be drawn from the graphs you have constructed? Mention particularly the effect on the dryness fraction for the steam after full expansion in the turbine.

198 Regenerative Heating

13.3 Repeat the calculations of Question 13.2 for isentropic efficiencies for expansions of (a) 0.95, (b) 0.90, (c) 0.85 and (d) 0.80. In each case the compression work can be assumed to be sufficiently small to neglect. Display your results in graph form.

What conclusions can be drawn about the practical efficiency acceptable in a particular case if the cost of the turbine rises as the square of its efficiency value?

13.4 The compression work transfer has been neglected in Questions 13.1, 13.2 and 13.4. Calculate this work transfer for each case and show that it is indeed negligible by comparison with the expansion work transfer.

13.5 A Rankine system works between the pressures 40 bar and 0.04 bar. Superheat is added to 600 °C but there is no reheat. There is one feed heat stage. Calculate the dryness fraction of the steam on full expansion through the turbine, the thermal efficiency and specific steam consumption of the cycle. The compression work transfer can be neglected and the turbine can be assigned ideal efficiency ($y = 1$). Compare your results with those for Question 13.2.

13.6 Repeat the analyses of Question 13.5 for a higher pressure of (a) 60 bar, (b) 100 bar and (c) 150 bar. Construct a graph to show the effect of boiler pressure on the working of the cycle.

13.7 The Rankine system works between the pressure range 150 bar and 0.04 bar with superheat to 600 °C. There are (a) two, and (b) three feed heat stages added. Calculate the effect of the further feed heat stages on the working characteristics of the cycle. The expansion work transfer can be supposed to be perfect and the compression work transfer can be neglected: there are no further losses in the system.

13.8 Repeat the calculations of Question 13.7 for (a) four, (b) six, (c) eight and (d) ten feed heat stages. Use your calculations to comment on the statement that modern power plant design often favours eight feed heaters for an optimum design.

13.9 It is suggested that the thermodynamics of a cycle with feed heat would be improved if a final feed heat line were added at the turbine entrance carrying steam from the boiler directly to the feed line, as shown in Fig. 13.14. In this way the feed water could be raised to the saturation liquid temperature corresponding to the boiler pressure involved and so reduce the required heat input. Show that the thermodynamics of the cycle would be unaffected by this additional feed line, the effect being simply to reduce the mass of working fluid in the cycle for a given boiler steam output. (Additional heat input to the feed water can be made in a practical plant by extracting enthalpy from the boiler flue gases using an economiser: this is explained in Chapter 14, Section 14.2.1.)

A cycle with no superheat and no reheat has a boiler pressure 100 bar and a condenser pressure 0.04 bar. There is a single feed heat line at 4 bar. Show that the thermal efficiency 0.432 and the specific steam consumption is 3.934

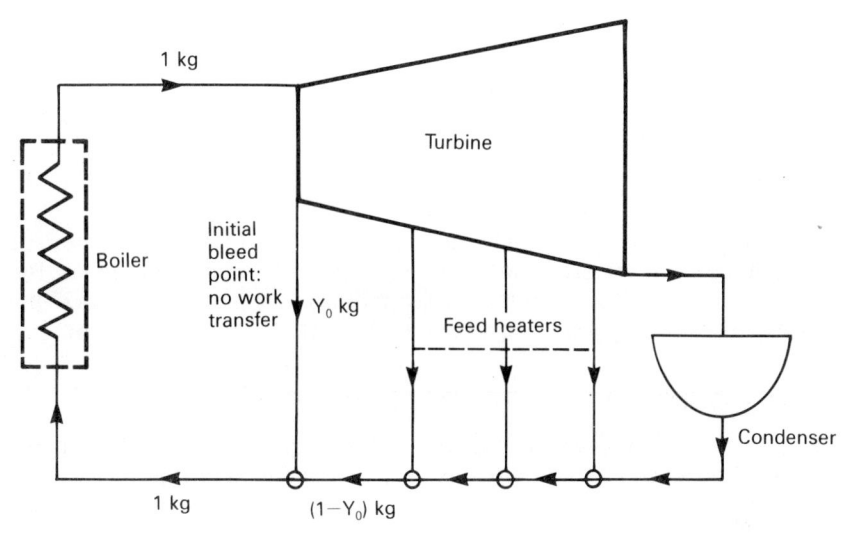

Fig. 13.14 The flow diagram for a feed heat system, the initial bleed point extracting steam from the turbine enetrance before work transfer occurs.

kg/kWh, with a feedline flow of $Y_1 = 0.232$ kg and a final dryness fraction $x = 0.645$. A further feed line is added at the turbine entrance (so the bled steam in this feed line has done no work). Show that the thermal efficiency and dryness fraction are unchanged but that the new $ssc = 6.332$ kg/kWh. Explain this result in terms of the T–s diagram.

13.10 The availability function B was defined in the text (Section 13.6.2) as $B = h - T_1 S$, where T_1 is some lower datum temperature for the system. The Second Law of Thermodynamics states that the entropy of a closed system either remains unchanged during a reversible change or increases if the change is not reversible. Substantiate the statement that, while the total energy of a system will remain constant during a thermodynamic change (First Law), the availability will either remain constant (reversible change) or will decrease (irreversible change). In the real world, therefore, where a change of state is associated with an increase of entropy, the change will reduce the availability for repeating the process. Consider the value of this statement as an alternative expression of the Second Law. (It was such ideas as this that, in the 19th Century, led Clausius to introduce the concept of entropy.)

14. THE FULL STEAM SYSTEM

Various modifications designed to improve the thermodynamic characteristics of the simple Rankine cycle have been considered separately in the last four chapters: it is now time to collect these analyses together and consider their combined effects in practical terms for application to a full steam plant.

14.1 Comments on the Simple Rankine Cycle

The application of the simple Rankine cycle to practical power production presents contradictory requirements. The higher thermal efficiencies (from Fig. 11.3 this means a little in excess of 0.4) and low specific steam consumption (some 3 kg/kWh) require a boiler pressure in the range 100–180 bar (pressure ratio about 4000). The dryness fraction then hardly exceeds 0.70. The higher dryness fractions (in excess of 0.88) on the other hand, require pressure ratios less than 10, and often less than 5 (see Question 11.10) high efficiency, low specific steam consumption and high dryness fraction taken separately require different cycle conditions.

There is the further disadvantage that the water must undergo a phase change (liquid/vapour) during the cycle and the latent heat of vaporisation is relatively high. Here heat is added at constant temperature. Indeed, the heat transfer for the liquid–steam evaporation region is less than that required to reach the boiling temperature only for boiler pressures above about 100 bar. Although such pressures are attractive on this count the dryness fraction falls and will be as low as 0.495 at the critical point pressure when the latent heat of vaporisation vanishes.

The dryness fraction is increased to some extent by the introduction of superheat, but substantial improvement can only be made by incorporating a reheat stage (see Chapter 12). The thermal efficiency is also improved by superheat (it may be either increased or decreased by reheat depending on the circumstances) but the major influence on the thermal efficiency is feed heating. The quantity of heat received by the cycle is determined by the temperature of the water leaving the last feed heater (but see Section 14.2).

14.2 Using the Flue-Gas Availability

Up to this point nothing has been said about the combustion process that heats the water. This is, in reality, an integral part of the plant dynamics and its working features are designed to tailor the requirements of heat reception in a particular case.

A work cycle has a characteristic lower limit for the temperature of heat reception and the furnace gases cannot be rejected at a lower temperature. This temperature becomes higher as the heat reception approaches isothermal: as an example, for the ideal case of the Carnot cycle (where the heat reception is at a single temperature) the temperature of waste gases would

14.2 Using the Flue-Gas Availability

need to be at least equal to this highest temperature of the cycle. The Rankine cycle is less restrictive since it receives heat over a wide temperature range. It is realised that the efficiency with which the fuel can be utilised is strongly dependent on the working characteristics of the cycle which it is to drive. The design of the boiler unit plays a vital role in determining the overall working thermal efficiency of the complete plant.

The boiler tubes are contiguous with the furnace in a way that depends in detail upon whether the fuel is coal or oil: in the former case the coal will be pulverised and so injected into the furnace as a 'fluid' with the air needed for combustion. The combustion region itself is held suspended in position by the entry fuel/air flow, the ash falling away and the hot gas drawn up the flue by convective flow. For a hot gas discharging to the atmosphere the temperature difference between the furnace and the atmosphere would in principle supply the buoyancy force necessary to produce the flue flow, but such a condition would reject an uneconomical amount of heat to the environment. In practice, at the actual rejection temperature of the gases, such natural convection is usually insufficient and appropriate pumps provide a forced convection to drive the flue gases through the chimney.

The temperature distribution within the furnace/flue system is obviously not uniform, and it is necessary to arrange the boiler tubing so that those regions that require the highest temperatures are in the hottest parts of the furnace. Ideally, the most effective thermodynamic use of the fuel is achieved when all the enthalpy of combustion has been transferred to the water so that the full availability provided by the fuel has been used. The flue gases which finally leave the boiler system are then close to atmospheric temperature. This is possible up to a few degrees in principle for the simple Rankine cycle which accepts heat over the full temperature range from the condenser to the boiler pressures. But the actual temperature difference, determined by the condenser pressure, has a lower limit of some 10 °C above that of the environment set by the rate at which heat must be rejected by the cycle itself. This limit is raised substantially when feed heating is used because then the lowest temperature for heat reception into the cycle is raised, in practice close to that for vaporisation. The lowest temperature that the flue gases can have on leaving the steam tube system is then relatively high. Because feed heating is an important component of modern power plant this reduction of overall plant efficiency can be serious.

This is alleviated to some extent because the temperature of the flue gases must be above the lowest dew point of the product gases of combustion. This limit is set by the composition of the rejected gases and so depends on the conditions of combustion in the furnace (see Section 14.5).

Two particular mechanisms allow a proportion of the residual availability of the flue gases to be utilised, increasing the overall thermal efficiency of the plant.

The Economiser

It was seen in Chapter 13 that the temperature of the feed water in a regenerative system can be raised arbitrarily close to the corresponding vaporisation temperature by increasing the number of bleed points, but in practice there are limits (financial and constructional) to the number of such points. It might be tempting to add a final bleed point taking a quantity of

202 The Full Steam System

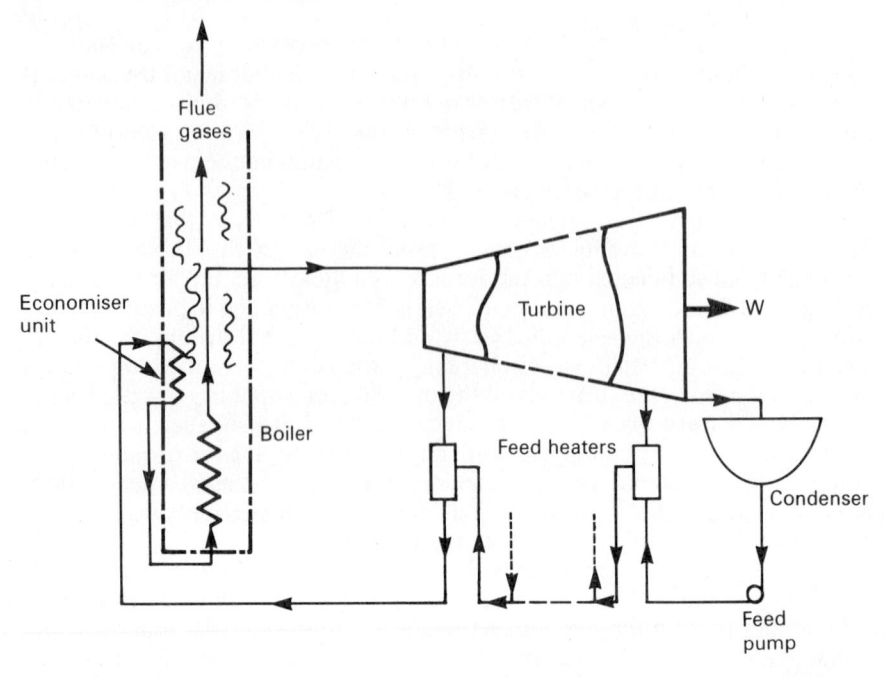

Fig. 14.1 Showing the fluid circuit for a cycle with feed heating and an economiser.

steam directly from the boiler before even reaching the turbine inlet but this will not, in fact, achieve a thermodynamic improvement of the cycle (see Question 13.8). If the system was ideal this would be the end of the possible improvement but this is not so in an imperfect world. The flue gases will still contain enthalpy above the lowest level of the cycle even after leaving the superheat and reheat sections, and some of this now waste enthalpy can be transferred to the feed chain as an energy bonus.

After leaving the turbine section but before entering the boiler, the feed heat tubes are passed through the flue gases for a final heat transfer (see Fig. 14.1) to raise the feed temperature. This system of tubes is called the economiser. The heat reception by the economiser is independent of the cycle heat transfer and so increases the effective thermal efficiency of the cycle. There is no effect on the specific steam consumption of the cycle. Although the economiser can properly be considered as part of the feed heat system it could be introduced even when no feed heaters are included. If there is no feed heating the temperature of the water entering the economiser will be low and there can be no pretence at achieving any level of thermodynamic efficiency for the heat transfer.

The Effectiveness of the Economiser/Feed Heat System
The economiser acts effectively as the final (high temperature) stage of the feed heat system although it does not involve bled steam and provides no work output. Where the economiser is included as part of the feed heat system, the distribution of n bleed points down the turbine will then portray a constant temperature interval starting from the economiser, there being

effectively ($n+1$) heaters in the line. The temperature range covers that from boiler to condenser.

Suppose there are n regenerative heater stages and the difference of the specific enthalpy between two consecutive bleed points is dh: the total enthalpy difference between the boiler and the condenser is dH. The ratio

$$F = \frac{\text{enthalpy change across the turbine}}{\text{enthalpy change from boiler to condenser}}$$

is called the effectiveness of the feed heat system. For the n bleed points of the feed system $F = n\text{d}h/\text{d}H$: including the economiser d$H = (n+1)$dh. Then $F = n/(n+1) = 1/(1+1/n)$. As $n \to \infty$, that is the number of feed heat points becomes indefinitely large, $F \to 1$ and the feed heat line accepts the maximum increment of enthalpy between the condenser and the boiler: the feed water is now delivered to the boiler at the vaporisation temperature corresponding to the particular boiler pressure used. For a finite number of heaters, $F < 1$ (as we would expect), and the enthalpy increment in the feed water is incomplete. As an example, for $n = 10$, $F = 0.909$: nearly 91% of the theoretical maximum enthalpy transfer to the feed water is achieved with 10 feed heaters plus an economiser. The feed system is expensive to build and the fewer heaters the better from the constructional point of view. If only eight feed points are included with an economiser, $F = 0.889$ and this is usually accepted as satisfactory in practice. It is surprising to notice that even $n = 1$ gives $F = 1/2$, showing the importance of regenerative heating in the energy budget of the cycle.

The Air Pre-Heater

Even after the economiser has extracted enthalpy from the flue gases the temperature is usually well above the dew points of the constituents so more enthalpy can be extracted from the gases. This is achieved by pre-heating the air entering the boiler using an air pre-heater.

Three conditions are necessary for the combustion of hydrocarbon fuels: (i) ignitable fuel; (ii) an appropriate quantity of air to form a stochiometric fuel/air mixture and (iii) a temperature (the combustion temperature) sufficient for the mixture to ignite. The combustion temperature for coal and oil is of the order of 400 °C and the air entering from the atmosphere is only at about 20 °C, some 5% of this temperature. However, the need to heat the air during the combustion process itself can be avoided if it is pre-heated in the furnace flue during its passage to the combustion chamber. This is done by passing the air through a bank of tubes in the upper part of the flue, where the flue gases have their lowest temperature. By raising the air to a high temperature before entering the combustion region an efficient combustion process is ensured without having to supply heat to maintain the combustion temperature.

14.3 The Total Plant: Overall Plant Efficiency

A modern steam plant is a very complicated construction with many components, of which the steady flow steam cycle is but one. This is shown schematically in Fig. 14.2. In assessing the overall efficiency of the system the

204 *The Full Steam System*

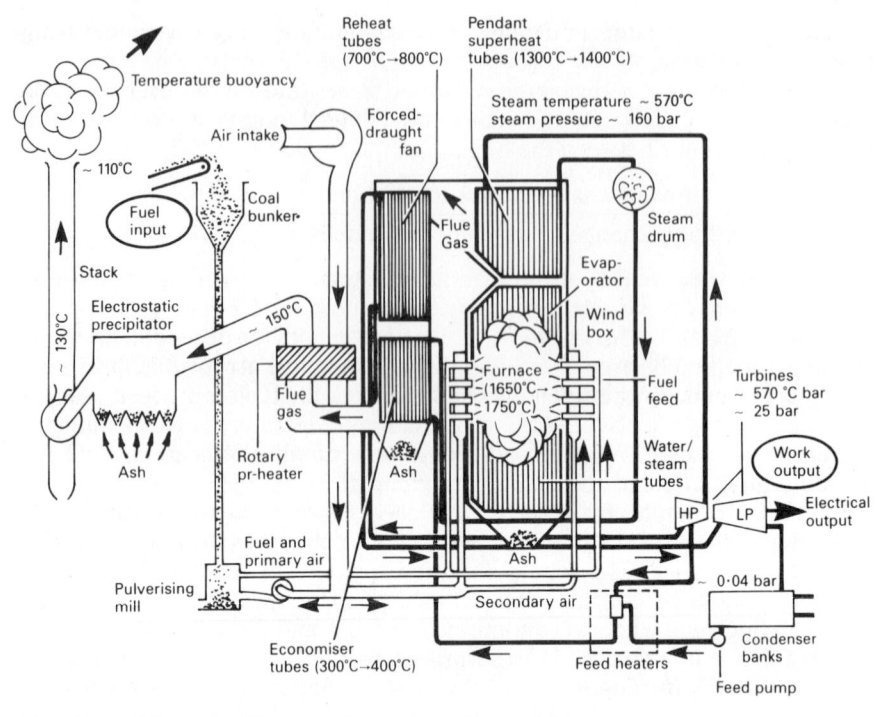

Fig. 14.2 Schematic diagram of a modern steam plant.

efficiency of the components other than the cycle must also be accounted for. We consider now in broad terms some of the criteria that are significant. We cannot quantify the analysis for two reasons: (i) this would involve a wide range of optimisation studies (best done using a computer) in which the various parameters are varied to achieve the most acceptable results, and we are here concerned with background principles, and (ii) it would be necessary to include the effects of the losses and inefficiencies of an actual plant for realistic results, and we are not concerned here with a specific case. Some possible plant designs are, however, considered in the Exercises that follows, to which the reader is referred. We might remember that the effective efficiency of the plant is to be measured ultimately in commercial terms based on the net profit from the energy supplied to the customer over two or three decades.

The Steam Cycle
We need the maximum thermal efficiency, lowest specific steam consumption and driest steam conditions on entering (and leaving) the turbine, consistent with the capacity of the heat source. For hydrocarbon fuels we can suppose the maximum temperature of the cycle to be set by the metallurgical properties of the components (particularly the steam tubes and turbine blades), both as regards temperature and pressure. This would not usually be the case for nuclear heat sources (see Chapter 16). Conditions will be chosen to be sub-critical (we will not consider the problem of the initial design

14.3 The Total Plant: Overall Plant Efficiency 205

specification, which includes thermodynamic considerations dominated by financial considerations and the availability of manufacturing skills).

Working Characteristics (see also Table 14.2)

The boiler pressure is chosen to be about 160 bar and the condenser pressure 0.04 bar. The accompanying low dryness fraction is partly overcome by introducing superheat (see Chapter 12). Moving near to the metallurgical limit, the superheat temperature will be taken as 500 °C. The maximum possible thermal efficiency requires regenerative heating (see Chapter 13) and eight feed heaters with an economiser (see Section 14.2) will be used. British practice has been to use one stage of reheat (see Chapter 11) to increase the dryness fraction (although some American designs use two stages).

The maximum efficiency is achieved when the feed bleed points are set along the turbine with approximately equal enthalpy differences between them (see Section 13.4), and this means approximately equal temperature differences as well. For the present case, with the highest pressure being 160 bar (347 °C dry saturated vapour), turbine bleed points could be taken at the pressures 110 bar (318 °C), 70 bar (286 °C), 44 bar (256 °C), 24 bar (222 °C), 11 bar (184 °C), 4.5 bar (148 °C), 1.3 bar (107 °C) and 0.07 bar (39 °C). The entropy on entering the turbine is 6.301 kJ/kg K: this corresponds to the dry saturation vapour temperature of steam at about 24 bar pressure (that is, about 1/7 of the boiler pressure). The single reheat could, therefore, be introduced immediately after the bleed point associated with this pressure although a lower reheat pressure (perhaps below 20 bar) might prove more effective. This question can be answered finally only by repeating the calculations for several reheat pressures.

The characteristics of this cycle are worked through by combining the arguments of Chapters 12 and 13. Without an economiser, the heat input is 1845 kJ/kg and the total turbine work produced is 962.7 kJ/kg. The corresponding thermal efficiency is 0.525 and the specific steam consumption is 3.74 kg/kJ. About 34 % of the input heat budget is regeneratively used to heat the feed water. The dryness fraction on leaving the turbine is 0.915: this is sufficiently high not to require a further reheat stage. Of the 1 kg of steam entering the turbine, 0.525 kg is expanded completely through to the exit, the remaining 0.475 kg being bled off at various stages during the feed heating. The volume flow leaving the turbine is then only about half that entering. The introduction of an economiser (Section 14.2), using flue waste heat, will raise the theoretical thermal efficiency to 0.572. The theoretical attractiveness of this cycle becomes clear when it is realised that the corresponding Carnot efficiency is 0.609.

Non-isentropic Behaviour

The expansion process will not be ideally efficient, as assumed so far, and this inefficiency must be accounted for. The high dryness fraction will help conditions in the turbine and it will probably be realistic to assume an isentropic efficiency of 0.89. This means that the expected thermal efficiency is lowered to 0.509 with the economiser, while the specific steam consumption is raised to 4.15 kg/kJ. These are still attractive values, but, of course, this is not the overall efficiency of the plant.

Other Losses

Not all the heat contained in the cycle provides useful work, and not all the work produced is available in the form of electrical power. Works auxiliaries (pumps, lighting etc.) will take about 1 % of the initial energy input and there will be radiation losses of perhaps 1–2 %. Again the combustion process will not be ideally efficient and a boiler efficiency can be assigned: typically, this might be expected to be close to 0.75 if the flue gas heat is included.

The Overall Efficiency

These various inefficiencies reduce the effective thermal efficiency of the plant taken as a single unit. It is not possible to make a general quantitative assessment of these factors but the data in Table 14.1 can be regarded as not untypical of the energy balance for a modern British power station, related to

Table 14.1. The Proportional Energy Budget for a Typical Modern Steam Plant.

Source	Energy Involved kJ	Percentage of Input Energy
Works auxiliaries	2.7×10^5	1
Regenerative heating	9.15×10^6	34
Boiler losses	2.9×10^6	11
Radiation losses	1.3×10^4	0.5
Turbo-generator and electrical losses	3.9×10^5	1.03
Condenser heat rejection	1.25×10^7	47.45

Sent out energy is 1×10^7 kJ/ton coal: overall efficiency = 38.5 %

Table 14.2. Working Characteristics for Some Electricity Generating Fossil Fuel Plants in the UK.

[PF = pulverised fuel (coal), OF = oil fuelled, and PF/OF means there is a choice. (a) year of first commissioning; (b) electrical energy offered to the grid; (c) thermal efficiency of the total plant.]

Station	Year (a)	Type of Firing	Steam Pressure bar	Temp. °C	Elect. Genertd. GWh (b)	Thermal Efficy. (c)
Cottam	1970	PF	159/38	566/566	1.33×10^4	36.06
Drax	1985	PF	166	568/568	2.03×10^4	37.08
Fawley	1971	OF	159	5348/538	2.24×10^3	35.18
Fiddler's Ferry	1973	PF	165/41	568	1.10×10^4	35.02
Ferrybridge B	1959	PF	110	530/513	5.88×10^2	30.71
Ferrybridge C	1968	PF	166	568/568	1.11×10^4	35.56
Grain	1984	OF	160/39	538/538	1.65×10^3	33.80
Ince B	1982	OF	166/40	541	1.08×10^2	26.62
La Collette (Jersey)	1973	OF	62	482	1.91×10^2	26.93
Stella South	1956	PF	66	496	3.02×10^2	23.84
Wakefield B	1957	PF	66	496	1.61×10^2	23.45
West Thurrock	1966	PF/OF	162/29	566/538	4.12×10^3	31.22

burning 1.016×10^3 kg (1 ton) of coal with calorific value 2.625×10^7 kJ (2.584 kJ/kg). With a thermal efficiency of 38.5%, the work output is 8.22×10^6 kJ/tonne; the energy lost or ejected is 1.803 kJ/tonne (that is rather more than twice as much). The overall efficiency of the plant (to be distinguished from the cycle efficiency) is about 38.5%. Data for twelve working fossil fuelled stations in the UK are collected in Table 14.2. The year of commissioning covers a range from 1956 to 1985 and there is a range of generating output. The effects of irreversibilities become relatively less for greater output and the move over this period towards this, and so more efficient stations overall, is evident. Corresponding data for nuclear plant are contained in Table 14.1.

14.4 Power and Heat Extraction Systems

In industrial processes it is often required to provide simultaneously both a substantial power output and a heat output, the heat output being in the form of hot water (called processed water). The conventional power plant can be modified to provide the heat output as well as the power output. For this purpose a bleed point is taken from the turbine, independent of the feed heat system, to lead off turbine steam at an appropriate temperature and pressure. This provides the required auxiliary heat by way of a heat exchanger, and accommodates the provision of the heat output while utilising the full temperature range between the metallurgical and atmospheric limits. The flow diagram for a simplified system, without the feed heat lines, is sketched in Fig. 14.3. The steam condensed in the process of heat exchange is returned to the boiler. The turbine arranged for this process is usually referred to as an extraction or pass-out turbine. This system shows important flexibility, changes in the heat and work outputs being easily made by varying the quantity of extracted steam. For a large system the loss of power need not be great.

Where the heat and power outputs are more modest, a substantial steam flow may be inappropriate. In this case the condenser of the steam cycle can be used to provide the processed water as shown in Fig. 14.4. The system is called a back-pressure turbine plant. The condensed steam is passed back to the boiler. The temperature of the processed water will generally be low.

Other variants are possible depending on the particular requirements and such systems are becoming more interesting as energy conservation becomes more important. All these plants are called collectively total heat plant. There is a potential value in such systems, and particularly in extraction systems, for power stations near cities where the processed heat can be used to warm the surrounding houses and factories. This has often been discussed in the UK but the economics of such schemes have always been marginal in the past. The move to larger generating units, to use the immediate thermodynamic advantages that such plants provide, has an environmental impact and large plant is best located away from populated areas. This however acts against the domestic use of processed water though the situation may well be different in the future, where a move to smaller power units is possible.

208 *The Full Steam System*

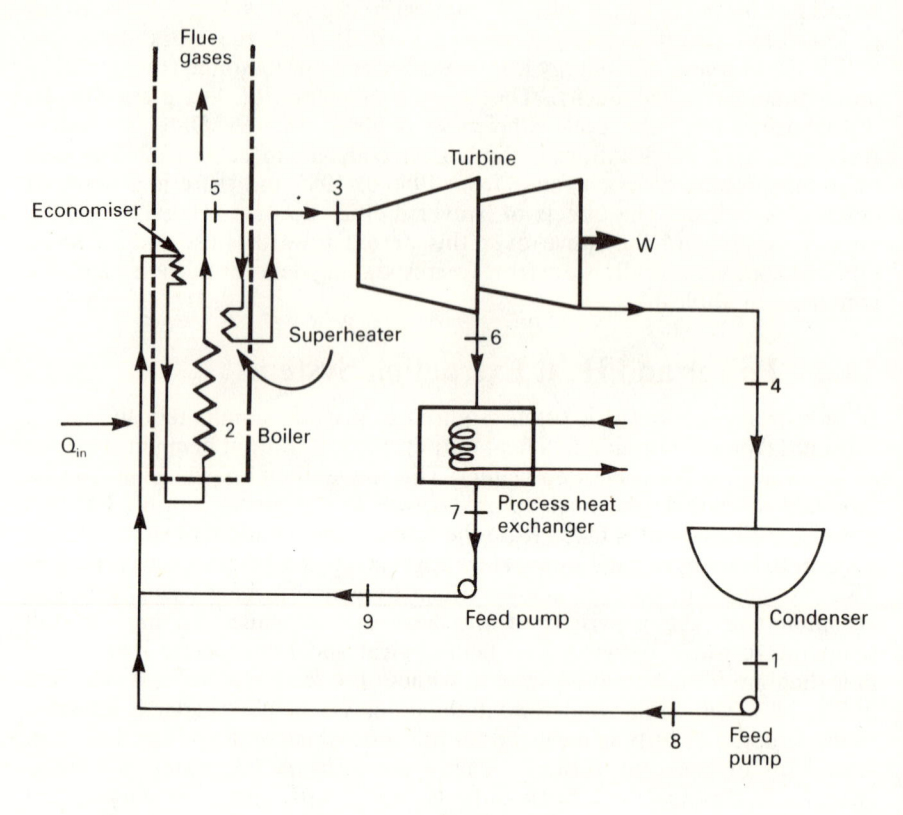

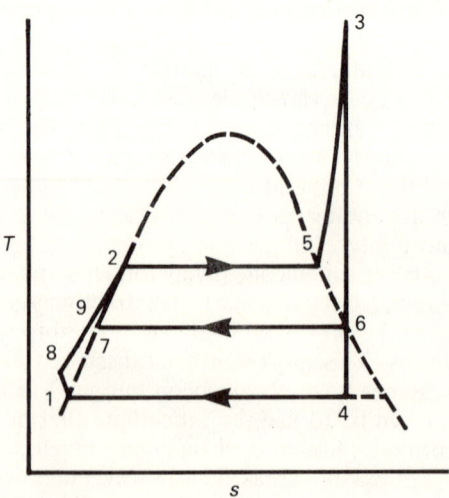

Fig. 14.3 The flow diagram and temperature-specific entropy diagram for an extraction turbine system with superheat and an economiser.

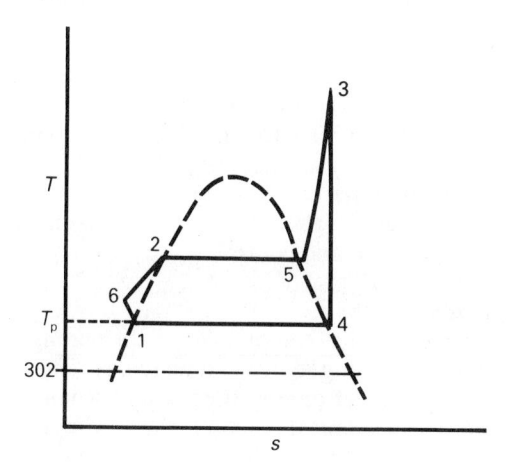

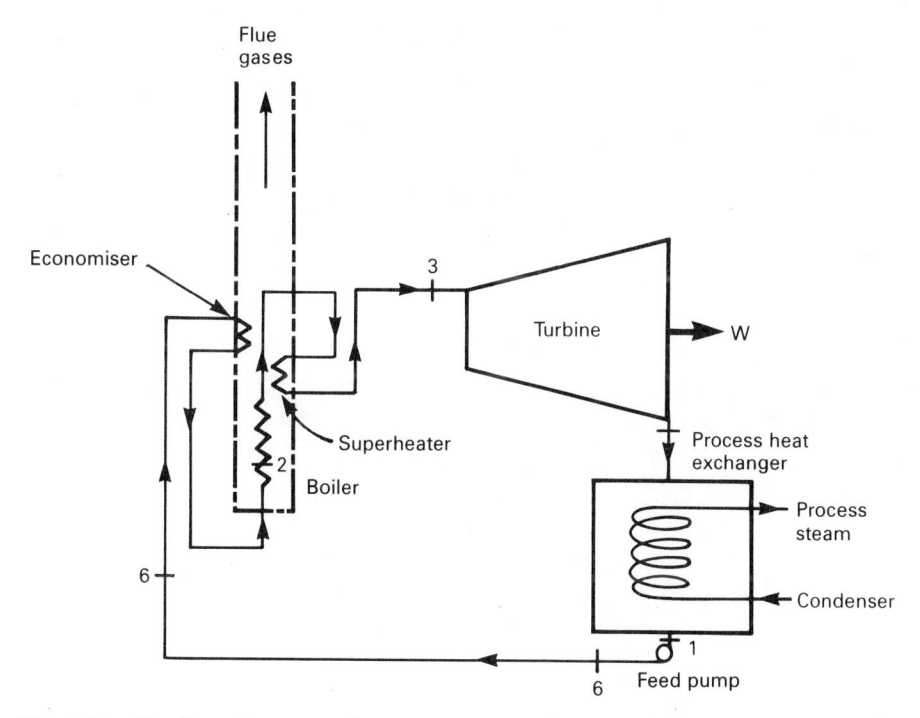

Fig. 14.4 The flow diagram and temperature-specific entropy diagram for a back-pressure turbine plant. T_p is the temperature of the process steam.

14.5 Combustion Products and Their Control

Although they fall outside our strict area of interest, the problems of controlling and dispersing the variety of products of furnace combustion must be accounted for in the practical working of the plant. The flue gases are corrosive and cannot be discharged to the atmosphere immediately after leaving the air pre-heater becuase they contain products of combustion harmful to the environment. Whereas the overall design of the plant is aimed primarily at thermal efficiency, the nature of the combustion products influences its strategic operation and control.

Impurities in the Fuel

Fossil fuels were formed by the compression and heating of once living materials (plant or machine) over geological time scales. The energy obtained by burning them is the chemical energy they stored when alive and which resulted from the solar photochemical reactions then. Although all fossil fuels have some chemical constituents in common (for instance carbon, oxygen and hydrogen) there are important variations of other elements arising from the different environments in which the living materials formed. For coal, these impurities come from the mineral content of the original plants and the soil in which the plants grew. For oil, bacterial action has supplemented the normal mineral content of the plankton-like source micro-organisms. The main impurities for these two fossil fuels are listed in Table 14.3.

Table 14.3. The Main Impurities in Coal and Oil Fuel.

Element	Impurity (% mass parts per million)	
	Coal	Oil
Silicon	2.6	0.03
Aluminium	2.5	0.07
Sulphur	1.5	2–4
Iron	1.3	0.01
Nitrogen	1.2	1.5
Calcium	0.95	0.1
Magnesium	0.37	0.01
Chlorine	0.34	0.03
Potassium	0.23	0.05
Sodium	0.15	0.05
Vanadium	0.15	0.15
Phosphorus	0.10	0.004
Cadmium	0.01	0.003
Copper	0.003	0.000 2

The coal used to fire power stations contains between 15 and 25 % mineral impurity. The oil fuel, which is the (thick black) residue of refinery distillation, has a comparable impurity level but with some 10 % being hydrocarbons of very high molecular weight including asphaltenes characterised by extensive polycyclic structures.

Features of Stack Emission

Flue gases are introduced to the atmosphere through a chimney stack, usually about 200 m in height, and there are four main regions for the dispersion. In the first, covering a radius of some 2 km from the stack, the gases have a high thermal buoyancy due to their high temperature (above 110 °C) and are swept high above the ground by the local wind. In the second region, where the plume buoyancy and internal turbulence have decayed, the gas distribution is controlled by atmospheric turbulence. The gas region widens and its lower edge moves downwards to meet the ground at a distance of between 5 and 20 km (depending on various factors, but particularly the weather) from the stack. Dry deposition of material on the ground takes place from this region onwards. In the third region, which spreads out to 100 km or more downwind of the stack, the vertical distribution of materials gradually disappears as the flue gases become fully mixed with the atmospheric constituents. The top boundary of this layer remains at a fairly constant height of about 1 km and its internal features are now independent of the particular stack characteristics. The fourth region extends these conditions outwards, even over intercontinental distances and the gas is now controlled primarily by meteorological factors. This region is the seat of many chemical transformations, including those forming acid rain.

Flue Gases from Coal

A typical 1000 MW coal fired power station, working continuously, will consume some 11 000 tonnes of coal and 100 000 tonnes of air per day. The coal is milled (see Fig. 14.2) into a fine powder (the particle diameter is usually much less than 300 μm) and is fed into the furnace through about 30 individual burners arranged geometrically. Some 3000 tonnes of ash is produced per day containing a range of the heavier mineral materials which will not burn (in the form of hollow spheres of diameter about 100 μm), together with other fused refuse such as unburnt fuel and soot. There is little commercial use for this ash although it has been used to a small extent as a filler for paint. The ash collects in the furnace and economiser hoppers but most is recovered by the precipitator. A small quantity (perhaps 5 tonnes per day, i.e. 1800 tonnes per year) escapes the cleaning stages and is discharged to the atmosphere. Small quantities of various inorganic salts and minerals, particularly sodium sulphate, adhere to the ash and are discharged into the atmosphere.

The flue gases, at the other end of the process, contain about 23 000 tonnes of carbon dioxide (CO_2) per day (8.4×10^6 tonnes per continuous year) and 3600 tonnes of water vapour per day (1.3×10^6 tonnes per continuous year). These are formidable quantities of pollutants. But a critical impurity is sulphur. This is present in the coal both as organic sulphur compounds and as iron sulphide (pyrites), and combustion with air produces sulphur dioxide (SO_2). Perhaps 1 % of the SO_2 is oxidised further to form SO_3: this combines with water vapour to form sulphuric acid (H_2SO_4) in the furnace gases which pass on through the superheater, reheater and economiser. The quantity of H_2SO_4 present is neutralised to a large extent by the alkaline ash but some is inevitably left to corrode the various metal surfaces. The SO_2 is the major impurity and the station will produce some 320 tonnes per day, or about 1.2×10^4 tonnes per continuous year.

Modern high temperature flames also oxidise the nitrogen present to form nitrogen oxides, particularly nitric oxide (NO). Hydrochloric acid (formed from sodium chloride and other salts) and nitric acid will also form, providing further corrosive agents in the flue gases, ultimately to be dispersed into the atmosphere. Some 3000 tonnes of NO will be emitted per 24 hour period.

Flue Gases from Oil

The residual fuel oil is injected into the furnace as a fine spray together with air. Again, sulphur is the major impurity and SO_2 is present in the flue gases in quantities fully comparable to those in coal. Again, sulphuric acid is formed but this now assumes a greater hazard than for coal because there is no alkaline ash to neutralise it. Carbon is the major component of the solid emission although there is also a range of partially burned hydrocarbons and asphaltenes. As oil becomes more scarce, and lower quality oils must be used for power production, so the proportion of unburned impurity materials in the oil flue gases can be expected to increase.

Pollution Control

The problem of the control of pollution is closely linked to that of attaining the optimum conditions for combustion in the furnace for the chosen fuel. This is a highly complicated matter that we cannot pursue here but it is useful to make a few comments.

The essential requirement, whether for coal or oil, is to maintain the stochiometric fuel/air ratio against variations of the fuel supply. For coal the difficulty is that of maintaining an even distribution of pulverised coal to all the burners simultaneously: up to a 50 % variation of the air/fuel ratio at an individual burner is common, although this ratio is more constant for the furnace as a whole. For oil the difficulties are clogging of the spray nozzles and maintaining an even mix of oil and air.

In either case, too little air leads to only partial combustion and the production of excess quantities of impurity gases, especially SO_2 and SO_3. Too much air reduces the thermal efficiency of the furnace and produces overheating of the flue gases.

The variations in composition of the flue gases that result from these problems present problems for pollution control. It will probably always be necessary to have some form of chemical cleaning of the flue gases, particularly for the SO_2. The elimination of the sulphur emission can be achieved, for instance, by the Babcock-Hitachi process which is becoming much used. The flue gases are passed through limestone which interacts chemically with the SO_2 to form calcium sulphate ($CaSO_4$). With water this becomes gypsum, a material used to make plasterboard for the building industry. A more efficient way of removing impurities may be to modify the furnace combustion itself, for instance by introducing a fluidised bed combustion system, but this is outside our present concern.

14.6 Comments on Metallurgical Limits

The metallurgical limit has been referred to a number of times in our arguments and it is important to be clear why this arises. Although the word limit might imply a sharp upper temperature boundary this is not, in fact, the case. Rather, there is a somewhat diffuse region of high temperature where difficulties that already exist at lower temperatures increase and become unmanageable e.g. heat transfer. The study of conditions in the limit is more important in modern plant because, with increasing boiler temperature and pressure, the traditional over-design of old is no longer possible: modern furnaces work close to the physical limit of the materials involved.

Furnace Heat Transfer
The furnace temperature is high (between 1600 and 1750 °C in places) but the temperature of the water being heated is below 360 °C even at 180 bar (only about 1/4 of the furnace temperature). Heating during furnace combustion is by radiation (ideally from incandescent carbon particles which burn away after a very few seconds—the residence time in the furnace). In all furnaces, the water tubes are unevenly heated and often receive heat on one side only. The tubes are packed closely round the furnace, often well away from the vertical and with significant bends. The water inside is a mixture of condensed water and forming steam and a wide variety of condensed water/steam mixtures is possible. Ideally, the steam should be evenly distributed but instead it often forms 'plugs' or sometimes regions which disturb the overall heat transfer. This condition is particularly significant in the so-called once through boiler where water at the boiler inlet is converted directly to dry saturated steam continuously through the boiler tubes. An alternative is the circulation boiler where the water is recirculated through the furnace, the steam being extracted through cyclones and collected in a steam drum. Here problems of instability in the circulation can leave individual tubes devoid of water and there can be other fluid flow difficulties even though the steam will have a density as high as 205 kg/m^3 at about 180 bar (only 1/4 that of the condensed water itself). The situation can be made worse by even small quantities of impurities in the water. These effects become particularly significant at high water tube pressures where the tube thickness is an important factor, and at the associated high steam temperatures.

Ancillary Heat Transfer
The heat transfer in the superheater, reheater and economiser is steam to fluid, radiation not now being significant. The steam temperature is comparable to that of the flue gases (superheated steam at 500 °C is maintained by flue gases at perhaps 1300 °C while reheat to 500 °C involves flue gases at about 800 °C). The water tubes are now heated at all angles but can suffer corrosion and deposition on the outside and impurity deposition (leading to capillary boiling or the so called 'wick-boiling' within the scale) on the inside. These are particularly important in the superheater section where the pressure is high and the metallurgical properties of the tube material are especially significant.

Limits Are Set
These various effects, of metal creep and heat transfer, become the more critical as the temperature and pressure increase. The furnace conditions also provide a greater range of impurity aggregates in the flue gases as the combustion temperature increases. The result is an adverse boiler efficiency and water tube failures. An optimum maximum temperature with the present

214 The Full Steam System

technology and materials is about 580 °C while the optimum upper pressure is rather less than 180 bar. It is this duo that is meant when the metallurgical limit is referred to.

14.7 The Usefulness of Exergy in Plant Assessment

We have seen in Section 2.5 that the availability function allows an estimate to be made of the maximum work, W_M, within a thermodynamic system operating between particular thermodynamic states, the changes being entirely reversible. If the final state is the environment the availability function is the exergy. The components of the plant can be treated as steady flow systems, and their behaviour analysed using the steady flow availability B between the entry and exit states. The total work availability of the combination is deduced by combining the individual values, the final state being that of the environment. A real system, however, is subject to the effects of irreversibilities and the actual work W that it provides will be less than the ideal between the same initial and final states. A prediction of how much less is given by the difference $W_M - W$, the effectiveness of the plant to produce work is sometimes expressed by the ratio W/W_M.

The specific steam consumption for a steam plant is given by $1/W > 1/W_M$ and this quantity determines the size of the boiler required in each case. The work effectiveness for a particular system can be interpreted alternatively as the ratio ssc(ideal)/ssc(actual) < 1, giving a numerical estimate of the boiler size relative to the ideal. The size is increased by the presence of irreversibilities in the system, including any inefficiencies in the boiler–combustion system itself.

The work output and the boiler characteristics for a real plant can be compared with the ideal cases in a quantitative way through the exergy. This information is useful for monitoring the behaviour of an operating plant; alternatively, for a new plant, it provides a quantitative framework for the comparison of alternative designs in the selection of the final circumstances.

The precise details of the application of these ideas depend upon the particular plant involved, and its proposed usage (for instance, the proportion of part load). Although the detailed assessment must depend on the particular plant circumstances, we can make some general comments on factors affecting the thermodynamic characteristics of plant in general terms. We use a modified Rankine steam plant as the basis for discussion.

Reference to Fig. 12.13, with superheating to 500 °C and one stage of reheat (usually sufficient to provide appropriate dryness of the steam), shows the general trends of the dependence of thermal efficiency, specific steam consumption and dryness fraction on variations of the boiler pressure p_b. The condenser pressure is kept constant at a level (0.04 bar in this case) largely controlled by the magnitude of the temperature gradient that must be maintained with the environment to ensure adequate heat transport for rejection. The thermal efficiency increases with p_b although after about 160 bar further increase is slow; the specific steam consumption has a minimum value at about 150 bar and the dryness fraction continually decreases. A compromise acceptable overall for the magnitudes of these various quantities is met for a boiler pressure between 150 and 160 bar. The boiler temperature then is close to 600 °C.

The effects of non-ideal components and the addition of the irreversible losses through heat conduction and pressure losses in the various tubes, pipes and ducts can now be made. The process is repeated for variations of the boiler pressure, reheat conditions, regenerative conditions and component characteristics, comparing the calculated values with the deduced exergy at each stage. The result is a set of overall preliminary designs from which a final selection can be made. Two central criteria for the selection are likely to be thermal efficiency of the total plant and the specific steam consumption of the boiler components. These various possibilities can, of course, be costed in financial terms through materials and times.

Nowadays it is required as part of the design that the plant shall cause less than a prescribed level of environmental pollution and this will require the appropriate removal of a range of sulphates, oxide gases and volatiles from the flue gases before they enter the atmosphere. There is a cost factor here which has only recently begun to be accounted for at the design stage. Flue gas pollution is less for lower boiler pressures and temperatures. The tendencies over the last two or three decades have been towards higher boiler temperatures and pressures (to increase the thermal efficiency especially) but there could well be a move in the opposite direction once anti-pollution stages are included in the initial costings. These features can also be accounted for through considerations of the exergy, although such applications essentially involve chemical analyses which are not our present concern.

These arguments can, of course, be applied equally well to combined plant or to other systems and form the framework for modern design activities.

14.8 Towards a Better Working Fluid

We have seen that the working characteristics of the cycle depend in important ways upon those of the furnace but they also depend on the properties of the working fluid. Steam is not by any means the best example. There would be enormous benefits from finding a more thermodynamically satisfactory working fluid and it is interesting to list the properties that would be sought.

Constant Temperature Furnace

The Rankine cycle fired by a fossil fuelled furnace provides a constant temperature heat reception for the cycle. While the steam cycle (explained in the previous chapters) works well enough in practice, it would be considerably simplified if a working fluid were available with more advantageous properties than steam. Heat reception over a wide range of temperatures is a feature to retain and regenerative heating with an economiser is thermodynamically essential, but superheating and reheating are unsatisfactory features that could be eliminated. It would also be desirable to work at lower boiler pressures (at most one or two tens of bar rather than 160 bar as for steam) and the elimination of the condenser vacuum would remove various practical difficulties. The reduction of the mass flow for steam would also reduce the size of the boiler and turbine and so reduce the capital cost of the plant.

A fluid able to achieve these requirements can easily be described. The mass flow, for a specific heat transport, decreases as the density increases so the new fluid will need a density similar to that of water or higher (which is

216 The Full Steam System

some 10 times that of steam at 100 bar). It would be an advantage if its viscosity was low to reduce the pump power required to drive the flow.

Conditions at the high temperature end would be simplified if the new fluid had a critical temperature substantially higher than the metallurgical limit, with a critical pressure which is relatively low (say a few tens of bar). This is opposite to water where the critical pressure is relatively high and the critical temperature much lower than the metallurgical limit.

The majority of the heat acceptance must be at high temperature (this is achieved for water by superheating). This would be possible if the specific heat capacity of the fluid were very high (so the saturated liquid line is very steep) and the latent heat of vaporisation large. The fluid would be readily raised to its boiling point at a relatively low pressure close to the metallurgical limit and further enthalpy would be added to provide dry saturated vapour. Coupled with a high critical temperature for the fluid, the need for superheating would be eliminated altogether. With regenerative heating, heat reception would be an essentially isothermal process and the thermal efficiency would be high. The other feature of superheating is the increase of the dryness fraction. If the dry saturated vapour line was steep, a high dryness fraction (in excess of 0.9) would follow easily without the need for reheating. Also, the saturation entropy at the entrance to the turbine would be high, thus providing a substantial work transfer from the cycle.

The condenser vacuum would not have a counterpart in the new fluid cycle if the fluid triple point involved a temperature only slightly above 29 °C and a pressure slightly greater than atmospheric. Heat rejection to the environment would then be controlled by a small positive pressure and air would not leak into the system were the condenser seal to be imperfect anywhere. The hypothetical T–s diagram for this new fluid is sketched in Fig. 14.5. With these properties the plant could be very simple. The fluid vapour would pass directly to the turbine, returning to the boiler from the condenser through a

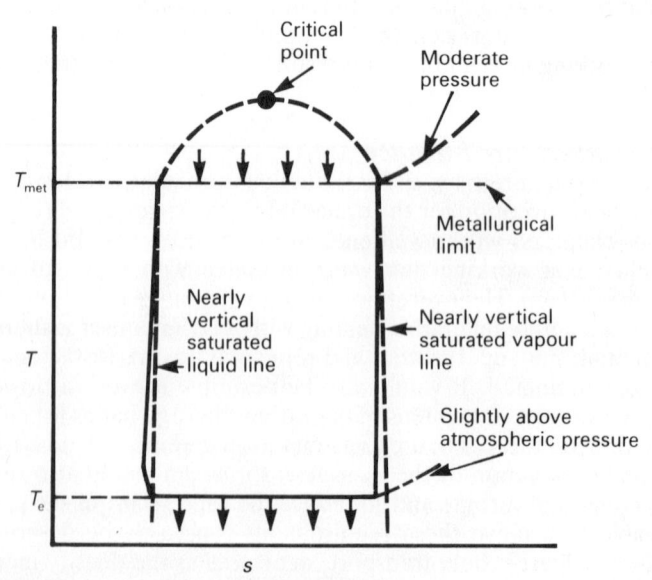

Fig. 14.5 The temperature-specific entropy diagram for a good working fluid. T_{met} is the temperature of the metallurgical limit; T_e is the temperature of the environment.

cascade feed heat system. There would not be superheat or reheat tubes in the furnace and no condenser vacuum to maintain. The piping and pump systems would be easily maintained, and cheap and simple to construct. It is necessary to remember that the new fluid would itself need to be cheap and simple to produce, plentiful, non-toxic, chemically stable and largely non-corrosive. Whether such a fluid could exist within the established laws of physics is something the reader must consider.

Varying-Temperature Heat Source
The requirements for the working fluid are different for a combined cycle if the heat reception by the Rankine component involves heat transfer from another fluid (a gas cooled reactor is an example, with heat transfer from CO_2). Now the second fluid will lose heat during the heat exchange process (because it is passing heat to the Rankine cycle) and the temperature will fall continually across the heat exchanger. The temperature of the Rankine fluid will correspondingly increase as it passes through the heat exchanger.

As the temperature of the other fluid decreases, that for the Rankine cycle fluid increases so the fluids must be set in counter flow. To achieve the best thermodynamic conditions the match between the temperatures of the two fluids at each point should involve infinitesimal temperature differences. A rising temperature at constant pressure is a characteristic of a superheated fluid so the new fluid now would need properties quite different from those for a constant temperature heat source. Because superheating is to be achieved at a moderate pressure it is necessary for the critical temperature to be much greater than the metallurgical limit but with a low critical pressure. The density should be high to provide sufficient mass flow without high fluid speeds. Again, the triple point pressure should be slightly above atmospheric to eliminate the need for a condenser vacuum. As for the case of the constant temperature heat source, the specific heat capacity of the liquid phase should be very large and the dry saturation line should fall sharply with increasing entropy.

14.9 Summary

1. The various modifications to the Rankine cycle considered separately in earlier chapters are now brought together in a systematic way to form a complete work transfer system.
2. The cycle thermal efficiency must, in this way, be combined with the efficiencies of the other components to provide an overall efficiency for the plant.
3. The enthalpy of the flue gases is important in allowing the introduction of an economiser as an additional feed heat stage, and pre-heating the furnace air is important to increase the efficiency of the combustion process.
4. The composition of the flue gases arising from combustion, and their effects on the plant, are briefly reviewed.
5. Problems of the release of flue gases to the atmosphere are briefly considered.
6. Comments are made about the properties that would be shown by a working fluid better than any known at present. It is realised that such fluids may well not exist.

218   *The Full Steam System*

7. Concern here is not only with the thermodynamics of the cycles, but with the ultimate efficiency and efficacy of the plant. This is to be judged in terms of the recurring costs (including interest payments and capital repayments) on the capital plant and the money raised by the sale of the work output of the system over an appropriately long life span.

14.10   Exercises

14.1   For a simple Rankine cycle, the enthalpy required to vaporise water at the boiler pressure B is $dh_B = h_{gB} - h_{fB}$, while the enthalpy required to raise the water from the condenser temperature to the liquid saturation temperature for boiling is $dh_H = h_{fB} - h_{fC}$, where h_{fC} refers to the condenser temperature. Using the steam tables, plot the ratio $dh = dh_B/h_H$ for boiler pressures between 20 bar and the critical pressure.

Repeat your calculations including the enthalpy h_s for superheating the steam to 500 °C. Plot the ratio $dh_s = [dh_B/(h_H+H_s)]$ over the same range of boiler pressures. Comment on your results.

14.2   It is intended to design a steam turbine power plant with a boiler pressure of 100 bar and a condenser pressure of 0.04 bar. There is superheat to 500 °C but no economiser. A feed heat system involving two turbine bleed points is to be used and a dryness fraction in excess of 0.90 is required at the turbine exit. Show that the bleed pressures should be 26 bar and 2.8 bar for the best thermodynamic performance. If reheat is made at the pressure appropriate to a saturated liquid, show that: (a) the dryness fraction at the turbine exit is 0.912; (b) the work output of the plant is 1411 kJ/kg; (c) the heat input is 1569 kJ/kg; (d) the thermal efficiency is 0.455 and (e) the specific steam consumption is 2.551 kg/kWh. (It is assumed that the expansion process is ideal and that there are no other pressure or heat losses in the system.)

14.3   The steam plant in Question 14.1 is modified so that the reheat pressure is 20 bar, the cycle remaining otherwise the same. Confirm that the thermal efficiency is now 0.495 and the specific work output is 2.478 kg/kWh. Why is this modification unacceptable using the criteria of Question 14.1? Would the design be acceptable if the fluid temperature after reheat was raised to 600 °C? What then would be the dryness fraction at the turbine exit?

14.4   The cycle described in Question 14.1 has an economiser added which brings the feed water to 98 % of the saturation liquid temperature for 100 bar boiler pressure, but is otherwise the same. What now are the working characteristics of the plant?

14.5   Write an account of the pollution associated with the raw burning of pulverised coal and of the measures (both mechanical and statutory) that are necessary to reduce it to very low levels. What action is needed in the longer term to make the effects of global coal-fired energy production appropriately clean over all work output levels? What ancillary effects would your proposals be likely to produce?

15. BINARY AND COMBINED CYCLES

Up to this point we have considered gas and steam cycles separately but advantages can result from combining them. There will generally be more than one vapour involved. These matters form the subject of the present chapter.

15.1 The Binary Vapour Cycle

The simple Rankine cycle using dry saturated steam approaches the thermodynamic ideal of isothermal heat acceptance through the addition of regenerative feed heating (Chapter 13). The heat rejection is maintained isothermal by the condenser vacuum. The dryness fraction is, however, too low, except perhaps at the lowest boiler pressures, and one stage of reheat (Chapter 12) is usually required to overcome this in practice: this, unfortunately, destroys the isothermal heat acceptance although the effect may be relatively small.

The maximum temperature attainable in a cycle is determined by the critical temperature of the working fluid and for water this is low in relation to the metallurgical limit. No fluid is known which (aside from being economically and physically acceptable) has on the one hand a critical temperature above the metallurgical limit and, on the other, has a triple point near the atmospheric temperature. Because the overall thermal efficiency is controlled by the maximum temperature of the cycle it is necessary to raise this temperature for steam if high efficiencies are to be achieved.

This can be done by combining the steam cycle with a second cycle, using a separate fluid with a critical temperature above the metallurgical limit, to span the temperature range between that of the steam cycle and the metallurgical limit itself. The heat rejected by the second cycle provides the heat acceptance for the steam cycle. Because the thermal conditions in the second cycle are in this sense superior to those in the steam cycle, we will refer to the second cycle as the top cycle. Such a combination of two vapour cycles is called a dual vapour cycle, or more specifically a dual vapour/steam cycle when steam is involved.

With reversible feed heating in both the top and steam cycles the heat acceptance for the combination is made isothermal by the top cycle at essentially the metallurgical limit while the heat rejection is made isothermal by the steam cycle at the atmospheric limit (about 10 °C above the mean temperature of the environment). The resulting intermediate thermal efficiency of the dual cycle can be very close to the ideal for the temperature range involved.

In practical application the steam cycle will need to be supplemented by a reheat stage to achieve an acceptably high dryness fraction. Some thermal

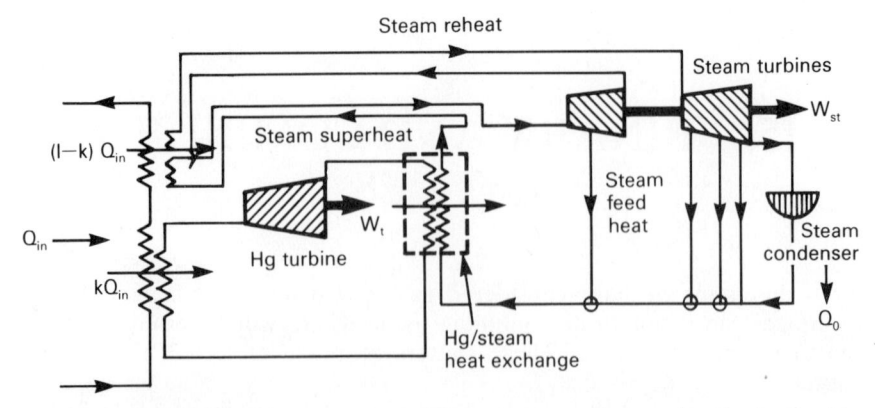

Fig. 15.1 The mercury/steam system. The mercury cycle has the simple Rankine form: the steam cycle has superheat, reheat and four stages (schematically displayed) of feed heat. An economiser system can be added but is not shown in the diagram.

efficiency will, in consequence, be lost and superheating may be introduced into the steam cycle. The further heat input necessary for these additional steam stages is provided by the flue gases of the boiler supplying the top cycle. The lay out for such a system is shown in Fig. 15.1.

15.2 Thermal Efficiency of the Dual Vapour/Steam Cycle

The conditions within the dual vapour cycle are readily set down, both in principle for an ideal cycle and for a real system.

Ideal Cycles

There is no superheating or reheating and each cyle has entirely reversible regenerative feed heating using an indefinitely large number of turbine bleed points so that heat acceptance and rejection are achieved isothermally (see Fig. 15.2). The expansion and compression work transfers are perfect and there are no losses in the system. The steam cycle uses 1 kg of dry saturated steam which is heated by the heat rejected by m kg of fluid in the top cycle. The heat received by the top cycle from a boiler is written Q_t and the heat Q_1 is rejected. This rejected heat is received by the steam cycle and is just sufficient to bring the water to the dry saturated state at a chosen pressure. The steam cycle rejects heat Q_o. The top cycle provides the net work output $W_t = (Q_t - Q_1)$ while the steam cycle provides the work output $W_s = (Q_1 - Q_o)$.

The thermal efficiency of the dual cycle is written η; η_t is the efficiency of the top cycle and η_{st} that of the steam cycle. Then we have

$$\eta = 1 - Q_o/Q_t \quad \eta_t = 1 - Q_1/Q_t \quad \eta_{st} = 1 - Q_o/(Q_1 + Q_h)$$

which are readily combined to give the relationship

$$(1 - \eta) = (1 - \eta_t)(1 - \eta_{st}) \tag{15.1}$$

15.2 Thermal Efficiency of the Dual Vapour/Steam Cycle

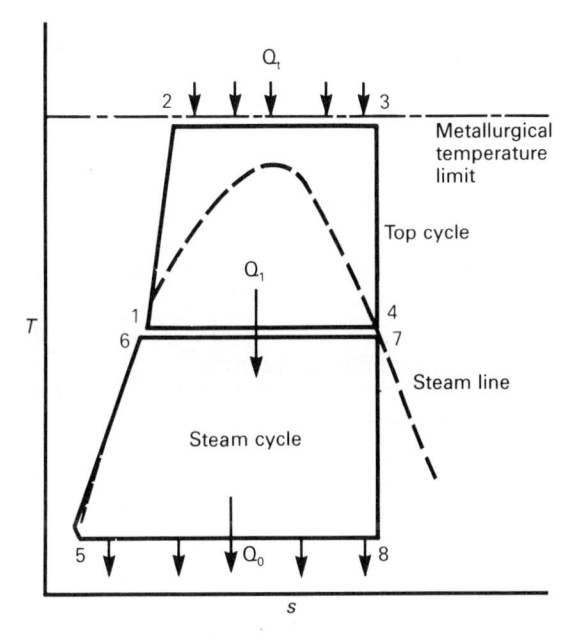

Fig. 15.2 The T–s diagram for the ideal dual vapour/steam cycle. Both the top and steam cycles have infinite feed heater arrays.

It is seen that the condition for $\eta > \eta_{st}$ is that $(1 - \eta_t) < 1$, and this condition always applies.

Practical Cycles
In practice there must be a finite number of feed heaters: an appropriate dryness of the steam at the turbine exit will be important and will involve additional heating. The heat transfer for these components will be obtained from the flue gases of the top cycle. The T–s diagram is shown in Fig. 15.3. With these additions to the cycles, the analysis for the ideal cycle must be modified to take account of the additional heat acceptance.

Let Q_{in} be the heat transfer into the binary cycle of which the fraction k enters the top cycle and the fraction $(1 - k)$ is available as ancillary heat to the steam cycle, as shown in Fig. 15.3. The arguments for the ideal cycle are recovered for $k = 1$.

For the Top Cycle
The heat input is
$$Q_t = kQ_{in}$$
while the net work output is
$$W_t = Q_t \eta_t = k\eta_t Q_{in}$$
The heat rejected is
$$Q_1 = Q_t - W_t = Q_{in}k(1 - \eta_t) \tag{15.2}$$

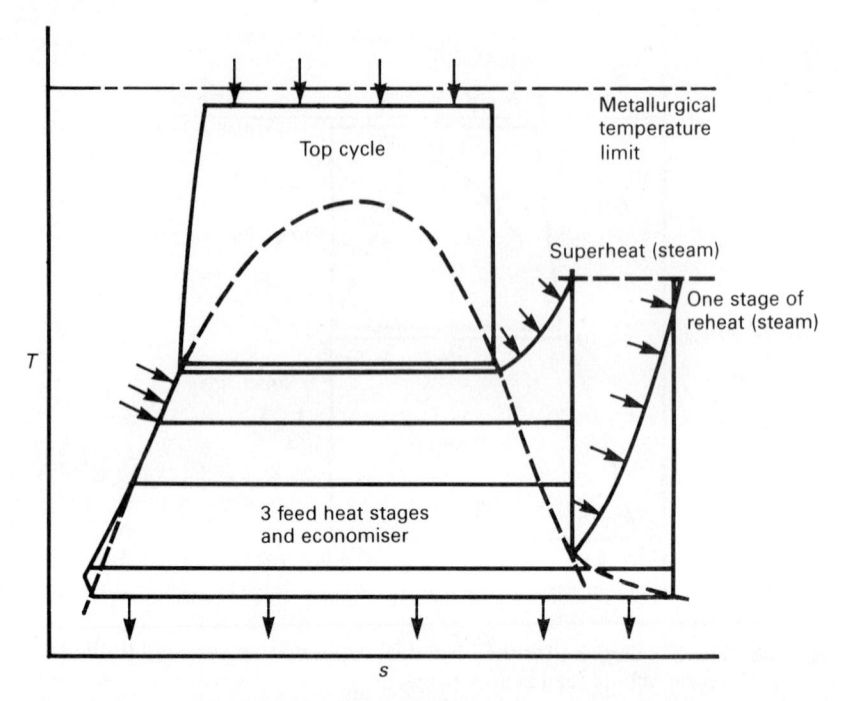

Fig. 15.3 The T–s diagram for a practical cycle. The top cycle has the simple rankine form. The steam cycle has feed heat with an economiser, superheat and one stage of reheat. The heat acceptance is indicated.

For the Steam Cycle
The heat input into the cycle is, using eqn. (15.2),

$$Q_{st} = Q_1 + (1-k)Q_{in} = Q_{in}(1-k\eta_t) \qquad (15.3)$$

The work output is

$$W_{st} = Q_{st}\eta_{st} = Q_{in}\eta_{st}(1-k\eta_t)$$

The heat rejected is, therefore, using eqn. (15.3)

$$Q_o = Q_{st} - W_{st} = Q_{st}(1-\eta_{st})$$
$$= Q_{in}(1-k\eta_t)(1-\eta_{st})$$

Combining eqns. (15.2) and (15.3), the thermal efficiency of the complete dual cycle is

$$\eta = 1 - Q_o/Q_{in} = 1 - (1-k\eta_t)(1-\eta_{st})$$

This expression is rewritten in the symmetrical form

$$(1-\eta) = (1-k\eta_t)(1-\eta_{st}) \qquad (15.4)$$

This reduces to the form of eqn. (15.1) when $k = 1$, that is when all the heat input reaches the top cycle. It is seen from eqn. (15.4) that η becomes greater the nearer k approaches unity. It seems, therefore, that the most efficient dual cycle is obtained by introducing the greater proportion of the incident heat to the top cycle. It is, in consequence, important that the steam cycle should include an effective feed heat system.

The mass flow in the top cycle is matched to that in the steam cycle by making the condition $Q_1 = Q_v$ where Q_v is the vaporisation energy of the steam at the chosen pressure (which is $(h_7 - h_6)$ in Fig. 15.2). For practical heat flow between the two cycles it is usual to maintain a temperature difference of 15 °C or so across the heat exchanger.

The dual cycle can appear very attractive in the light of eqns. (15.1) and (15.4) since feed heat can be applied to the top cycle as well as to the steam cycle, and a total thermal efficiency very close to the ideal is possible.

15.3 The Mercury/Steam Cycle

From the thermodynamic point of view mercury is a very suitable working fluid for the top cycle. Its boiling point is 234.3 °C at 1 bar and at the turbine metallurgical limit its vapour pressure is lower than 30 bar. This means that it can be expanded from near the metallurgical limit without superheating. It cannot, however, be condensed at the atmospheric rejection temperature without using a very high vacuum because the vapour pressure and specific volume are then extremely low, requiring a condenser of enormous size. It is therefore sensible to use a steam cycle for the heat rejection.

The temperature for heat acceptance by the mercury is to be about 600 °C (metallurgical limit): from the mercury tables we find $T = 604.6$ °C at 24 bar, an agreeably low pressure which will be accepted. Heat rejection is at a temperature marginally higher than the maximum for the steam cycle, which is supposed to be for dry saturated steam at 100 bar, for which the temperature is 311.2 °C. It is found from the mercury tables that mercury at the pressure 0.6 bar has the temperature 231.6 °C and this will be accepted as the rejection temperature here. The steam condenser pressure is 0.04 bar.

To illustrate the structure of the cycle it will be supposed that both the steam and mercury cycles have the simple Rankine form without feed heat, reheat or superheat. The T–s diagram is shown in Fig. 15.2. Consider the mercury and steam cycles separately.

The Mercury Cycle
Using the notation of Fig. 15.2,

$h_1 = 44.85$ kJ/kg Hg $h_2 = 80.75$ kJ/kg Hg   (mercury tables)
$h_3 = 368.6$ kJ/kg Hg $s_3 = 0.4839$ kJ/kg Hg

Also $s_f = 0.1078$ kJ/kg Hg and $s_{fg} = 0.4869$ kJ/kg Hg
This gives $x_4 = 0.772$.
Because $h_{fg4} = 0.7724$ kJ/kg Hg we find

$h_4 = 271.2$ kJ/kg Hg

From these values it follows that

$Q_t = (h_3 - h_1) = 323.7$ kJ/kg Hg
$Q_{out} = (h_4 - h_1) = 226.3$ kJ/kg Hg
$W_m = (h_3 - h_4) = 97.4$ kJ/kg Hg

224 Binary and Combined Cycles

The thermal efficiency and specific mercury consumption (*smc*) are

$$\eta_m = 97.4/323.7 = 0.301$$
$$smc = 3600/97.4 = 36.96 \text{ kg/kWh} \tag{15.5}$$

The Steam Cycle Alone
From the steam tables

$$h_5 = 121 \text{ kJ/kg H}_2\text{O}, \quad h_6 = 1408 \text{ kJ/kg H}_2\text{O}$$
$$h_7 = 2725 \text{ kJ/kg H}_2\text{O}, \quad s_7 = 5.615 \text{ kJ/kg K H}_2\text{O}$$

After the expansion

$$x_8 = 0.645 \quad \text{and} \quad h_8 = 1690.3 \text{ kJ/kg H}_2\text{O}$$

This gives

$$Q_o = (h_8 - h_5) = 1569.3 \text{ kJ/kg H}_2\text{O}$$
$$W_{st} = (h_7 - h_8) = 1034.7 \text{ kJ/kg H}_2\text{O}$$
$$Q_1 = (h_7 - h_5) = 2604 \text{ kJ/kg H}_2\text{O}$$

The thermal efficiency and specific steam consumption are

$$\eta_{st} = 2604/1034.7 = 0.397$$
$$ssc = 3.479 \text{ kg/kWh} \tag{15.6}$$

Matching the Cycles
It is necessary to ensure that the heat rejection by the mercury cycle equals the heat reception *for vaporisation* $Q_v = (h_7 - h_6)$, by the steam cycle. The mass flow of mercury must be adjusted for this purpose. Suppose z kg of mercury are required to vaporise 1 kg of steam, that is $zQ_1 = Q_v$. With $Q_v = 1317 \text{ kJ/kg H}_2\text{O}$ (steam data)

$$z = 1317/226.3 = 5.820 \text{ kg Hg/kg H}_2\text{O}$$

The Dual Cycle Efficiency
The combined work output $W = W_{Hg} + W_{st}$

$$W = 5.820 \times 97.4 + 1034.7 = 1601.5 \text{ kJ/kg H}_2\text{O}$$

The heat input to the system is in two parts. For the mercury cycle

$$Q_t = kQ_{in} = 5.820 \times 323.7 = 1883.7 \text{ kJ/kg H}_2\text{O}.$$

The additional heat in the steam cycle is required to bring the liquid to the vaporisation temperature,

$$(1-k)Q_{in} = (h_6 - h_5) = 1287 \text{ kJ/kg H}_2\text{O}$$

Therefore

$$Q_{in} = 1883.7 + 1287 = 3170.7 \text{ kJ/kg H}_2\text{O}$$

This gives the thermal efficiency of the total cycle

$$\eta = 1601.5/3170.7 = 0.505 \tag{15.7}$$

The fraction k is derived from $(1-k)Q_{in}/kQ_{in} = 1287/1883.7$ from which it follows that $k = 0.5941$. Of the heat entering the dual cycle, about 59%

enters the top cycle and 41 % enters the steam cycle. With k known, and using eqns. (15.5), (15.6) and (15.7), it is a simple task to confirm that the condition (15.4) is indeed obeyed.

The dual vapour cycle shows a substantial increase in thermal efficiency over the steam cycle alone. Further examples of dual vapour cycles are considered in the Exercises, Section 15.7. Account is also taken there of expansion isentropic inefficiencies and of inefficient heat exchange between the mercury and steam cycles, neglected so far (Question 15.6).

A few dual cycles have been built to operate on these lines, particularly in the United States, but none have been built since about 1950 and there seem to be no plans for more. Dual cycles achieved overall plant efficiencies of nearly 40 % at a time when steam plant alone was showing overall efficiencies of about 30 %. Since then higher steam plant efficiencies have become standard (primarily due to the ability to use higher boiler pressures) and the superiority of the dual vapour plant has largely disappeared at the present time because the complexity of the mercury cycle makes it far less attractive to build and operate than the relatively simple steam plant. This philosophy may change with a greater awareness of energy conservation, requiring improved thermal efficiency, and with the possible move to smaller plant.

15.4 Low Heat Cycles

It was seen in Section 14.2 that the lower limit for the temperature of the flue gases which leave the cycle is fixed by the lowest cycle temperature. For an efficient feed heat plant this temperature is closely that of steam vaporisation, and the introduction of an economiser and air pre-heater does not reduce this temperature to the regions of the dew points of the flue gases. Proposals have been made, from time to time, to use the availability that still remains in the flue gases on leaving the cycle as a source of enthalpy for a cycle accepting low grade heat. The essential difficulty here again is the working fluid. Ammonia is a possibility from a thermodynamic point of view but its physical properties are not satisfactory. Other possibilities included freon hydrocarbons, and some preliminary work was done to this end some years ago, but the harmful effects of freons on the atmosphere (and particularly on the atmospheric ozone) now rule them out entirely. The future for the thermodynamic use of low grade heat does not seem bright but the possibility should always be kept in mind.

15.5 A Combined Cycle

Instead of using a second work cycle to supply the heat to vaporise the feed water, a completely independent heat source can be used that does not necessarily involve a vapour. Such a system is often called a combined cycle. We shall be considering one case of this in Chapter 16 where the nuclear reactor is the source of heat but here we can note that a gas turbine is sometimes used for this purpose. Such plant is sometimes called Stag or Cogas plant, and appears very attractive for marine use and for generating base power electricity (see Chapter 7, Section 7.6).

The gases leaving a gas turbine usually have a temperature of about 350 °C and this is suitable for heating water to form steam. With an upper

226 Binary and Combined Cycles

temperature of perhaps 830 °C, the temperature ratio is about 2.3. From the steam tables it is found that the specific enthalpy of water must be raised by some 3250 kJ/kg to reach a superheat temperature of 500 °C at 100 bar and this must be provided by the rejected heat from the gas turbine. For a gas turbine with one reheat and one intercooling stage and with heat exchange, we see from Fig. 7.12, Chapter 7 that the thermal efficiency of the gas cycle $\eta \sim 0.65$ for a temperature ratio of 2.3 and a pressure ratio of 3, assuming that the system is ideally efficient. From Fig. 7.13 the specific work output W_s is found to be closely 300 kJ/kg. Using these figures the heat Q_o rejected by the gas cycle is $Q_o = W((1/\eta) - 1) = 215.4$ kJ/kg. The mass flow m of gas to provide the entropy required to fire the steam cycle is then $m = 3250/215.4 \sim 15$ kg gas per kg of steam. The mass flow is increased if account is taken of the gas cycle inefficiencies or if the gas cycle (as would be more likely) is more elementary in not having intercooling or reheat. The mass flow is reduced if the steam cycle has regenerative heating: if the feed heat system is ideally effective the gas mass flow is $m \sim 8.2$ kg gas per kg of steam. Eqn. (15.1) will apply to this case. The gas turbine exhaust gas is usually rich in oxygen and this could be used by burning additional fuel to heat the steam.

15.6 Summary

1. The evaporation of the working fluid consumes energy which could be derived from a separate heat source. This leads to the concept of the binary vapour/steam cycle in which a second (top) cycle provides heat to the steam cycle.
2. Mercury is taken as the 'top' working fluid and the properties of the mercury/steam cycle are considered.
3. The possibility of a 'bottoming' cycle is also considered, to use the waste flue enthalpy from a steam cycle. It is recognised that achieving the required degree of availability from a satisfactory working fluid raises difficulties.
4. The combined cycle is introduced in which the heat for the evaporation of water is obtained altogether separately from a fluid cycle.

15.7 Exercises

15.1 A mercury/steam dual cycle has *both* the separate mercury and steam cycles as simple Rankine cycles without superheat, reheat or feed heat. The mercury cycle works between the pressures 24 bar and 0.10 bar, while the steam cycle works between 40 bar and 0.04 bar. Show that:

(a) the thermal efficiency of the mercury cycle is 0.382
(b) the thermal efficiency of the steam cycle is 0.363
(c) the mass flow of mercury is 8.296 kg per kg of steam
(d) the thermal efficiency of the total dual cycle is 0.543
(e) the work output of the dual cycle is 2032.5 kJ/kg H_2O
(f) 74.17 % of the heat input to the dual cycle passes through the top cycle
(g) that the eqn. (15.4) of the main text is satisfied.

15.2 The expansion and heat exchange inefficiencies of the mercury/steam dual cycle of Question 15.1 are now accounted for. The mercury turbine has the isentropic efficiency $y_{Hg} = 0.75$, the steam turbine has the isentropic efficiency $y_{water} = 0.90$ and the heat exchanger linking the mercury and steam cycles has the effectiveness 0.85. Show that:

(a) the thermal efficiency of the mercury cycle is 0.287
(b) the thermal efficiency of the steam cycle is 0.327
(c) there is a flow of 7.186 kg mercury per kg steam
(d) the thermal efficiency of the dual cycle is 0.464
(e) the work output of the dual cycle is 156 kJ/kg H_2O
(f) 71.32 % of the heat input to the dual cycle passes through the top cycle
(g) eqn. (15.4) of the main text is satisfied.

What modification to the heat exchange between the two cycles might be made were this design to be built for practical use?

15.3 The dual cycle of 15.1 applies again, without inefficiencies, but now the mercury (top) cycle has two stages of feed heat added. The steam cycle remains as a simple Rankine cycle, without feed heat. Show that:

(a) the most effective bleed point pressures for the mercury cycle are at 7 bar and 1 bar
(b) the thermal efficiency of the mercury cycle is 0.397
(c) there is a flow of 9.375 kg mercury per kg steam
(d) the thermal efficiency of the dual cycle is 0.551
(e) the work output of the dual cycle is 2099.3 kJ/kg steam
(f) 74.62 % of the heat input to the dual cycle passes through the top cycle
(g) eqn. (15.4) of the main text is satisfied.

15.4 A dual cycle is constructed from a mercury cycle and a steam cycle. The mercury cycle has a simple Rankine form without feed heat and works between the pressures 24 bar and 0.10 bar. The steam cycle has dry saturated steam at 40 bar passed to a turbine without superheat: after expansion it is condensed at 0.04 bar. There are two stages of regenerative feed heating and an economiser in the steam circuit which brings the water to the liquid saturation temperature. There is no steam reheat. All the components behave as if ideal and there are no pressure losses in the system. The compressor work can be neglected. Show that:

(a) the most effective steam bleed pressures are 11 bar and 1.3 bar
(b) the thermal efficiency of the steam cycle is 0.400
(c) the thermal efficiency of the dual cycle is 0.606
(d) the total work output is 1867.1 kJ/kg steam
(e) 90.06 % of the heat input passes to the top cycle
(f) the formula (15.4) of the main text is satisfied.

15.5 The dual system of 15.4 is modified so that *both* the mercury cycle and the steam cycle each have two stages of feed heat, the other conditions remaining the same. Show that:

(a) the flow of mercury now is 9.375 kJ/kg steam
(b) the thermal efficiency of the dual cycle is 0.615

228   Binary and Combined Cycles

(c) the top cycle receives 90.27 % of the heat input
(d) the work output of the cycle is 2840.1 kJ/kg steam
(e) the equation (15.4) of the main text is satisfied.

15.6 Compare and contrast the working characteristics of the ideal dual cycles treated in Questions 15.1, 15.3, 15.4 and 15.5. Substantiate the statement in Section 15.2.2 of the main text that it is sufficient and desirable in practice to construct the mercury cycle with a simple Rankine form (without feed heat) when it is used as the top cycle in a dual mercury/steam cycle.

15.7 The steam cycle in the dual cycle of Question 15.4 has superheat added, the steam being brought to 500 °C. The two feed heaters remain in the steam cycle and there are no feed heaters in the mercury cycle. The expansions are ideal in both the mercury and steam turbines and the compression work can be neglected in each cycle. Remember that it was found in Question 15.1 that the thermal efficiency of the mercury cycle is 0.382. Show that:

(a) the thermal efficiency of the steam cycle is 0.427
(b) the thermal efficiency of the dual cycle is 0.590
(c) the work output of the dual cycle is 2195.9 kJ/kg steam
(d) 74.49 % of the heat input passes through the top cycle
(e) the dryness fraction of the steam at the turbine outlet is 0.828
(f) the formula (15.4) of the main text is satisfied.

15.8 The dual cycle of Question 15.7 has reheat added, the steam being brought back to the initial superheat temperature of 500 °C. It is required that the dryness fraction should be not less than 0.95. Show that a reheat pressure of 4 bar will satisfy this requirement. The dual mercury/steam cycle system remains otherwise unaltered. Show that:

(a) the thermal efficiency of the steam cycle is 0.437
(b) the thermal efficiency of the dual cycle is 0.575
(c) the work output of the dual cycle is 2475.4 kJ/kg steam
(d) 64.48 % of the heat input passes through the top cycle
(e) the relation (15.4) is satisfied.

What is the final dryness fraction of the steam at the outlet to the steam turbine?

15.9 Using the arguments and notation of Section 15.2.2 of the main text, the heat required by the steam cycle for superheat, reheat and for the economiser is denoted by $(1-k)Q_{in}$. The heat required to vaporise the water to dry saturated conditions is denoted by Q_1. Introduce the new symbol s for the ratio

$$s = (1-k)Q_{in}/Q_1$$

Show that

$$1/k = [s(1-\eta_t) + 1]$$

and that eqn. (15.4) of the main text can be rearranged to become

$$(1-\eta) = k(s+1)(1-\eta_t)(1-\eta_{st})$$

Confirm that this expression reduces to the form (15.1) of the main text when all the heat input is accepted by the top cycle alone (that is $k = 1$). Confirm that the various dual cycles treated in these Exercises satisfy this relationship.

15.10 A combined cycle consists of a gas turbine cycle as the top cycle, combined with a steam cycle.

The gas cycle is of the closed type using helium ($\gamma = 1.66$) as the working fluid. It has a heat exchanger but no reheat or intercooling. The high temperature is 850 °C and the temperature ratio is 2.3. The pressure ratio is 3.

The steam cycle has a high pressure of 40 bar and a low pressure of 0.04 bar. The dry saturated steam is superheated to 500 °C. There are two feed heaters and reheat at 4 bar to the initial superheat temperature (refer to Question 15.8).

The heat rejected by the gas turbine provides all the heat input for the steam cycle. All components can be assumed ideal, the compression work of the steam cycle is negligible and there are no other pressure losses in the system. Find:

(a) the thermal efficiency of the gas cycle
(b) the gas flow per kg steam
(c) the thermal efficiency of the combined cycle
(d) the work output of the combined cycle.

Show that the combined cycle satisfies eqn. (15.1) of the main text.

15.11 A coal-fired steam plant, with superheat to 568 °C, reheat to the same temperature and full regenerative feed heating with an economiser, has a boiler pressure of 165 bar and a condenser pressure of 0.04 bar. Its thermal efficiency is $\eta_{st} = 0.395$ and the specific steam consumption is 2.70 kg/kWh. It is proposed to convert the plant into a combined cycle by introducing a natural-gas fired gas turbine system on the front end of the steam plant. The existing thermodynamic features of the steam plant are to remain unaffected.

If the thermal efficiency of the gas turbine system is $\eta_g = \beta\eta_{st}$, show that eqn. (15.4) of the main text will apply in the form

$$(1 - \eta) = (1 - \eta_{st})(1 - \alpha\eta_{st})$$

where $\alpha = \beta k$, and η is the total thermal efficiency of the combined cycle.

Deduce that $\eta > \eta_{st}$ if $\alpha > 0$. What is the value of β for the maximum possible theoretical increase in the thermal efficiency of the steam plant? What would be a realistic value for α in this case and what then would be the value of η? Consider the practical implications of this result, taking account of the constructional and economic features of the gas turbine side necessary to approach it. What modifications to the thermodynamic characteristics of the existing steam plant could be advantageous in increasing the overall thermal efficiency? What effect would this have on the work output of the combined plant?

15.12 The requirement to use energy resources to the maximum advantage while reducing environmental pollution to the minimum places strong restrictions on the processes of energy production. In particular, it is desirable that the power plant (whether steam alone or as a dual or combined cycle) should have a thermal efficiency of at least 0.40 from coal pile to busbar. Consider the implications of these requirements for the design of a base power plant. What are the implications of the size of plant for your arguments?

16. NUCLEAR POWER CYCLES

The nuclear power plant differs from the fossil fuel plant in its energy source. The work transfer has still, so far, been achieved using a steam cycle (although other possibilities exist) and the arguments of the previous chapters are immediately available for application to this new case.

For the gas and vapour cycles powered by a fossil fuel the furnace conditions are to a large measure open to choice within the constraints of the materials forming the furnace. This is not necessarily the case for a nuclear heat source because the safe operation of a nuclear reactor can place constraints upon its working that affect the characteristics of the associated work cycle in a critical way. For this reason the work cycle is secondary to the nuclear heat source. For the reactor systems (thermal reactors) that have operated commercially so far the temperature of the reactor core must be held low (less than 700 K), so that the (secondary) work cycle depending on it functions under conditions that might be regarded as primitive by modern fossil fuel standards. The power characteristics of the reactor are still sufficiently impressive, however, to provide a highly attractive heat source overall provided the associated work cycle is chosen appropriately.

Although it strictly falls outside our range of discussion, we shall comment briefly on the nuclear reactor heat source for power cycles. The characteristics of several operating power plants will be considered as examples of the steam conditions to be expected generally in current nuclear plant. The time scale for new developments seems likely to extend to the further rather than the nearer future so we will not consider the various hypothetical alternatives to steam as the working fluid.

16.1 The Chain Reaction

The nuclear reactor relies for its energy on the disintegration (fission) of atomic nuclei (principally of two isotopes of uranium) by collision with neutrons. The technology is more recent than for fossil fuels, being less than 40 years old.

The atomic structure of matter was finally demonstrated (through studies of the Brownian motion) during the first decade of the present century. The nuclear model of the atom was developed by Rutherford at about the same time. The atom is composed of a central nucleus, containing most of the mass of the atom, and is surrounded by orbiting electrons. The nucleus is composed of protons (carrying a positive electrical charge) and neutrons (without a charge). The neutron is only slightly heavier than the proton and for many purposes the two can be regarded as of equal mass and are collectively called nucleons: the mass of the nucleon is taken as 1.6×10^{-27} kg. For comparison, the mass of the electron is 9.1×10^{-31} kg. These masses are unimaginably small but macroscopic matter contains a large number of atoms: the number per litre (the Avogadro or Loschmidt number) is 6.3×10^{27}. The number of protons and neutrons in the nucleus of an atom varies across the Periodic Table. There are about equal numbers of each for the lighter elements but there are about twice as many neutrons as protons in the nuclei of the heavier elements.

The electron carries a negative electrical charge ('the charge on the electron' = 1.6×10^{-19} coulomb) of the same magnitude as that carried by the positively charged proton. The atom as a whole is electrically neutral so the number of electrons orbiting the nucleus is equal to the number of protons in the nucleus (the atomic number). This is the nuclear model. The simplest case is the most abundant isotope of hydrogen, in which the nucleus contains one proton only and there is one orbiting electron. The nucleus of the other isotope (heavy hydrogen or deuterium) contains a proton and a neutron. The correctness of the nuclear atom model was demonstrated for hydrogen by N. Bohr who showed in 1913, using the principles of quantum mechanics that then were new, that the spectral structure of atomic hydrogen can be accounted for to a remarkable accuracy using the simple nuclear atomic model.

That the nuclei of atoms can be disintegrated by collision with energetic (fast) neutrons beamed at them was established during the 1930s. The composition of the nucleus is generally changed in the process to become that of a different chemical element and energy is released. The energy that must be given to the neutrons for this to be achieved is comparable to that released by the disintegration. There is no possibility of using this process as a source of energy. The mechanism would be capable of practical applications, however, if (a) the energetic disintegration could be accomplished by neutrons of very low energy (say comparable to that of the atoms and molecules of the surroundings at room temperature or about 0.025 ev = 4×10^{-21} J), (b) the nucleus/neutron collision occurred by chance, and sufficiently often, with neutrons and atoms colliding in a random fashion, and (c) if a further neutron is produced at each disintegration ready for a further encounter with another nucleus. Such a process could release a substantial excess of energy in a self-sustaining way, forming a 'chain reaction'.

Such a chain reaction was established in 1938 when the disintegration of the uranium isotope ^{235}U was first discovered using neutrons of low energy. In this case, the neutron has an energy comparable to the random energy of the atoms at ordinary (room) temperature: neutrons with about this energy are called thermal neutrons and each has an energy about 0.025 ev = 4.0×10^{-21} J, while the energy released in each nuclear disintegration is about 200 MeV = 3.2×10^{-11} J. Each disintegration also produces, on average, rather more than 2 neutrons. This mechanism is the key to the development of the nuclear reactor. The energy released during fission appears in the form of the kinetic energy of the various resulting atomic particles. Most of this kinetic energy is absorbed by neighbouring atoms through various interactions and is converted into random energy, that is heat. It is this heat that is available for use as the heat source for a work cycle.

The Second World War (1939–1945) saw an enormous advance in the application of fission to military uses, energy release occurring almost instantaneously in atomic weapons. This involves fuel extremely rich in ^{235}U, while the isotope of plutonium (^{239}Pu) behaves in a similar way and is also involved.

Once the war was over the information about nuclear disintegrations so obtained was available for the different problem of applying the fission process to a new nuclear technology for peaceful uses. In such applications the energy must be released over a long period of time, and under strictly controlled conditions, with much lower proportions of ^{235}U or ^{239}Pu. The United Kingdom, the United States and the Soviet Union each began independent nuclear reactor programmes to achieve these aims, which have led to the modern nuclear industry.

In each of these developments, the nuclear reactor forms a combined cycle (see Section 14.6) with a steam (Rankine) cycle, the nuclear reactor replacing the furnace of the fossil fuel plant. The first nuclear reactor to produce electricity for a commercial grid was opened at Calder Hall, Cumbria, in 1959. This was barely twenty years after the special features of uranium fission had been discovered, a very remarkable speed of development.

16.2 Two Types of Reactor

The nature of the reactor heat source is controlled by the energy of the neutrons sustaining the chain reaction. When released during the fission process the neutrons have an energy of about 4.2×10^{-13} J ($= 2$ MeV). For a given fissile fuel, the chain reaction requires neutrons of a particular energy. A chain reaction can be sustained in all fissile materials for neutron energies in excess of 1 MeV. Some nuclei, and particularly ^{235}U and ^{239}Pu, are also very susceptible to fission at thermal energies, when the neutrons have been slowed down to have the same energy as the atoms of materials at ordinary (room) temperatures. The individual fission processes are particularly uncomplicated for fission either with high energy neutrons or with nuetrons of thermal energy, but for neutron energies in between, the behaviour of the fission process depends critically on the precise neutron energy involved, and the fine engineering control necessary for safe application of the fission process proves very difficult. Fortunately, the fission energies associated with these intermediate neutrons are a small proportion of the whole.

In practice, the nuclear furnace is designed to work either using all neutron energies including the highest (the fast neutron reactor, or fast reactor) or with thermal neutron energies only (the thermal reactor). In the former case, neutrons can be used fresh from fission; in the latter case, the neutrons must be slowed down (using a moderator) to thermal energy levels before they can trigger a chain reaction. It will be realised that the characteristics of the reactor will be different in the two cases and this must affect the steam cycle, but it is not relevant for our present purpose to consider the details of these reactor heat sources further.

For neutrons at either high or thermal energy, the nuclear fuel (usually uranium) must be enriched by the addition of a percentage of ^{235}U or ^{239}Pu to maintain the chain reaction. For the thermal reactor an enrichment of up to 3 or 4 % is usual (depending on the particular design); for the fast reactor the percentage enrichment needs to be in the range 20 to 50 %.

16.3 Thermal Reactors

The heat generated in the reactor is removed by a coolant, which is then transferred to the steam cycle for work transfer. The coolant can be a gas (gas

16.3 Thermal Reactors

cooled reactor) or a liquid (for instance water, perhaps under pressure to prevent boiling). For reasons of safety, the coolant is not used directly to drive an expansion turbine but the heat content is transferred, through a heat exchanger, to a secondary fluid for this purpose. This secondary fluid has so far always been water to be converted to steam but this need not be so in the future. The particular form a thermal reactor takes depends on the cooling mechanism used to remove the heat produced in the reactor fuel core.

The Gas Cooled Reactor

Reactor developments in the UK have, until now, been centered on the use of a gas as the reactor coolant, and carbon dioxide has been used for this purpose. Gas cooled reactors thus involve a heat exchanger transferring heat from the CO_2 reactor coolant to the water in the separate steam cycle. The heat transfer to the water is not at constant temperature, the CO_2 loses enthalpy as the water gains it (see Section 14.6.2).

The Dual Pressure Cycle

The differing thermal properties of CO_2 and water make thermodynamic reversibility in the heat exchanger (linking the reactor to the steam cycle) difficult to achieve even approximately. Whilst the temperature of the CO_2 decreases through the exchanger (see Fig. 16.1) that of the steam rises to the boiling point and there stays constant during vaporisation, though a further limited rise would follow superheating. Reversibility would require the water

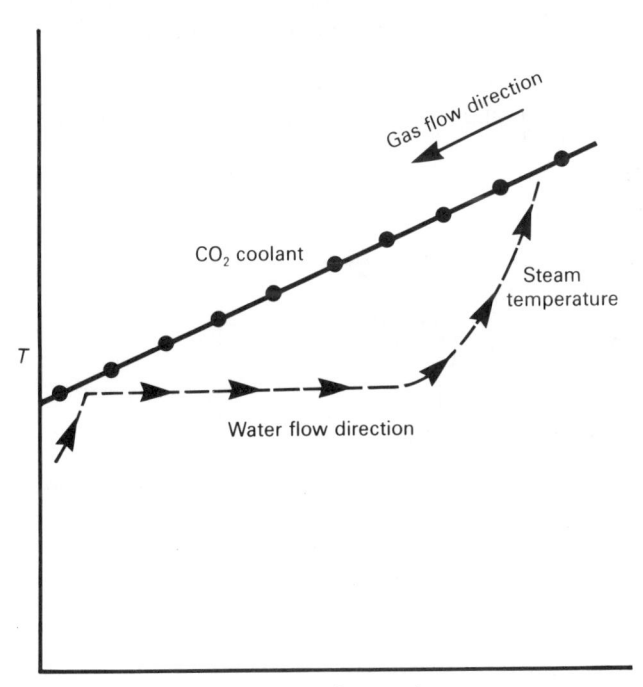

Fig. 16.1 Showing the nature of the heat transfer between the reactor gas coolant and the water to form superheat steam.

234 Nuclear Power Cycles

temperature to rise steadily in sympathy with the fall in the CO_2 temperature. A matching of the two temperature profiles could be approached by presenting the water-side to the gas-side through a number of stages (in principle, through an indefinitely large number of stages) differing by the water pressure. The highest water pressure is matched to the highest gas temperature, the lower pressure stages matching the falling gas pressure distribution through the heat exchanger. This is shown in Fig. 16.1.

For a finite number of pressure stages the match between water and gas temperatures is not close. In practice, the heat transfer is affected by heat conduction in the boundary layers so a finite temperature difference (perhaps of the order of 17 °C) is necessary to achieve adequate heat conduction. This smallest temperature difference is achieved at certain points, called pinch points. Elsewhere the temperature difference is greater.

Although thermodynamically there should be many different pressure cycles for the heat transfer process, in practice only two were used in the early gas cooled reactors. As an example, Bradwell uses two stages at pressures of 48 bar and 12 bar, with two thirds the steam being generated at the higher pressure, providing steam at a common temperature of 350 °C at the rate of 391 kg/s.

Higher Gas Pressure

The thermal conditions in the reactor gas-side of the heat exchanger depend on the gas temperature and the mass flow. Improvements in the reactor fuel element design and the introduction of pre-stressed concrete pressure vessels for the reactor have allowed the reactor thermal conditions in British gas cooled reactors to be continually improved. As the CO_2 temperature and pressure are raised the need for multi-pressure stages is eliminated. This is the case for the station built at Wylfa (see Fig. 16.2) and commissioned in 1973.

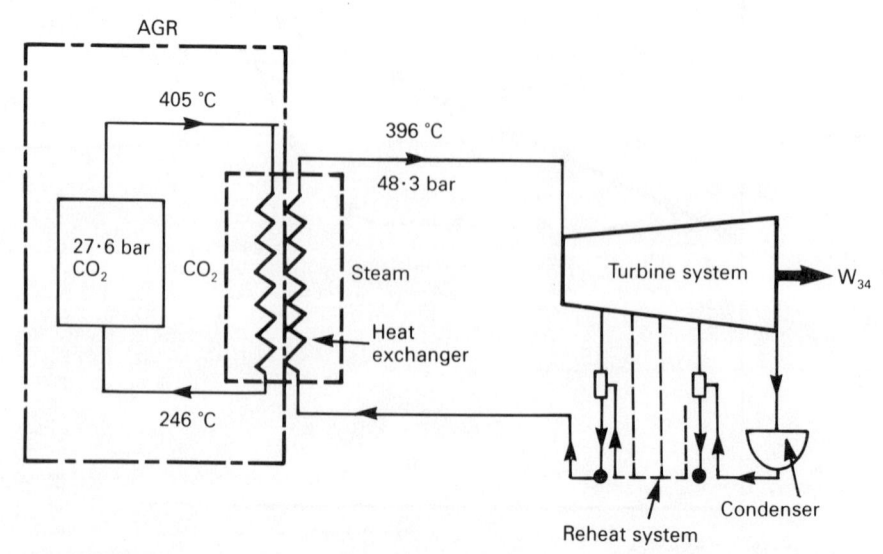

Fig. 16.2 Schematic representation of the Wylfa Nuclear Plant using the advanced gas cooled reactor.

The reactor is cooled by CO_2 gas at a pressure of 27.6 bar. The gas temperature leaving the reactor is 405 °C and it enters at 246 °C, giving a heat exchanger temperature drop of 159 °C. Steam leaves the heat exchanger at a temperature of 331 °C and a pressure of 39 bar which is a superheating of about 70 °C. The steam is reheated in a steam-steam system and there is a regenerative feed heat system raising the feed water to about 130 °C on entry to the heat exchanger. The work output is 1180 MWe. This gives a cycle thermal efficiency of slightly more than 0.30 although the overall efficiency of the plant is rather lower (see Table 16.1).

Table 16.1. Steam and Efficiency Data for the Nuclear Power Stations of the UK Commissioned Between 1959 and 1985.

Station	Year	Boilers kg/s	Steam pressure bar	Steam temp. °C	Elect. GWhe[a]	Efficiency %
Calder Hall	1959	400	15.2/3.3	320/190	1.77×10^3	22.6
Chapelcross	1960	404	15.5/3.5	329/185	1.76×10^3	18.5
Berkeley	1962	413	20/4	319/316	1.26×10^3	21.78
Bradwell	1962	391	48(H)/12(L)	350	2.00×10^3	23.93
Hunterston A	1964	476	40/11	385/364	2.65×10^3	24.83
Dungeness A	1965	686	97/39	393/395	3.23×10^3	28.05
Hinkley Pt. A	1965	472	43/12	350/336	4.21×10^3	24.66
Trawsfynydd	1965	672	48/16	340/335	3.30×10^3	24.38
Sizewell	1966	739	45	387	2.87×10^3	37.71
Oldbury	1968	504	26	350	3.45×10^3	27.69
Winfrith	1968	149	62	281	0.43×10^3	32.00
Wylfa	1973	1261	39	331	7.75×10^3	26.33
Dounray PFR	1975	271	160	538	9.40×10^2	42.00
Hunterston B	1977	828	167/41	541	8.84×10^2	37.99
Hinkley Pt. B	1978	994	152/36	498	7.88×10^3	37.63
Dungeness B	1985	660	153/38	551/364	2.80×10^3	32.94

[a] Gigawatt hour electric.

The gas cooled reactor has been a particular development in the UK and it has the great advantage of reactor safety should anything go wrong. The move to the Advanced Gas Cooled reactor (AGR) has been bedevilled by development problems and this has led to delays in commissioning and consequently escalating costs. But the steam conditions are closely similar to those of the most modern fossil plants Hunterston B, Hinkley Point and Dungeness B (Table 16.1). Proposals for the further development of the AGR programme are likely now to be curtailed, if not abandoned, due to recent government policies on energy.

The Water Cooled Reactor

Water is an excellent moderator and coolant for the reactor and the water cooled reactor has been extensively developed, particularly in the United States, France and the Soviet Union. The important limitation for such reactors is the critical temperature for water, 374 °C. The details of the reactor depend upon whether the water is allowed to boil.

The Pressurised Water Reactor (PWR)

If the water is not to boil in the reactor its temperature must not exceed 374 °C, and the pressure must be greater than the saturation pressure. This type of reactor/steam system has been developed particularly in the United States and what is virtually a standard design has been achieved. This is sketched in Fig. 16.3. The water temperature is about 330 °C (~40 °C below the critical) and the pressure 153 bar (20 bar above the saturation pressure) with the exit and inlet water temperatures at the reactor at 295 °C and 330 °C respectively. Steam is generated at 73 bar with a temperature of 290 °C which means without superheat, while a feed heat system returns the steam to the heat exchanger at a temperature of 232 °C, giving a moisture fraction of the steam in the heat exchanger of less than 0.003. The power of the reactor is 3800 MW giving an output of 1300 MWe.

This reactor is probably the most widely used in the world. The greatest advantage of this design lies in the moderating properties of water, which allow for a compact reactor and so a small plant overall. The capital cost is therefore low in comparison with the AGR, and this offsets the rather poor thermal efficiency of the steam cycle due to the low temperature of the steam.

The small size of the PWR makes it very suitable for marine propulsion and small propulsion plants have been developed for this purpose. It is particularly attractive for submarine propulsion because, unlike the chemical reactions required for fossil fuels, the chain reaction does not need oxygen to maintain its activity. Small plant have also been used for orbital space craft and there would seem to be a great future for such applications.

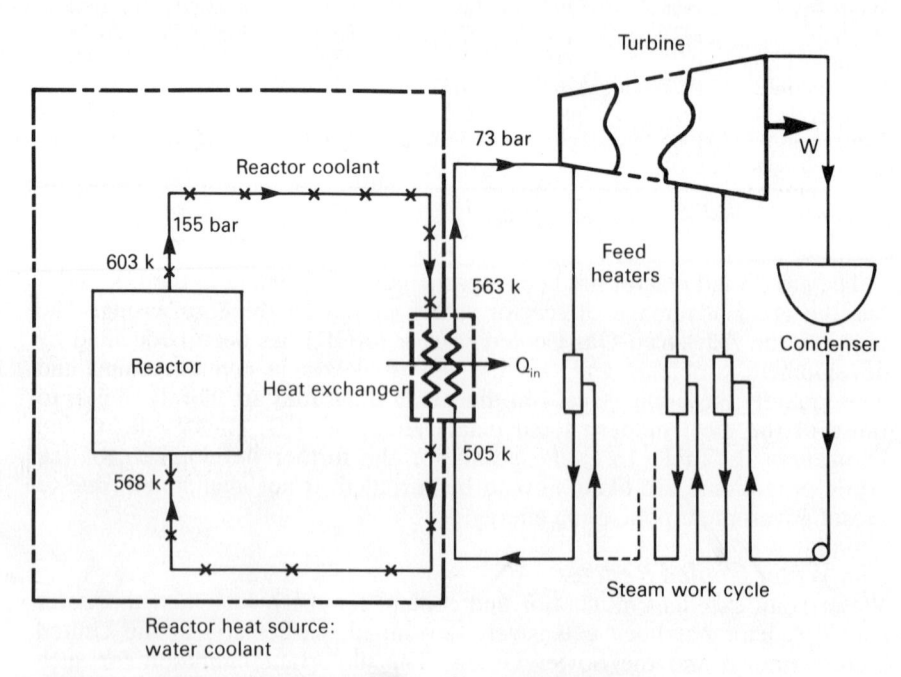

Fig. 16.3 The flow diagram for the pressurised water reactor. The coolant in the reactor circuit is water under high pressure to prevent boiling.

16.3 Thermal Reactors

The Boiling Water Reactor (BWR)
The reactor pressure is reduced substantially if the coolant water is allowed to boil. This is even more true if the steam at the reactor is dried and expanded directly through the turbine from the reactor without using an intermediate heat exchanger. This direct-cycle type of reactor has been developed in the United States, and the Dresden power station is a modified form involving a dual pressure system sketched in Fig. 16.4. The steam conditions at the outlet of the reactor are 288 °C and 73 bar: steam returns to the reactor inlet at 260 °C giving a temperature drop of about 30 °C across the heat exchanger. In the dual pressure steam system one line provides steam to the turbine at 70 bar pressure and 285 °C while the other is at 36 bar pressure and 244 °C. Both steam lines have regenerative heating and water is returned to the low pressure heat exchanger at 207 °C.

The steam conditions are seen to be generally similar to those for the PWR, which are not as advanced as for the AGR or for the fossil fuel equivalents. The lower water pressure in the reactor of the BWR reduces still further the capital cost of the plant. There is perhaps still some doubt about the boiling coolant flows but indications are that they can be safely controlled. Water in the turbine is linked directly to the reactor but radioactive contamination of the turbine, which could arise from impurities in the water being made radioactive in the reactor.

The Heavy Water Reactor (HWR)
Water is composed of two atoms of hydrogen and one of oxygen. Naturally occurring hydrogen is composed of two isotopes (as mentioned in Section 16.1) one, deuterium, with twice the mass of the natural isotope, though each variant has a single positive nuclear electric charge and so a single orbiting

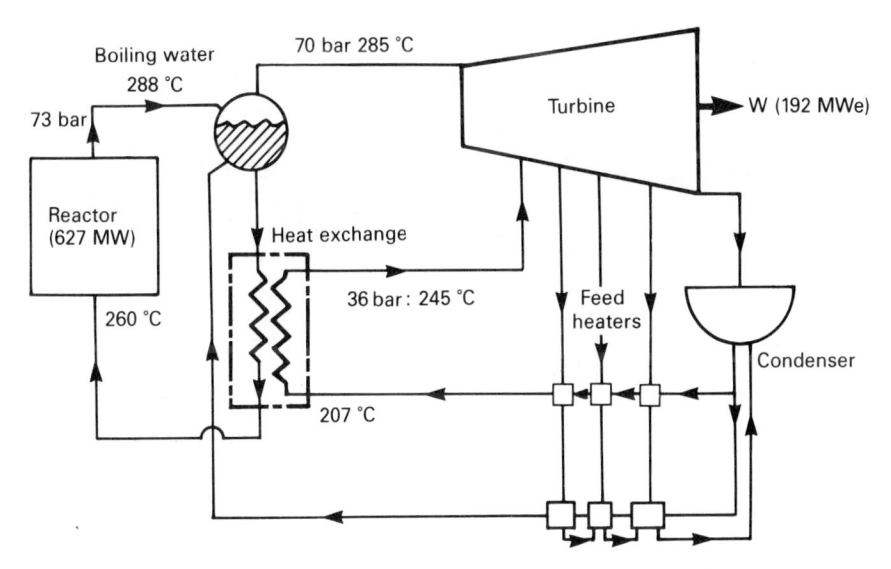

Fig. 16.4 The flow diagram for a dual pressure direct cycle boiling water reactor (after the Dresden 1 design in the USA).

238 Nuclear Power Cycles

electron. They are chemically indistinguishable but 99.985 % of any natural sample of hydrogen is normal while 0.015 % is the heavier deuterium. Consequently, water itself shows these two proportions of ordinary water (H_2O) and heavy water (D_2O with deuterium (D)). Heavy water is an excellent moderator and coolant and has the advantage that a chain reaction can proceed with natural uranium (which contains only 0.715 % ^{235}U) without the addition of any more (that is, without enrichment). Heavy water is expensive because the collection process is difficult and lengthy but the enrichment of natural uranium (which is avoided) is also costly. Heavy water reactors have been developed particularly in Canada under the name Candu (from *Can*adian *D*euterium *U*ranium) and their general form, based on the reactor at Pickering, Ontario, is shown in Fig. 16.5. The heavy water is used as the moderator and coolant and is heated typically to a temperature of 295 °C at 100 bar. The heat is transferred from the heavy water at high pressure through a heat exchanger (called a calandria) to ordinary water at 41 bar pressure with a steam temperature of 251 °C (no superheat). This water passes through the turbine system to provide the work transfer of 515 MWe. A feed heat system allows the water to return to the calandria at a temperature of 171 °C.

As in the PWR and BWR, the steam conditions for the HWR are not comparable with those of modern fossil fuel plant, but it is the overall cost of producing the electrical output that is ultimately important and the HWR appears competitive in these terms.

Other Possibilities

There are many other designs on the drawing board for new approaches to the production of power by nuclear reactors, particularly without using steam. They are unlikely to be developed in practice in the near future, due to present world economic circumstances, but may well be used in the future. We will not consider these here.

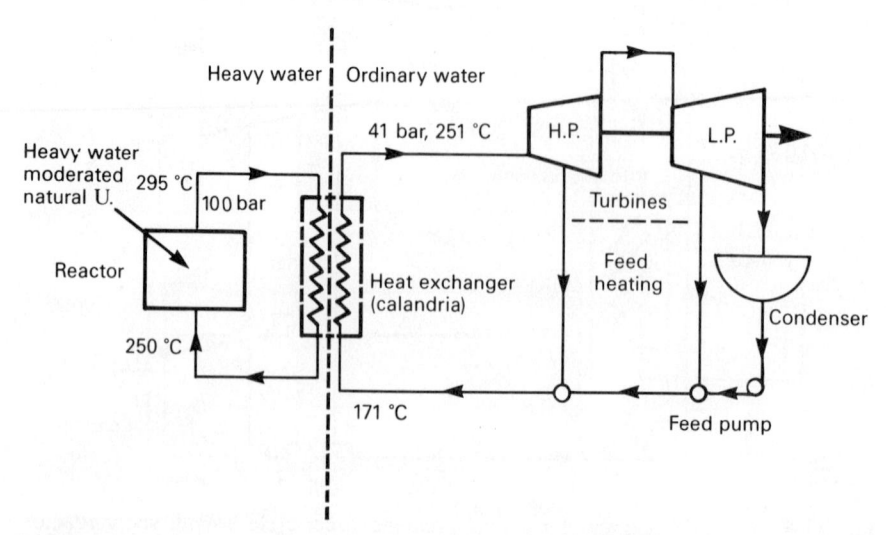

Fig. 16.5 The flow diagram for the heavy water reactor (Candu).

16.4 Reactors Using Fast Neutrons

The steam conditions available from fast reactors are quite different from thermal reactors. All the fast neutron energies are used in heat production but the uranium fuel must be heavily enriched by between 25 and 50 % of ^{235}U or ^{239}Pu, to maintain the chain reaction. Such reactors have been developed in several countries and the British Prototype Fast Reactor (PFR) at Dounray (see Table 16.1) is perhaps not untypical of those in other countries as well.

If the large quantity of heat produced in the reactor core is to be removed efficiently a coolant fluid of low Prandtl number must be used to ensure that sufficient of the thermal boundary layer is removed with the main stream speed outside the velocity boundary layer. In practice this means a liquid metal, and sodium or a sodium/potassium mixture is used for this purpose. The system is shown schematically in Fig. 16.6. The coolant temperature at the outlet from the reactor is high, entering the sodium/water heat exchanger at about 590 °C and leaving at 370 °C. Steam is generated at 160 bar and 565 °C (so giving about 300 °C superheat) and there is reheat and feed heat: the system is comparable to the best modern fossil fuel practice and achieves an overall efficiency of 42 %.

16.5 Final Comment

Nuclear systems have now reached a stage where they form a mature component of world energy production. They will continue to play an increasing part in this as oil becomes more scarce, and so more expensive, particularly for countries where coal is not plentiful or where mining is expensive. The world reserves of uranium are vast (even greater than coal) and thorium (^{232}Th) can also be used to provide the fissile fuel ^{233}U which is similar in many ways to ^{235}U and ^{238}U in its interaction with neutrons. Nuclear reactors differ from fossil furnaces in that they can breed their own

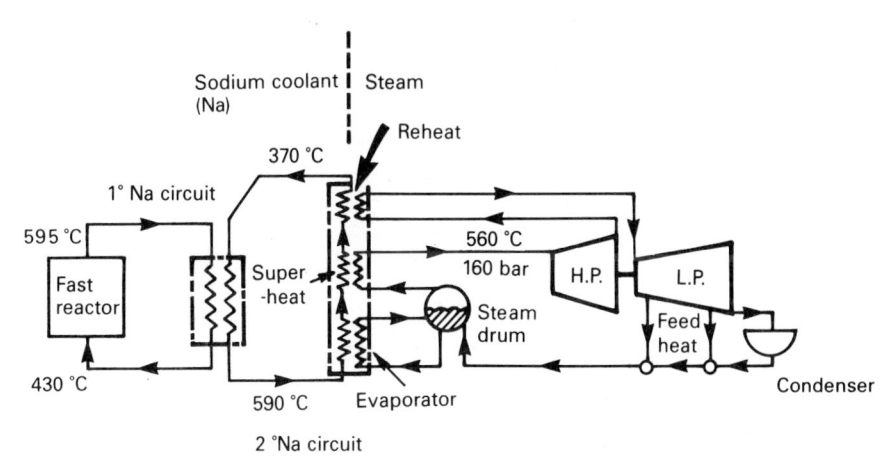

Fig. 16.6 The flow diagram for the fast reactor system (prototype) showing the primary (1°) and secondary (2°) sodium coolant circuits and the full steam system.

fuel, converting a proportion of low grade uranium to isotopes that are fissile with thermal neutrons. The energies involved are high and the fuels must be handled with far greater care than fossil fuels. All fuels present dangers but this should not mean they cannot be used: technology must be developed to allow them to play their proper place in the energy budget of the industrialised world.

16.6 Summary

1. Nuclear transformations release energy which can be used to obtain work transfer.
2. The neutron chain reaction provides large quantities of energy at continuous working in excess of one year without recharging.
3. Two ranges for neutron energy are of special technological interest: one is at high energies (>1 MeV) and the other at low (thermal) energies about 0.024 eV.
4. This has led to the development of two types of reactor, the fast reactor where the full range of neutron energies is used and the thermal reactor where only the thermal neutron energies are involved.
5. The thermal reactor has been developed particularly because it requires less advanced technology than the fast reactor.
6. The steam conditions for the thermal reactor are determined by the coolant used for the reactor, but are generally less attractive than for fossil fuels.
7. The best steam conditions have been achieved with CO_2 gas as the reactor coolant (AGR), but technical difficulties have made such stations expensive to build. Pressurised water as a coolant (PWR) provides more elementary steam conditions but does not involve particularly advanced technology and such plant is cheap to build and run.
8. Another system uses heavy water as the coolant (HWR) and has the advantage that the nuclear fuel does not require enrichment. However, the associated steam cycle is not very efficient.
9. The best steam conditions, comparable to those for modern fossil fuelled plant, are associated with fast reactors.

INDEX

Adiabatic
  conditions 9, 28, 43
  equation for a gas 28
  heat transfer 179
  reversible conditions 28
  work transfer 29
Air pre-heater 203
Aircraft propulsion 108
Availability 18
  flue-gas 200
  functions 21, 181, 199
  restrictions on 21

Berthelot equation 35
Binary cycle 219
Bleed point locations 187
  optimum 189
Boiling water reactor 237

Carnot
  engine 14, 30
  cycle 15
    with gas 21
    with steam 131
      characteristics 136
  inefficiencies 136, 147, 148
  non-ideal components 36
  numerical values, gas 32
  performance coefficient 18
  refrigerator 18
  temperature scale 17
  thermal efficiency 17, 30
Cascade, feed heating 190
Chain reaction 230
Characteristics
  Carnot cycle with gas 30
    with steam 134
  dual mercury/steam 224
  dual vapour/steam 221
  Joule cycle 66
  Joule cycle with intercooling 64
  Joule cycle with isothermal
    compression 74
  Joule cycle with reheat 61
  Joule cycle with reheat and
    intercooling 76
  Rankine cycle 143
    with feed heat 192

    with reheat 167
    with superheat 159
    with superheat, reheat and feed
      heat 205
Clausius equation 35
Closed, cycle 2
  system 1
Cogad 107
Cogas 107
Cogod 107
Cogos 107
Combined cycle 225
Combustion products 210
  control 211
Compression stage 1
Compression, isothermal 73
Complete Joule cycle 91
  system 97
Compressor work transfer 142
Cycle
  Carnot 15
  closed 2
  complete 91
  diesel 112
  dual pressure 233
  Joule 41
  multi-stage Joule 70
  nuclear 230
  open 2
  Otto 116
  Rankine 140
Cyclic change 11
  irreversible 5
  reversible 5

Diesel, principle 112
  cycle 112
  thermal efficiency 114
  work output 115
Dieterici equation 35
Diagram
  Mollier 128
  p-T for steam 125
  p-V for steam 126
  T-V for steam 128
Diffuser 8
Dual pressure cycle 233

Dual transformation   19
　vapour/steam cycle   220

Economiser   201
　effectiveness   202
Efficiency
　Carnot   17
　diesel   112
　isentropic   36
　simple Rankine   140
Energy
　blood point balance   186
　budget, steam plant   206
　equation   6
　free   19
　internal   6
Engine, reversible   27
Enthalpy   7
　exchange   41
Entropy   5, 9
　gas   9
　general   10
　heat exchange   10
Equation, Berthelot   35
　Clausius   35
　energy   6
　ideal gas   8
　van der Waals   36
Equation of state
　empirical   35
　explicit expressions   34
　ideal gas   8, 21
　　validity of   27
　real gas   33
Equlibrium, thermal   3
Ericson cycle   110
Energy   181
　plant assessment   214
Expansion stage   1
Extraction systems   207

Feed heating   177
　practical implementation   182
　using cascade   190
　with superheat   192
　with superheat and reheat   208
　without superheat   183
First law of thermodynamics   5
Flow process   2
Flue gas   211
　availability   200
Free energy   19
　Gibbs   20
　Helmholtz   20
Fuel impurities   210

Furness heat transfer   213

Gas, constant   8, 27
　cooled reactor   233
　cycles   14, 25, 41, 56, 81, 91
　maximum work output   58
　flow, Carnot   31
　with intercooling   63
　with reheat   56
　with reheat and intercooling   67
Gibbs free energy   20

Heat   4
　and power   209
　conservation   5
　exchange, Joule cycle   81, 85
　extraction systems   207
　　Joule cycle, complete   97
　　Joule cycle with heat exchange   96
　　Joule cycle with irreversibilities   85
　　total system   100
　pump   18
　transfer   4
　　furnace   213
　　optimum for feed heating   189
　with reheat   91
　with intercooling   96
Heavy water reactor   237
Helmholtz free energy   20

Ideal gas   27
Isentropic efficiency   36, 51
Isothermal compression   73
Industrial gas cycle   109
Intercooling   63
Internal energy   6
Irreversibility, effects   11
Irreversible change   5
Isothermal compression   73

Joule, cycle   41
　characteristics   44
　effects of irreversibilities   51
　experiment   9
　h-s diagram   43
　intercooling characteristics   66
　maximum specific work output   49, 54
　multi-stage cycles   70
　p-V diagram   43
　　specific work output   58
　　thermal efficiency   59
　reheat characteristics   58
　reheat with irreversibilities   60
　thermal efficiency   46
　T-s diagram   42

Index

with heat exchange 81
with intercooling 63
with irreversibilities 51
with isothermal compression 43
with reheat 57, 78
  work output 59
with reheat and intercooling 67
working characteristics 44

Low grade heat cycle 225

Marine gas units 107
Mean receptor temperature 149, 154
Mercury/steam cycle 223
Metallurgical limit 212
Mole 8
Mollier diagram 128
Mixed cycle 118
Multi-stage gas cycles 70

Non-flow process 2
Non-steady flow cycles 112
Nozzle 7
  efficiency 7

Open, system 1
  cycle 2
Otto cycle 116

Perpetual motion
  first kind 6
  second kind 12
Potential, thermodynamic 19
Pressurised water reactor 236
Process
  flow 2
  non-flow 2
Propulsion
  aircraft 108
  marine 107
Power plant, steam 204
  working characteristics 205

Rankine cycle 140
  characteristics 141
  comparison with Carnot cycle 147
  compressor work transfer 142
  effect of boiler pressure 144
  irreversibilities 148
  supercritical conditions 169
  with reactors 233
  with reheat 160
Reactor 232
  boiling water 237
  fast 239

  gas cooled 233
  heavy water 237
  pressurised water 236
  water cooled 235
Refrigerator 13
  Carnot 18
  effectiveness 18
Regenerative heating (*see* feed heating) 177
Reheat, Rankine cycle 160
  criterion 161
Reversible change 5

Second law of thermodynamics 11
Specific gas flow, Carnot 31
Specific work output
  Carnot 31, 38
  Joule cycle with intercooling 66
  Joule cycle with reheat 59
  Joule cycle with reheat and intercooling 68
  maximum, Joule cycle 49
Stability, thermodynamic 20
Stack emission 211
Stage, compression 1
  expansion 1
State, thermodynamic 2
  of system 2
Steady flow energy equation 6
Steam
  full system 200
  Mollier diagram 128
  other thermodynamic variables 127
  properties 123
  p-T diagram 125
  p-V diagram 125
  tables 127
  total plant 203
  triple point 125, 126
  T-s diagram 128
  T-V diagram 123
Stirling cycle 40
Supercritical conditions 169
Superheating 154, 158
  effects of boiler pressure 169
  effects of irreversibilities 156
  effects on thermal efficiency 165
  numerical example 158
  with regenerative heating 183
    with reheat 165
  work of compression 159
System, availability 18
  open, closed 1
  state 2

Temperature 3
  scale 3
Thermal
  equilibrium 3
  reactor 232
Thermal efficiency
  Carnot cycle, with gas 30, 36
    with steam 133
  diesel 115
  dual cycle 220
  Joule cycle 45
  Joule cycle with reheat 59
  Joule cycle with intercooling 66
  Joule cycle with reheat and
    intercooling 68
  Joule cycle with isothermal
    compression 45
  Joule cycle, multistage reheat and
    intercooling 79
  Joule cycle with heat exchange 84, 86
  Joule cycle, complete 93, 95
  Otto 118
  mixed constant pressure/constant
    volume 118
  Rankine cycle
    supercriticial 171
    supercritical with reheat 173
    with steam 143
    with superheat 155
    with reheat 163
    with superheat and reheat 165
    with feed heating 195
Thermodynamics 1
  bleed point conditions 187

consequences 12
first law 5
irreversibility 11
second law 11
Thermodynamic, potentials 19
  state 2
Throttle 7
Temperature, mean receptor 149
Total, cycle 91
  heat 6, 7
  plant 203
  system 100
Transformation, dual 19
Triple point 125

van der Waals equation 35
Virial, expansion 33
  coefficients 33, 36

Water cooled reactor 235
Work 4
  maximum 19
  output 3
    Carnot 38
    Diesel 115
    Otto 118
  ratio, Carnot 31
  transfer 4, 6

Work transfer
  adiabatic 29
  isothermal 30
Working fluid 3
  improved 215